SOIL SOLUTION CHEMISTRY

SOIL SOLUTION CHEMISTRY
Applications to Environmental Science and Agriculture

JEFFREY D. WOLT

JOHN WILEY & SONS, INC.

New York / Chichester / Brisbane / Toronto / Singapore

Library of Congress Cataloging in Publication Data:
Wolt, Jeffrey D.
    Soil solution chemistry : applications to environmental science
  and agriculture / Jeffrey D. Wolt.
      p.   cm.
    ISBN 0-471-58554-8 (cloth)
    1. Soil solutions—Research—Methodology.   2. Soil chemistry—
  —Methodology.   I. Title.
  S594.W7   1994
  631.4′ 1—dc20   94–10959

Printed in United States of America

10 9 8 7 6 5 4 3 2 1

To Evelyn

CONTENTS

Preface   xi

Symbols   xiii

    Constants / xvi
    Defined Quantities / xvi
    Conversions / xvi
    SI Units / xvii

Periodic Table of the Elements   xviii

1 Chemical Distribution in Soil Environments   1

    1.1   The Soil Environment / 1
    1.2   Soil as a Chemically Reacting System / 2
    1.3   Soil Solution as a Hub Predictor of Soil Processes / 4
    1.4   Soil Solution Definitions / 4

2 The Soil Solution   6

    2.1   Historical Perspective / 7
    2.2   Concepts of Soil Solution / 13

3 Chemical Statics and Dynamics Applied to Soil Solution   22

    3.1   Limitations to Equilibrium and Kinetic Approaches / 22
    3.2   Chemical Equilibrium and Kinetic Models Compared / 24
    3.3   Statics in Review / 27
    3.4   Dynamics in Review / 47

4 Master Variables   **74**

    4.1 Soil pH / 74

    4.2 Ionic Strength / 85

    4.3 Electrical Potential / 86

    4.4 Suspended Colloidal Materials / 92

5 Obtaining Soil Solution: Laboratory Methods   **95**

    5.1 Operational Needs and Practical Constraints / 95

    5.2 Centrifugation / 104

    5.3 Column Displacement / 109

    5.4 Pressure Membrane Extraction / 114

    5.5 Saturation Extracts / 116

    5.6 Water Extracts of Soils / 118

    5.7 Complexation and Exchange Techniques / 118

    5.8 Summary / 120

6 Obtaining Soil Solution: Field Methods   **121**

    6.1 Common Attributes of Lysimetric Methods / 122

    6.2 Monolith Lysimeters / 126

    6.3 Filled-In Lysimeters / 129

    6.4 Tension Lysimeters / 130

    6.5 Ebermayer Lysimeters / 135

    6.6 Watershed and Field-Scale Sampling of Soil Solution / 139

    6.7 Summary / 141

7 Soil Solution Composition   **144**

    7.1 Total Ion Composition of Soil Solutions / 144

    7.2 Ion Speciation in Soil Solution / 153

    7.3 Computational Modeling of Soil Solution Composition / 165

8 Quantity-Intensity Relationships   **169**

    8.1 Ratio Law / 170

    8.2 Ion Exchange / 172

    8.3 Chemisorption / 174

9 Mineral Stability and Pedogenesis   **181**

    9.1 Ion Activity Products / 182

    9.2 Stability Diagrams / 186

    9.3 Rates of Mineral Weathering / 202

10 Chemical Availability   **209**

  10.1   Bioavailability / 209
  10.2   Geochemical Availability / 217
  10.3   Biogeochemical Availability / 218

11 Soil Solution Aluminum   **220**

  11.1   Aluminum Distribution in Soil Solution / 221
  11.2   Measurement of Solution Aluminum / 224
  11.3   Partitioning Soil Solution Aluminum / 225
  11.4   Aluminum Bioavailability / 227
  11.5   Aluminum Geochemistry / 236

12 Trace Metals in Soil Solution   **246**

  12.1   Soil Biogeochemistry of Trace Elements / 246
  12.2   Speciation of Trace Metals Arising from
         Waste Disposal / 247
  12.3   Complexation Reactions of Trace Elements / 250
  12.4   Cadmium / 255
  12.5   Micronutrient Cations / 261
  12.6   Trace Inorganic Ligands / 268
  12.7   Technetium / 271

13 Dissolved and Colloidal Organics   **275**

  13.1   Reactive Natural Organic Chemicals in Soil Solution / 275
  13.2   Metal–Humic Matter Interactions / 281
  13.3   Quantifying Dissolved Organic Carbon Effects on Soil
         Solution Composition / 285
  13.4   Dissolved and Colloidal Organic Carbon Effects on
         Bioavailability and Transport / 289

14 Xenobiotics in Soil Solution   **293**

  14.1   Speciation and Complexation in Soil Solution / 294
  14.2   Sampling and Analytical Approaches / 295
  14.3   Availability / 303
  14.4   Environmental Chemistry / 306
  14.5   Summary / 309

References   **311**

Index   **339**

PREFACE

Soil chemistry has evolved from an agronomic focus to be broadly applied in a number of disciplines. For this larger body of soil chemistry practitioners, soil solution chemistry deserves particular attention as a predictive and diagnostic approach for elucidating bioavailability, mobility, and geochemical cycling of chemicals in soil.

Applying soil solution chemistry to problem solving in agronomy, environmental science, and geochemistry increasingly interests students, researchers, and consultants concerned with chemical retention and reactivity in soil environments. Considerable research contributions to soil solution chemistry, especially relating to soil environmental chemistry, have occurred in the past decade. Reducing this body of research to practice will assure continued growth of soil solution chemistry as a powerful approach for monitoring, assessing, and interpreting chemical processes occurring in soil environments. This book—devoted to the theory, method, and application of soil solution chemistry—was written with this objective in mind.

Although earlier texts—most notably, those of Sposito and Lindsay and the compilation of literature edited by Elprince—have addressed theory of soil solution chemistry, no textbook or reference has addressed both the applied theory and methodology of soil solution chemistry. *Soil Solution Chemistry: Applications to Environmental Science and Agriculture* updates the field with an emphasis on applied problem solving. The book is comprised of 14 chapters in four sections overviewing soil solution chemistry (Chapters 1 and 2), outlining theory and methodology (Chapters 3 to 6), addressing general applications (Chapters 7 to 10), and discussing specific applications of broad interest to environmental scientists and agronomists (Chapters 11 to 14). Special emphasis is given to environmental science, as this is where the greatest potential for future application of soil solution chemistry will occur.

In writing this text, I have drawn on experiences gained in instructing and advising students attempting to apply soil solution chemistry to the resolution of problems relating to chemical fate and behavior in soils. Traditional texts in soil chemistry take an inside out look at topics such as bioavailability, mobility, and geochemical cycling by emphasizing the soil solid phase and the implications of solid-solution phase relationships to environmental availability of chemicals. Because the environmental availability of chemicals relates mostly to the occurrence and distribution of chemicals in the soil aqueous phase (soil solution), I take the approach in this text of emphasizing the direct analysis and interpretation of soil solution composition in order to gain insights as to chemical reactivity and availability in soils. This very different focus will be useful to soil chemistry educators in advanced courses seeking alternative ways to discuss soil chemistry. Additionally, educators in allied disciplines (environmental science, chemical and environmental engineering, geochemistry, ecology) desiring a textbook for teaching soil environmental chemistry may find this approach useful in their particular areas of disciplinary expertise.

I am very grateful to those who have given their time to review and comment on portions of this book. In particular, I would like to thank Frank Sikora for review of Chapter 3, N.V. Hue for review of Chapters 4 and 5, Imo Buttler for reviews of Chapters 2 and 6, John Graveel for review of Chapter 13, and Ron Turco for review of Chapter 14. These various reviewers have greatly aided in clarifying the presentation and ferreting out mistakes in the text. Mistakes and ambiguities that remain are of course my own, but I thank these reviewers for trying to keep them to a minimum. Finally, I am especially grateful for the unwavering support of my wife, Evelyn.

Jeffrey D. Wolt
Indianapolis, IN

SYMBOLS

Symbol	Definition	Section
A	Helmholtz free energy	3.3.1
A	Debye-Hückel parameter	3.3.6
A	Arrhenius equation frequency factor	3.4.6
A	reaction affinity	3.4.7
AR	activity ratio	8.1
a_i°	ion size parameter	3.3.6
A	activity	2.1.3
A_S	electrolyte activity	3.3.5
A_+, A_-	single ion activity	3.3.5
$A_{\pm}$	mean ionic activity	3.3.5
α	Elovich pre-exponential factor	3.4.4
α	gamma function exponential factor	3.4.4
β	extended Debye-Hückel parameter	3.3.6
β	Elovich exponential factor	3.4.4
β	gamma function pre-exponential factor	3.4.4
CEC	cation exchange capacity	5.1.1
C	concentration	3.2.1
$C_{\pm}$	mean ionic concentration	3.3.5
C_0	bulk phase concentration	2.2.1
C	influx concentration	3.2.2
Δc_p	change in heat capacity at constant pressure	3.3.3
χ	mole fraction	3.3.5
DOC	dissolved organic carbon	4.4
$D_w(\theta)$	diffusivity function	6.1.1

Symbol	*Definition*	*Section*
d_{ex}	effective thickness of the diffuse double layer	2.2.1
d_q	heat absorbed	3.4.7
$d\sigma$	entropy production	3.4.7
δ	correction for deviation of chemical potential from ideality	3.3.5
E	internal energy	3.3.1
E	electric potential	3.3.5
E*	activation energy	3.4.6
E_H	equilibrium redox potential	4.3.1
E_j	junction potential	4.1.2
E_m	membrane potential	4.1.2
EC	electrical conductivity	4.2.1
ε	dielectric constant	3.3.6
F	Gibbs free energy	3.3.1
$\mathcal{F}$	Faraday constant	3.3.5
f	fugacity	3.3.5
f	empirically derived activity coefficient	3.3.6
g	solid phase activity coefficient	9.1.2
γ	fugacity or activity coefficient	3.3.5
$\gamma_{S^+}, \gamma_{S^-}$	McInnes activity coefficient	3.3.5
γ_+, γ_-	single ion activity coefficient	3.3.5
$\gamma_\pm$	mean ionic activity coefficient	3.3.5
γ'	molar activity coefficient	3.3.5
H	Enthalpy	3.3.1
h	pressure head	6.4.1
I	ionic strength	3.3.6
IAP	ion activity product	3.3.5
j	number of reactive sites	13.2.2
K	equilibrium constant	3.2.1
K	selectivity coefficient	8.1
K^c	conditional equilibrium constant	3.3.6
K_a	acid dissociation constant	3.4.2
$K(\theta)$	unsaturated hydraulic conductivity	6.1.1
K_d	distribution coefficient	8.3
K_{ex}	exchange coefficient	8.1
K_f	Freundlich coefficient	8.3.1
K_G	Gapon coefficient	8.1
K_H	Henrys law constant	3.3.5
K_K	Kerr coefficient	8.1
K_{KO}	Krishnamoorthy and Overstreet coefficient	8.1.1
K_l	Langmuir coefficient	8.3.2
K_M	Michaelis-Menten constant	3.4.4
K_O	overall binding constant	13.2.2

Symbol	*Definition*	*Section*
K_{OS}	outer sphere association constant	7.2.2
K_p	partition coefficient	8.3.3
K_S	Monod constant	3.4.4
K_{sp}	solubility product	9.1
K_V	Vanselow coefficient	8.3
K_W	dissociation constant of water	4.1.2
k	rate constant	3.2.2
k	salting coefficient	14.2.4
k_i	interchange rate constant	7.2.2
k'	pseudo first-order rate constant	3.4.2
k_{app}	apparent distribution coefficient	14.4.1
k_{ex}	exchange rate of water	7.2.2
μ	chemical potential	3.3.5
μ	Monod growth rate	3.4.4
μ_{max}	Monod maximum growth rate	3.4.4
μ^{el}	electrochemical potential	2.1.3
μ°	standard state chemical potential	2.1.3
μ_S	electrolyte chemical potential	3.3.5
μ_+, μ_-	single ion chemical potential	3.3.5
$\mu_{\pm}$	mean ionic chemical potential	3.3.5
n	number of moles of particles	3.2.1
n	Freundlich constant	8.3.1
ν	stoichiometric coefficient	3.3.5
ν	formation function	13.2.2
θ	soil water content	2.2.2
P	pressure	3.3.1
pe	negative logarithm of electron activity	4.3.4
pH	negative logarithm of A_{H^+}	4.1.1
p_sH	negative logarithm of C_{H^+}	4.1.1
p(k)	probability density function	3.4.4
Q	reaction quotient	3.2.2
R	gas constant	2.1.3
R	resistance	4.2.1
r	rate of solution transfer	3.2.2
r	reaction exchange rate	3.4.7
r_m	radius of influence	6.4.1
S	entropy	3.3.1
T	temperature	3.3.1
$t_{1/2}$	half-life or half-time	3.4.2
V_{max}	Michaelis-Menten maximum rate	3.4.4
$\underline{V}$	partial molar volume	4.1.2
v	velocity	3.2.2
v_0	initial rate of reaction	3.4.4

Symbol	Definition	Section
ξ	extent of reaction	3.3.5
y_S	empirically derived activity coefficient for a nonpolar molecule	14.2.4
ψ_m	water matric potential	2.2.2
ψ_w	water potential	2.2.2
Z	number of collisions per second	3.4.6
z	valence	2.2.1
$\emptyset$	electrostatic activity coefficient	2.2.1

CONSTANTS

Symbol	Definition	Value
$\mathscr{F}$	Faraday constant	$= 96{,}490$ C mol^{-1}
R	Gas constant	$= 8.314$ J mol^{-1} deg^{-1}
		$= 0.082057$ L atm mol^{-1} deg^{-1}
		$= 1.987$ cal mol^{-1} deg^{-1}
	Ideal gas molar volume (T = 273.15 deg and 101.33 kPa)	$= 22.4$ L mol^{-1}

DEFINED QUANTITIES

Symbol	Definition	Value
ρ_w	Density of water (T = 273.15 deg and 101.33 kPa)	$= 1.00$ kg L^{-1}
	Standard pressure	$= 1$ atm
		$= 101.33$ kPa
	Standard temperature	$= 298.15$ deg^{-3}

CONVERSIONS

Physical Quantity	SI unit	Equivalent unit
entropy	4.184 J mol^{-1} deg^{-1}	$= 1$ cal mol^{-1} deg^{-1}
pressure	101.33 kPa	$= 1$ atm
		$= 1.013$ bar
temperature	273.15 deg	$= 273.15 + $ °C

SI UNITS

Physical Quantity	Unit	Symbol
conductance	seimens	$S = AV^{-1}$
electric charge	coulomb	$C = A$ sec
electric current	ampere	A
electric potential	volt	$V = W A^{-1}$
energy, work, heat	joule	$J = N$ m
force	newton	$N = kg$ m sec^{-2}
length	meter	m
mass	kilogram	kg
power	watt	$W = J$ sec^{-1}
pressure	pascal	$Pa = N$ m^{-2}
quantity	mole	mol
time	second	sec
temperature	kelvin	deg or K
volume	liter	L

Periodic table of the elements

New notation → (top row: 1 … 18)
Previous IUPAC form →
CAS version →

1 / IA	2 / IIA	3 / IIIA / IIIB	4 / IVA / IVB	5 / VA / VB	6 / VIA / VIB	7 / VIIA / VIIB	8	9 VIII	10	11 / IB	12 / IIB	13 / IIIB / IIIA	14 / IVB / IVA	15 / VB / VA	16 / VIB / VIA	17 / VIIB / VIIA	18 / VIIIA
1 H 1.0079																	2 He 4.00260
3 Li 6.941	4 Be 9.01218											5 B 10.81	6 C 12.011	7 N 14.0067	8 O 15.9994	9 F 18.9984	10 Ne 20.179
11 Na 22.9898	12 Mg 24.305											13 Al 26.9815	14 Si 28.0855	15 P 30.9738	16 S 32.066(6)	17 Cl 35.453	18 Ar 39.948
19 K 39.0983	20 Ca 40.08	21 Sc 44.9559	22 Ti 47.88	23 V 50.9415	24 Cr 51.996	25 Mn 54.9380	26 Fe 55.847	27 Co 58.9332	28 Ni 58.69	29 Cu 63.546	30 Zn 65.39	31 Ga 69.72	32 Ge 72.59	33 As 74.9216	34 Se 78.96	35 Br 79.904	36 Kr 83.80
37 Rb 85.4678	38 Sr 87.62	39 Y 88.9059	40 Zr 91.224	41 Nb 92.9064	42 Mo 95.94	43 Tc (98)	44 Ru 101.07	45 Rh 102.906	46 Pd 106.42	47 Ag 107.868	48 Cd 112.41	49 In 114.82	50 Sn 118.71	51 Sb 121.75	52 Te 127.60	53 I 126.905	54 Xe 131.29
55 Cs 132.905	56 Ba 137.33	57 La★ 138.906	72 Hf 178.49	73 Ta 180.948	74 W 183.85	75 Re 186.207	76 Os 190.2	77 Ir 192.22	78 Pt 195.08	79 Au 196.967	80 Hg 200.59	81 Tl 204.383	82 Pb 207.2	83 Bi 208.980	84 Po (209)	85 At (210)	86 Rn (222)
87 Fr (223)	88 Ra 226.025	89 Ac▲ 227.028	104 Unq (261)a	105 Unp (262)a	106 Unh (263)	107 Uns (262)a											

★ Lanthanide series

58 Ce 140.12	59 Pr 140.908	60 Nd 144.24	61 Pm (145)	62 Sm 150.36	63 Eu 151.96	64 Gd 157.25	65 Tb 158.925	66 Dy 162.50	67 Ho 164.930	68 Er 167.26	69 Tm 168.934	70 Yb 173.04	71 Lu 174.967

▲ Actinide series

90 Th 232.038	91 Pa 231.036	92 U 238.029	93 Np 237.048	94 Pu (244)	95 Am (243)	96 Cm (247)	97 Bk (247)	98 Cf (251)	99 Es (252)	100 Fm (257)	101 Md (258)	102 No (259)	103 Lr (260)

Note: Atomic masses shown here are the 1991 IUPAC values (maximum of six significant figures). a Symbols based on IUPAC systematic names.

SOIL SOLUTION CHEMISTRY

CHAPTER 1

CHEMICAL DISTRIBUTION IN SOIL ENVIRONMENTS

Chemicals occur naturally, or are deposited either purposefully or inadvertently, in soil environments. Clarifying chemical fate within, or routes of transport out of, the soil is essential to both agriculture and environmental science. Soil chemistry describes the fundamental processes of chemical transfer and transformation occurring among and within compartments of the soil environment. These processes are dominated by chemical reactivity in the soil aqueous phase—soil solution—as moderated by intimate interaction of soil solution with other compartments of the soil environment. Thus, soil solution chemistry provides a unique focus for elucidating chemical fate and behavior in the soil environment.

1.1 THE SOIL ENVIRONMENT

The dominant feature of the soil environment is the interaction over time of climate and biota on parent material in a landscape to produce a physicochemically and morphologically distinct entity. As typically conceived, *soil* is a naturally occurring three-dimensional body situated in the landscape, consisting of unconsolidated materials differentiated into horizons, and exhibiting the effects of biological activity and of physical and chemical weathering.[1] The boundary of soil with the atmosphere is distinct, whereas boundaries with water and geologic materials may be less distinct. *Soil chemistry* is the study of chemical reactions occurring in this unique environment. The precepts of soil chemistry, however, are frequently broadened to encompass the vadose environment and unconsolidated materials constructed or altered by man.

1

Soils vary laterally across landscapes and vertically within the soil profile as a consequence of the differential interaction—both in rate and intensity—of soil-forming factors (time, relief, parent material, climate, and biota). The soil chemist's understanding of "soil" is based on investigations either of subsamples of natural soil removed from the environment or from model systems comprised of isolated soil components, as when exchange on a synthetically prepared oxyhydroxide is used to model soil exchange reactions. Thus, the soil chemist's precepts of soil are limited to the extent that simplified models of "soil" are extrapolated to the natural soil environment.

1.2 SOIL AS A CHEMICALLY REACTING SYSTEM

Soils are chemically reacting systems; thus, applying static models without considering process dynamics (the efficiency and extent of reactions) may prove insufficient for describing chemical reactivity in soils (Richter, 1987). Designing and interpreting relevant experiments to model the soil environment must account for dynamic chemical processes as well as for the transfers between soil compartments at a scale of resolution appropriate for the processes considered. Soils are frequently modeled experimentally as systems exhibiting localized equilibrium, but the applicability of this approach is dependent on the spatial and temporal scale involved. The chemical processes leading to the genesis of a soil body exhibiting unique physicochemical and morphological features occur over spatial and temporal scales appropriate for the assumption of quasi-equilibrium; processes governing aqueous phase availability of nutrients or pesticides may not. The applicability of static versus dynamic approaches to description of chemical reactivity in soil systems is further described in Chapter 3.

Compartments

The soil environment is a heterogeneous[2] system where chemicals are distributed between complex, intimately interacting compartments (biotic, liquid, solid, and gas phases). Within soil compartments there is considerable complexity as well. Biotic compartments (plants, microbes, and soil fauna) act as surfaces for sorption and desorption, sinks for uptake and degradation, and sources for production of chemicals in the soil environment. The soil solid phase contains mineral, amorphous inorganic, and organic components that endue soils with unique features of chemical retention and reactivity. The soil liquid phase—soil solution—contains dissolved chemicals in free and complexed form as well as chemicals associated with colloidal particles. Chemical availability and mobility in soil are both dependent on chemical occurrence in soil solution. Thus, soil chemistry entails an understanding of the chemistry of soil solutions. It is useful to consider soil solution as the hub about which the other interacting compartments of the soil revolve (Fig. 1-1).

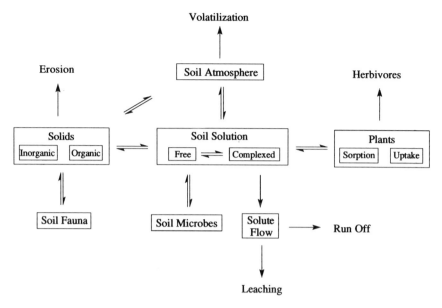

FIGURE 1.1   Interacting compartments of the soil, and routes of mass transfer out of the soil environment.

Transfers

Soils in nature are continuous flow reactors that can continuously exchange matter and thermal energy at their boundaries and, therefore, can only be completely characterized in dynamic terms. Describing chemical reactivity for flow reactors requires that transfers between compartments are sufficiently detailed to account for mass balance within the system. However, compartments of soil can be conceptually and experimentally isolated to yield simplified models of the soil environment amenable to description in static terms (Fig. 1-2). Stationary state thermodynamic equilibrium models where mass fluxes are zero, or steady state kinetic models where energy and mass fluxes are constant are useful descriptive approaches for soil systems.

Processes

Chemical reactivity in soils is dominated by aqueous phase processes—hydrolysis, hydration, carboxylation, oxidation-reduction, and solution. Additionally, the biochemical processes of synthesis and metabolism occur in biotic compartments of organisms intimately contacting the soil solution. Solution phase reactivity is moderated by partition of lipophilic molecules into biomass or soil organic matter, volatilization to soil atmosphere, or retention in the soil solid phase by either exchange or precipitation. The specific processes controlling chemical fate in the soil environment are not always evident; but the net effect of these various inter-

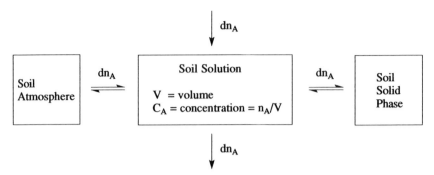

FIGURE 1.2 Simplified model of soil solution as an open system where mass of component A is exchanged with the environment and between soil phases. A steady state condition exists when net flux of A through the system is constant ($\Sigma dn_A = 0$); as fluxes between soil solution and the surroundings become infinitesimally small, the system approaches a stationary state and is closed with respect to mass transfer. [After Stumm and Morgan, 1981.]

acting soil processes is manifested in the *intensity factor*—the concentration of chemical of interest occurring in soil solution at any point in time.

1.3 SOIL SOLUTION AS A HUB PREDICTOR OF SOIL PROCESSES

Measuring chemical intensity in the soil environment is critical for understanding the processes of chemical decomposition active in soil weathering and for assessing the environmental availability—bioavailability, mobility, and geochemical cycling—of chemicals in soils. Soil solution chemistry focuses on the available fraction of chemical occurring in the soil environment at any point in time and is a useful predictor of mass transfer from the soil to other environmental compartments (Fig. 1-1). Thus, analysis of soil solution composition is frequently more instructive than is analysis of whole soils or soil extracts. Soil solution analysis yields information of both a static and dynamic nature, dependent on experimental design and modeling scale, for describing the nature, direction, extent, and rate of chemical reactions.

1.4 SOIL SOLUTION DEFINITIONS

Soil solution—the aqueous liquid phase of soil at field moisture contents—represents the medium where soil chemical reactions take place (Adams, 1974; SSSA, 1989). Soil solution is differentiated from soil water, the "moisture that fills the pores between soil particles" (Kohnke, 1968), in that it represents the aqueous phase of the soil at or below field capacity. Despite the simplicity of the foregoing definition of soil solution, a concept of soil solution is considerably more difficult to arrive at in practice because

1. Measurement of soil solution chemical components in situ is not possible.
2. Soil solution chemical composition varies with soil moisture content, especially in weakly buffered soils.

Consequently, soil solution is often operationally defined by the particular method whereby it is obtained. This is a considerable problem that has hindered understanding of soil solution chemistry, because what many researchers have termed "soil solution" may bear little resemblance to true soil solution as it occurs in natural soil. The conceptual definition of "soil solution" is discussed in Chapter 2, while operational definitions are described in Chapters 5 and 6.

NOTES

1. This definition reflects the definitions of soil given by Hilgard (1912), Marbut (1935), Joffe (1949), and Brady (1974).
2. The soil environment is heterogeneous in the sense that intensive properties (temperature, pressure, and chemical potential) vary throughout the soil body and cannot be represented by an average for a normally distributed set of values. Homogeneity can be assumed for most soil studies, but this assumption depends on the spatial and temporal scale selected. As one proceeds from point or site models of soil to field or regional models of soils, the assumption of homogeneity is less appropriate. See Richter, 1987 (p. 167) and Sposito, 1981 (p. 4).

CHAPTER 2

THE SOIL SOLUTION

Soil solution represents the dominant site of chemical reactivity in soils. It is the natural medium for plant growth, and it represents the chemical fraction immediately available in the environment. Therefore, the soil solution is broadly addressed in most aspects of soil chemistry. A key feature distinguishing soil solution chemistry from the approaches to understanding chemical reactivity in soils that focus on the soil solid phase is the emphasis on obtaining a model—conceptual, experimental, or computational—that realistically mimics soil solution in situ.

Soil solution chemistry provides useful approaches to characterizing soil chemical processes of importance to agricultural and environmental science, provided certain basic assumptions are met. First, if soil solution represents the natural medium for plant growth, then soil solution analysis allows for prediction of plant response to chemicals occurring in the soil environment. Second, if soil solution can be related to mobile water in the soil environment, then soil solution composition can be used to predict the forms and amounts of chemicals that may reach ground and surface water through transport from the soil environment. Third, if soil solution approaches a steady state relative to the soil solid phase, then soil solution composition can be used to predict solid phase components controlling chemical distribution in soil. Fourth, if soil solution composition controls bioavailability to soil microbes, soil solution monitoring with time provides information as to the concurrent processes of sorption and metabolism. The validity of these various assumptions is dependent on the way that soil solution is conceptualized and how that concept is translated into an operational method whereby soil solution can be obtained and its composition expressed in a meaningful way.

The greatest progress in soil solution chemistry has occurred in proving the veracity of obtaining and analyzing "unaltered" soil solution for diagnosis of plant growth response and for predicting solid phase control of soil solution composition.

6

This has occurred through the coupling of displacement and analysis of soil solution with consideration of ion speciation and complexation, and expression of soil solution composition in thermodynamic terms. In conjunction with nutrient culture studies, applied soil solution chemistry has proven a useful interface between soil fertility and plant physiology for the description of plant mineral nutrition and phytotoxicity. Applications of soil solution chemistry relative to chemical distribution in the solution phase are shared with the disciplines of geochemistry and aquatic chemistry. The use of applied soil solution chemistry for describing dynamic processes of solute flow and chemical metabolism is more recent and is restricted to relatively few examples.

2.1 HISTORICAL PERSPECTIVE

Soil solution chemistry emerged as a subdiscipline of soil chemistry in the late 1960s through the convergence of two avenues of research that had occupied soil chemists for several decades. The first line of research was investigation of soil solution as the natural medium for plant growth; soil scientists sought to relate diverse plant growth response across environments to nutrient and trace metal intensity in the soil environment. This led to the development and verification of methodology for obtaining and analyzing the "true" soil solution. The second was the application of the theory of electrolytic solutions to soil solutions. This approach allowed soil solution composition to be expressed in a meaningful manner, both conceptually and practically.

2.1.1 Early Chemical Edaphology

Some of the earliest fundamental studies of plant nutrition—the pot culture experiment of van Helmont in 1652 and the solution culture studies of Woodward in 1699—provided evidence as to the importance of soil solution as a nutrient source for plant growth. These studies, however, predated the development of a base of chemical science that would allow their correct interpretation. Beginning with the work of de Saussure, first published in 1804, and culminating with the experiments of von Sachs and Knop in the 1860s, the ability of plants to gain nutrition from solution culture was established (Epstein, 1972; Russell, 1973). This, coupled with increased understanding of the chemical nature of soils and its relationship to plant growth, led to the late nineteenth century recognition of soil solution as the natural medium for plant growth.

Schloesling in 1866 initiated the first compositional analysis of "unaltered" soil solution. He obtained soil solution from soil at field moisture contents (19.1% moisture) using a displacing solution of carmine-colored water. This allowed him to discriminate the soil solution from the moving front of displacing solution as it filtered through the soil (Russell, 1973). This procedure, with slight modification, remains one of the most reliable techniques for obtaining intact soil solution.

In the United States, scientists influenced by the German school of agricultural chemistry—which emphasized chemical analysis of plants and soils for determination of crop nutrient needs and fertilizer formulation—instituted studies to describe the chemical nature of soils as reservoirs of plant nutrients. Cameron and coworkers at the Bureau of Soils of the United States Department of Agriculture recognized the inadequacy of whole soil and soil extract analysis for description of plant nutritional analysis early in the twentieth century. These workers attempted to describe the "crop-producing power of soils" by analyzing water extracts of soils and soil solution displaced by centrifugation of saturated soils for Ca, K, and P content. Cameron (1911) made this conclusion from this early work: "Many attempts have been made to extract the solution naturally existing in soil and to analyze it. The results obtained have not been very satisfactory, owing mainly to the mechanical difficulties involved."

Despite the difficulties of obtaining, analyzing, and interpreting soil solution composition, the promise of this approach for unraveling questions concerning differential plant response across soil environments sustained intense research activities on soil solution chemistry that lasted through the 1930s.

2.1.2   Sampling the "True" Soil Solution

Empirical evidence early in the twentieth century indicated the inadequacy of water extracts of soil as appropriate models of soil solution. Table 2.1 presents selected data of Burgess for chemical composition of soil solution and water extracts of soils. The chemical intensity of electrolytes in soil solution is a function of various processes—solution of both freely soluble and sparingly soluble salts, ion

TABLE 2.1   Composition of Pressure-Displaced Soil Solution as Compared to That of 1:5 Soil–Water Extracts[a]

	Solute Concentration (μg/g water-free soil)					
	Ca	Mg	K	NO_3	PO_4	SO_4
	Hanford fine sandy loam					
Soil solution	35	8	7	51	0.3	23
1:5 soil–water extract	83	20	27	66	9	65
	Kimbal fine sandy loam					
Soil solution	26	5	4	25	0.2	21
1:5 soil–water extract	54	14	21	45	4	46
	Madera fine sandy loam					
Soil solution	54	18	13	90	0.7	35
1:5 soil–water extract	78	29	62	123	50	79

[a]P.S. Burgess, 1922, *Soil Science,* 14:191–216.

exchange, and sorption—for which dilution has an uncertain effect on the various anion and cation components of the aqueous phase. Therefore, water extracts of soil do not provide consistent models of soil solution composition. Despite extensive efforts to rationalize chemical composition of water extracts of soil with soil solution composition,[1] the approach was largely abandoned in favor of techniques seeking to obtain unaltered soil solution.

Morgan (1916) and Burgess (1922) summarized the various approaches for obtaining soil solution as drainage waters (lysimetry), soil extracts, artificial roots (suction lysimetry), centrifugation, column displacement, pressure displacement (including direct pressure and hydraulic pressure displacement), and filter absorption. The investigation of such a wide variety of techniques points to the difficulty in reliably obtaining "true" soil solution in a practical, useful manner. Assessing the reliability of techniques for obtaining soil solution generally involved showing that changes in such things as pressure, centrifugation time, or nature of displacing solution either had no effect on solution composition, or influenced solution composition in a predictable manner. Burgess, for example, found specific resistance and Ca and Mg content of soil solutions obtained by direct pressure to be unaffected by pressures up to 16,000 psi (111 MPa).

Demonstrating the veracity of a technique for obtaining "true" soil solution was more problematic, because direct measurement of soil solution composition was— and remains—impossible. Ischerekov in 1907 demonstrated constancy of soil solution composition by measuring total salt content of successive fractions of displaced soil solution. This approach has been adopted in most subsequent studies of soil solution displacement methodologies to indicate the veracity of the technique employed as a measure of "true" soil solution composition.[2] Burd and Martin (1923) demonstrated that successive increments of displaced soil solutions were largely invariant in composition (Table 2.2). Soil solution composition as determined by centrifugation and displacement techniques is compared in Table 2.3. One-to-one correspondence between differing techniques may indicate that the "true" soil solution has been sampled.

Constant composition for successive soil solution fractions obtained by column displacement, pressure extraction, or centrifugation and a good correlation between differing displacement techniques proves only that the techniques employed are reproducible and provide comparable models of soil solution—not that they obtain soil solution per se. Soil water is attracted to the soil matrix at increasingly greater energies proceeding from the center of water-filled soil macropores to the water films coating soil particles (Fig. 2-1). Techniques for obtaining soil solution commonly recover < 50% of soil water for soils at field moisture content. Thus, equating displaced soil solution composition with in situ soil solution composition is conceptually difficult, especially when the effect of an electric double layer may cause differential distribution of anions and cations through the soil aqueous phase. Consequently, various operational definitions and conceptual models of soil solution are given to justify differing applications of soil solution chemistry to studies of chemical reactivity in soils (see section 2.2, Concepts of Soil Solution). The veracity of any soil solution displacement technique for measuring "true" soil

TABLE 2.2   Constancy of Composition for Successive Increments of Soil Solution Obtained by Column Displacement[a]

Increment[b]	Specific Resistance	Ca	Mg	K	NO₃	Cl	PO₄[c]
	ohms			*μg/mL*			
1	103	1280	278	250	5356	720	
2	104	1340	284	228	5183	720	
3	103						
4	103						10.9
5	102						
6	103						
7	103	1360	278	244	4885	680	
8	108						
9	104						11.1
10	104						
11	106						
12	107	1260	226	276	4860	680	
Average	104	1310	267	250	5071	700	11.0

[a]Sandy loam soil at 7.3% moisture equivalent; Burd and Martin, 1923, *J. Agric. Sci.,* 13:265–295.
[b]Successive 5 mL increments of displaced soil solution.
[c]Composites of the third through sixth and the eighth through eleventh increments, respectively.

solution composition is validated by the predictive ability of the method used. Within this definition of veracity, soil solution studies prior to the 1960s largely failed because the methodologies employed were unable to yield data whereby chemical reactivity across environments could be usefully expressed. During this earlier period of intense interest in the soil solution, development and application of the theory of electrolytic solutions to soil systems provided the foundation for future success in soil solution chemistry.

2.1.3   Applying the Theory of Electrolytic Solutions to Soils

Debye and Hückel in 1923 successfully described the electrical potential associated with a charged sphere in solution and consequently formulated the *Debye-Hückel limiting law* which allowed for the computation of single-ion activity coefficients. Thus, the framework for the modern theory of electrolytic solutions was established whereby activity of ions in solutions could be described in quasi-thermodynamic terms. The eventual successful application of Debye-Hückel theory for expression of ion composition of soil solutions was proceeded by a period of research that related soil chemical phenomena to the chemistry of electrolytes and served to establish the conceptual utility of single-ion activities in soil chemistry.

By the 1930s, the glass electrode had been widely adapted for the measurement of soil pH. The commonly observed "suspension effect"—a higher pH is observed

TABLE 2.3 Comparison of Soil Solution Composition Obtained by Various Methods

	pH	EC	Ion Difference[a]	K	Na	NH$_4$	Ca	Mg	Al	NO$_3$	Cl	PO$_4$	SO$_4$
		dS m^{-1}	%						mmol L^{-1}				
Lucedale sandy loam (Rhodic Paleudults)[b]													
Column displacement	5.12			0.46		0.01	1.70	0.79	0.0015			0.0013	0.26
Centrifugation	5.02			0.45		0.01	1.58	0.75	< 0.0004			0.0001	0.25
Centrifugation with an immiscible liquid	5.04			0.43		0.01	1.68	0.75	< 0.0004			0.0001	0.25
Memphis silty clay loam (Typic Paleudalfs)[c]													
Vacuum displacement	5.91	1.19	4.8	0.616	0.152	0.035	5.10	0.884		9.48	3.34	< 0.002	0.272
Centrifugation	6.13	1.20	3.6	0.580	0.160	0.037	5.29	0.998		8.03	3.79	< 0.002	0.279

[a]Ion difference = |Σcations − Σanions|(100/(Σcations + Σanions), assuming insignificant hydrolysis or protonation to shift ion distribution.
[b]Adapted from F. Adams, C. Burmester, N.V. Hue, and F.L. Long, 1980, *Soil Sci. Soc. Am. J.* 44:733–735.
[c]Adapted from J. Wolt and J.G. Graveel, 1986, *Soil Sci. Soc. Am. J.* 50:602–605.

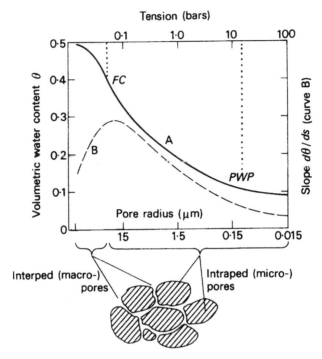

FIGURE 2.1 Idealized graphs of (*a*) the water characteristic function and (*b*) specific water-holding capacity for a clay soil [White, 1987]. The chemical potential of water attracted to soil particles (the matric potential, ψ_m) is expressed as tension—the negative pressure difference between soil water and free water.

for electrodes when immersed in the supernatant of a soil-water slurry than when immersed in a soil-water suspension—led to decades of investigations as to the nature of the added electrical potential generated from suspended soil particles. One consequence of these investigations was the consideration of electrochemical potentials and their expression across phases of varied composition. This led soil scientists to realize that for phases of varied composition—such as soil solution and solid phases—resolution of electrochemical potential into separate electrical and chemical terms was not possible within a thermodynamic context.

Concurrent with investigations relative to the suspension effect, a debate occurred over an issue of even broader practical concern: the nature of plant nutrient uptake by the plant root from the soil. The *contact-exchange theory*, promulgated by Jenny and coworkers in the late 1930s, maintained that plant-essential nutrients were obtained predominantly through root-mining of nutrient-rich zones—ion swarms—in close proximity to soil particles. Although the contact-exchange theory was dismissed as a dominant mechanism for explaining plant mineral nutrition, scientific debate of this issue again led to a consideration of electrochemical potential as a governing principle for description of ion activities in solutions (the ambient soil solution) and suspensions (water and colloids in close association at the surface of soil particles). In brief, the electrochemical potential of an ion in solution (μ_i^{el}) is comprised of chemical and electrical components:[3]

$$\mu_i^{el} = \mu_i^{\circ} + RT \ln A_i + RT \ln \emptyset_i$$

The inability to resolve electrical and chemical potentials of an ion in solution as separate terms has led some to dismiss the utility of ion activity (A_i) as a useful descriptor of ion availability in soils. Khasawneh (1971), however, summarizes the view shared by many soil solution chemists: "Fortunately, solutions extracted from soils at optimum moisture content represent solutions that are relatively far from charged soil surfaces, where the electric field of surface charges has dropped to near zero. Thus, the chemical composition of the soil solution, when expressed in ion activity units, is a measure of the electrochemical potential of ions in the soil-water system."

Both the concept of single-ion activities and the application of chemical potentials of ions divorced from their electrical effect fall outside of the strictures of thermodynamics. The adaptation by soil solution chemists of the concept of single-ion activities has been dominantly influenced by the strong empirical evidence of experimental utility; the concept yields diagnostically meaningful descriptions of the chemical and biological availability of chemicals in the soil environment.

2.2 CONCEPTS OF SOIL SOLUTION

The conceptual model of the soil environment introduced in Chapter 1 considers soil as an intimately interacting network of compartments exchanging mass and energy (see Fig. 1-1). In fact, the soil environment represents a continuum of phases exhibiting indistinct interfaces at the molecular level. Solutes in the soil aqueous phase may be associated with bound water at the surfaces of soil colloids, free water percolating through soil macropores, water in the free space of plant roots, or immobile water in soil micropores. Herein lies the fundamental conceptual difficulty in bridging models employed by the soil solution chemist with observations of chemical reactivity and transport observed in the natural soil environment: there is a distinct difference between what soil solution is in fact—the aqueous phase of the soil and its solutes—and what is termed "soil solution" in practice on the basis of conceptual and practical definitions. Conceptual clarity is therefore essential to the application and interpretation of soil solution chemistry as the approaches utilized and inferences drawn are strongly tinged by what is taken as "soil solution" in any given instance.

2.2.1 Effect of the Electrical Double Layer

Perhaps the most physicochemically unique feature of soils as chemically reacting systems is the manifestation of net negative charge by the soil solid phase. The accumulation of counter charge by a soil particle or surface in order to maintain electrical neutrality leads to a differential distribution of anions and cations in the surrounding soil solution. This effect may be described from Donnan equilibrium, but it is more completely defined on the basis of double-layer theory.[5]

When interpreted from an equilibrium perspective, the effect of a solid phase exchanger is the exhibition of an average electrical potential difference—the Donnan potential—for an ion passing between an exchange phase and a surrounding aqueous phase. The net effect for a soil solution is the prediction of a volume from which anions are effectively excluded—the Donnan free space—such that ion distribution is differentially manifested in the bulk soil solution phase and the soil solution within the Donnan free space. The degree to which components in the Donnan free space participate in a soil process of interest—nutrient availability, solute transport, biodegradation—and the degree by which a given soil solution displacement and analytical technique accommodates the effect of the Donnan free space determine the suitability of a given application of soil solution chemistry.

More realistically, the equilibrium balance between the countervailing effects of Coulombic forces distributed across a plane parallel to a charged soil surface and the osmotic potential of the surrounding soil solution describes the diffuse distribution of cations and anions in solution (Fig. 2-2). The combined Gouy-Chapman and Stern models describe the surface charge and compact (Stern) and diffuse layers of counter ions collectively known as the electrical or diffuse double layer (DDL). Of practical significance to the soil solution chemist is the effective thickness of the diffuse layer (d_{ex}, nm) according to

$$d_{ex} \cong \frac{2 \times 10^9}{z\sqrt{\beta C_0}}$$

(a)   **(b)**

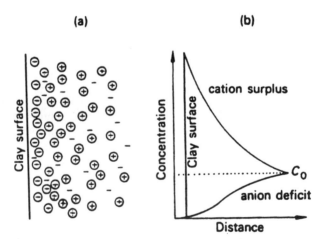

FIGURE 2.2   (*a*) Ion distribution at a negatively charged clay surface. (*b*) The concentration profiles associated with differential anion and cation distribution with distance from the clay surface [White, 1987].

where z is cation or anion valence, β is the DDL constant ($= 1.084 \times 10^{16}$ m mol^{-1} at 298.15°K), and C_0 is the electrolyte concentration of the bulk soil solution (mol m^{-3}). As evidenced by the example of Table 2.4, valence of the dominant solution phase electrolyte and ionic strength of bulk soil solution markedly effect d_{ex}. The effect of the DDL is more strongly manifested for dilute soil solutions.

As the operational method for obtaining soil solution is altered—as for example when forces and/or times of centrifugal displacement of soil solution are varied—the anticipation is that different proportions of the DDL will be sampled. The degree to which this is in fact observed influences the interpretative strength achievable for a given application of soil solution chemistry. Figure 2-3 illustrates the effective pore volume drained at midpoint as a function of midpoint driving pressure, along with the approximate volume of the DDL as a percent of pore volume, for centrifugal displacement of a Harwell silt loam.[6] At low midpoint driving pressures (≤ 0.5 MPa) complete drainage is limited to pore diameters > 750 nm and an inconsequential fraction of the DDL is sampled; thus, the soil solution obtained is representative of C_0—the bulk soil solution electrolyte composition. The fraction of the DDL that is sampled may increase with higher centrifugation speeds (upwards to 20,000 rpm in this example), longer times of centrifugation, or reduced soil water content. Soil pores, however, typically exhibit equivalent pore size diameters $> 1,000$ nm. Therefore, with the possible exception of high-speed centrifugal displacement and high-pressure vacuum extraction techniques, soil solution composition does not significantly sample the DDL and can be considered representative of homogeneously distributed electrolytes occurring in the bulk aqueous phase of the soil.[7]

2.2.2 Soil Solution and Soil Water

Definitions of Soil Water. What "soil solution" means in either a conceptual or operational context is determined by the fraction of soil water to which soil solution composition is related. Therefore, description of the portion of soil water sampled by a soil solution displacement method is important for interpreting soil solution composition and its relationship to solute transport and bioavailability. The

TABLE 2.4 Calculated Diffuse Layer Thickness for Different Concentrations of 1:1 and 2:1 Electrolytes

Electrolyte Concentration, C_0 (mmol cm^{-3})	Effective Diffuse Layer Thickness, d_{ex} (nm)	
	NaCl	CaCl$_2$
0.1	1.94	1
0.01	6.2	3.2
0.001	19.4	10.1

From R. E. White, 1987, *Introduction to the Principles and Practice of Soil Science,* 2nd ed., Blackwell Scientific Publications, Oxford.

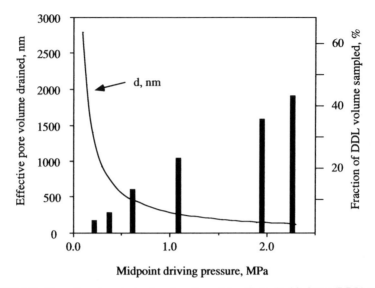

FIGURE 2.3   Pore diameters drained and portion of the dilute double layer (DDL) sampled as functions of variation in midpoint driving pressure for centrifugal displacement of soil solution.

water characteristic function is useful in this context, as it describes the relationship between the water *matric potential* (Ψ_m) and soil water content (Θ). As illustrated in Figure 2-4, the soil water characteristic function (or moisture characteristic) varies among soils. Dissimilarities in water characteristic functions among soils are especially apparent when soils vary markedly in texture or organic carbon content. The fraction of soil water sampled and the water yield obtained varies when the soil solution displacement procedure is applied to the same soil at different water content, or to dissimilar soils at the same water content.

The partial molar free energy or chemical potential of water (see section 3.3.5, Chemical Potential) is affected by pressure, temperature, soil water content, dissolved solutes, and gravity.[8] The potential energy of water is commonly described in terms of water potential (Ψ_w) which is the chemical potential of water expressed on a volume basis. The water matric potential is, thus, that component of Ψ_w attributable to the change in free energy of water that occurs as a result of soil water content (Θ). The matric potential is due to the combined effects of surface tension and surface adsorption of water within pores interacting with soil solids. Matric potential increasingly dominates Ψ_w as Θ decreases.

The water characteristic function is constructed by measuring Θ with change in tension placed on soil (a moisture release curve). The soil tension associated with Θ for a particular soil is equivalent to the Ψ_m counteracting the applied tension. The tension is inversely related to the soil pore radius of the largest pores holding water at that tension, when soil shrinking is negligible with drying (Fig. 2-1). The slope of the water characteristic function at any point ($d\Theta/d\Psi_m$) describes both the

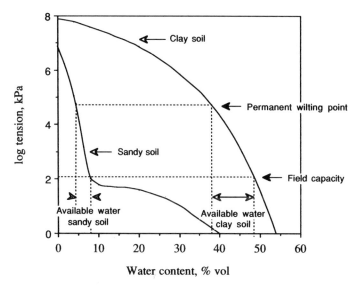

FIGURE 2.4 Relationship between soil water content, water tension, and available water for a sandy soil and a clay soil. [After Mengel and Kirby, 1979.]

water yield and the pore sizes drained for an infinitesimal change in tension. The greater the value of $d\Theta/d\Psi_m$, the greater is the volume of pores holding water within the range of $\delta\Psi_m$ (White, 1987). The relationship of pore radius to Ψ_m allows for description of soil water content in terms of soil pore size classes holding water against a given tension. A convenient, albeit somewhat arbitrary, grouping of soil pores into size classes is shown in Table 2.5 relative to the dominant physical processes with which water occupying these pores is associated. The utility in using pore size ranges to classify soil water is the ability to distinguish micro- and mesopore capillary water subject to diffusive flow through the soil profile from noncapillary *macropore water* that contributes to channelized or bypass flow in soils.

Water transport through the soil profile, however, is a dynamic nonequilibrium process, rather than a static process as implied through the use of pore size terminology. In some soils, continuous pores within the mesopore range have been observed to be important contributors to saturated flow of water. Thus, the static classification of water by pore size ranges can lead to misinterpretation regarding how initial soil moisture content and rate of water supply interact with pore size to influence bypass flow. This leaves case-specific direct measurements of water transport in soils as the only method whereby water transport can be unambiguously described.

However direct measurements of water transport, such as with breakthrough curves, do not lead to unambiguous interpretation as to the physical process of water flow. The concept of *mobile and immobile water* is employed to interpret breakthrough curves by using a variable to fix the proportion of soil water participating in the observed transport of a solute. Although widely used, the concept of

TABLE 2.5 Suggested Characteristics of Three Soil Porosity Classes[a]

Soil Porosity Class	Pressure Range, kPa	Equivalent Pore Diameter Range, um	Dominant Phenomena
Micro	< -30	< 10	Evapotranspiration; matric pressure gradient for water distribution
Meso-	-30 to -0.3	10 to 10,000	Drainage; hysteresis; gravitational driving force for water dynamics
Macro-	> -0.3	$> 1,000$	Channel flow through profile from surface ponding and/or perched water table

[a]R. J. Luxmoore, 1981, Micro-, meso-, and macroporosity of soil, *Soil Sci. Soc. Am. J.*, 45:671.

mobile/immobile water obscures the continuous nature of soil water as well as the effect of time on the distinction of water as mobile or immobile. Bouma and co-workers (Booltink and Bouma, 1991; Van Shipout et al., 1987) have described *internal catchment*—noncontinuous macropores within the soil which can act as internal soil reservoirs of infiltrating water as water percolates through soils—that accounted for $\approx 33\%$ of water balance in well-structured clay soil. The assignment of this portion of soil water as mobile versus immobile is dependent on time—as water percolates through soil, the internal catchment acts as a sink of immobile water during the initial stages of transport that can be remobilized at a later time.

Solute Transport in Relation to Soil Water. Bypass flow water exhibits short contact times with the soil matrix as compared to capillary water that is subject to diffuse flow through the soil profile; the contact time of internal catchment water in soil determines the degree to which it approximates diffuse versus bypass water. Laboratory soil solution displacement methods (Chapter 5) are typically designed to approximate diffuse water in quasi-equilibrium with the soil solid phase. Thus, the applicability of soil solution composition determined by such methods to problems of chemical transport is limited to the degree to which diffuse water is mobile in a given soil system. Soil systems at quasi-equilibrium represent an optimized case for retardation of solutes by soil because time is not a constraint to the degree of chemical reactivity between solutes and the contacting soil solid phase. Soil solution studies designed from a dynamic perspective can be used to gauge the rate-dependent nature of solution-phase/solid-phase reactions. Rates of reaction relative to rates of water flux through soil can then be used to quantitate the degree that solid phase reactivities may retard solute transport through the soil profile.

Field lysimetry (Chapter 6) results in sampling of soil water of uncertain origin. Lysimeter waters may represent combinations of bypass water originating from recent rainfall or irrigation events and diffuse or internal catchment waters exhibiting varied contact times with the soil matrix. Consequently, compositional analysis of lysimeter solutions and soil solution obtained by laboratory displacement may substantially differ. This has been demonstrated for soil solutions obtained from forested Typic Cryohumods by centrifugal displacement and from low-tension plate lysimetry (Zabowski and Ugolini, 1990). Displaced soil solutions exhibited distinct seasonal patterns of nutrient ion composition, whereas lysimeter soil solutions typically had lower nutrient ion compositions and did not exhibit seasonal variation. These differences were attributed to greater biotic influence on displaced soil solution, that sampled diffuse soil water, in contrast to lysimeter soil solutions, where relatively short residence times of mobile water did not reflect the influence of soil biota.

2.2.3 Operational Definition of Soil Solution

No one approach to obtaining soil solution is appropriate to all applications. Uncertainties regarding the physical origin of lysimeter water result in uncertain interpretations regarding chemical processes governing the chemical composition of lysimeter soil solutions. In contrast, uncertainties concerning the contribution of solutes in diffuse soil water—as measured by laboratory displacement of soil solution—result in uncertain conclusions regarding flux of solutes through the soil profile. Considerations of biologically consequential processes relating to plant nutrient availability, phytotoxicity, and soil metabolism are best related to chemical composition of diffuse soil water as reflected in composition of displaced soil solution. Considerations of chemical transport, however, are more clearly reflected in soil solutions obtained from lysimetry—provided the solutions obtained represent mobile water in the soil environment. "Soil solution" is, therefore, operationally defined by the methodology employed for its acquisition and by the analytical methodology employed, as will be discussed in subsequent chapters.

In Chapters 5 and 6, various methodologies for obtaining "unaltered" soil solution are described, each having inherent biases that influence the interpretation of the result obtained. So the soil solution chemist must clearly describe the methodologies and assumptions employed for an application involving displacement and analysis of soil solution. Conceptual vagueness concerning operational definition of soil solution remains as a primary limitation to the wider application of soil solution chemistry.

NOTES

1. For example, see the extensive discussion of Hoagland et al. (1920) which developed and justified the use of soil water extracts for assessment of plant nutrient status of California soils. The use of 1:1 and 1:5 soil-to-water extracts developed by Hoagland and coworkers continue, with modification, to find widespread use for assessing fertility status of arid region soils.

2. For a discussion of total salts as a measure of veracity of soil solution see F. Adams, 1974.

3. The term $\mu_i^\circ + RT \ln A_i$ describes chemical potential of the i-th component in terms of the standard state chemical potential (μ_i°) and activity (A_i). The electrical potential of the i-th component is described by the electrostatic activity coefficient ($\emptyset_i$) that corrects chemical activity for electrical effects.

4. This standard definition of "soil solution" is from the *Glossary of Soil Science Terms,* 1987, Soil Science Society of America (SSSA), Madison, WI.

5. Useful descriptions of double-layer theory and its application to soils are found in Singh and Uehara, 1986; and White, 1987, pp. 97–100.

6. The example presented here is developed from the data of Kinniburgh and Miles, 1983. The functional relationship of pore diameter drained to midpoint driving pressure is calculated from,

$$d = \frac{4\sigma \cos \Theta}{T_c \rho}$$

which describes capillary diameter (d, nm) as a function of surface tension ($\sigma = 0.007$ N m^{-1}), contact angle of wetting ($\Theta = 0°$), the capillary pressure (T_c, MPa) counteracting the centrifugal driving pressure, and the bulk phase density ($\rho = 1$). The effective DDL thickness (d_{ex}) was modeled for Harwell silt loam using the program GOUY (Burnette and Schwab, 1987) with the following inputs:

bulk phase concentration, $C_0 = 0.005$ mol L^{-1}

scaled midplane potential, $Y_b = 0.01$ (weakly interacting particles)

surface charge potential, $\Gamma_s = 1.67$ μmol_c m^2 (CEC = 100 cmol$_c$ kg^{-1} and

surface area = 6×10^5 m^2 kg^{-1})

The effective DDL thickness (d_{ex}) for the case of a monovalent ion is 29 nm. The fractional volume of DDL sampled as a percent of effective pore volume was calculated for cylindrical pores as:

$$\text{DDL fractional volume, } \% = [1 - (r - d_{ex})^2/r^2] \times 100$$

where r is effective pore radius.

7. Other factors relating to high pressure displacement probably have a more consequential influence on soil solution composition. The table below summarizes the effect of relative centrifugal force (RCF) on soil solution anion concentrations for a Typic Haplorthod Oa horizon; increased F$^-$, Cl$^-$, and SO$_4^{2-}$ in soil solution with increased RCF was attributed to release from destructive collapse of microbes and plant roots (Ross and Bartlett, 1990).

Soil Solution Anions Displaced with Increasing Speed of Centrifugation

Centrifuge RCF	F^-	Cl^-	NO_3^-	SO_4^{2-}
$m\ s^{-2}$		$\mu mol\ L^{-1}$		
390	9c*	126b	1040a	102b
880	53bc	208b	1046a	117ab
1560	36bc	240ab	1062a	127ab
2400	117a	374a	1056a	133a

*Values in a column followed by the same letter are not significantly different (Duncan's multiple range test, $P < 0.05$).

8. The chemical potential of water (μ_w) is described (Nye and Tinker, 1977) as

$$\mu_w = (\overline{V})_{T,P,n_i} dP - (\overline{S})_{T,P,\Theta} + \left(\frac{\delta\mu_w}{\delta\theta}\right)_{T,P,n_i} d\Theta + \left(\frac{\delta\mu_w}{\delta n_i}\right)_{T,P,\Theta} dn_i + Mgh$$

where μ_w is a function of

the change in partial volar volume ($\overline{V}$) with pressure (P)

the change in partial molar entropy ($\overline{S}$) with temperature (T)

the change in μ_w with respect to moisture content (Θ), the matric effect

the change in μ_w with respect to the number of moles of solute (n_i), the osmotic effect

the effect of gravity as described by the mole weight of water (M), the acceleration of gravity (g), and the change in height (h) from some reference height

Water potential (Ψ_w) is the volume-weighted chemical potential

$$\Psi_w = \frac{(\mu_w - \mu_w^\circ)}{(\overline{V})}$$

where μ_w° refers to the chemical potential of water in the standard state of pure water.

CHAPTER 3

CHEMICAL STATICS AND DYNAMICS APPLIED TO SOIL SOLUTIONS

The translation of chemical models to the environment is hampered by the complexity of natural systems, such as chemically and physically heterogeneous soil systems. A model, however, need not give one-to-one correspondence with natural systems to be useful. It is sufficient if a simplified model provides useful generalizations whereby the processes regulating the chemical composition of natural systems may be conceptualized.

Statics—the study of systems at equilibrium, and dynamics—the study of movements within systems or kinetics offer important, but differing, insights into soil chemical processes. The utility of either the static or dynamic approach to soil chemical reactivity requires that their limitations for given applications be clearly discerned.

3.1 LIMITATIONS TO EQUILIBRIUM AND KINETIC APPROACHES

For soils, chemical thermodynamic equilibrium is a valid model for the generalization of many processes governing chemical reactivity. The free energy concept of chemical thermodynamics allows for prediction of the nature and extent of processes that approach equilibrium. The soil is an open dynamic system in a state of chemical, physical, and biological flux; therefore, the visualization of soil systems as exhibiting quasi-equilibrium requires simplifying assumptions. The applicability of these assumptions is very much related to the scale of resolution required for addressing the hypotheses being tested.

This is exemplified by the case where a quasi-equilibrium approach is used to predict the solid phase mineral controlling a solution phase component in a soil B horizon. The B horizon is a zone of illuviation where mass accumulates with time from vertical fluxes of material from overlying soil horizons. Despite the dynamic nature of this soil compartment, the assumption of localized equilibrium can give insight as to processes occurring within the B horizon provided

1. Equilibrium is approached for the sorbed and solution phase distribution of the components of interest, relative to fluxes of these components through the B horizon and their distribution among other soil phases over the time scale of resolution considered.
2. There is a spatial homogeneity of the B horizon, or the soil subsample analyzed represents the average of normally distributed properties of the system.

When representative subsamples of the B horizon soil are homogeneous with respect to properties such as temperature, pressure, bulk density, and electrolyte content, the assumption of localized equilibrium holds and the analysis of displaced soil solution provides information translatable to the soil considered.

If the B horizon soils for this case exhibit similarity at the family level—if they are taxonomically similar in B horizon texture, mineralogy, and the climactic regime under which they have been weathered—the assumption of quasi-equilibrium is appropriate over the time scale of soil genesis (years to centuries). Over shorter time scales (seasons) the assumption may be invalidated by localized events. For example, spodic B horizons may exhibit control of soil solution Al through quasi-equilibrium precipitation–dissolution reactions governed by a soil solid phase oxyhydroxide mineral such as gibbsite; yet on a seasonal basis material fluxes through the B horizon due to a transient event, such as snowmelt, may result in short-term departures from control over soil solution Al by the soil mineral phase (Wolt, 1990).

Kinetic approaches, while conceptually more appropriate to dynamic systems, also may have limits in applicability to soils. First, soil systems frequently exhibit heterogenous kinetics and must be simplified by homogeneous kinetic models. Second, reaction rates for key processes of interest may be either too rapid or too slow to monitor experimentally, as is the case for ion exchange occurring at reaction times on a microsecond scale or mineral weathering reactions measurable only on an annual scale. Third, kinetic studies of soil reactions frequently lack physical (spectroscopic) evidence as to the occurrence of postulated reaction products or intermediates. Because of these limitations, dynamic studies of soil systems typically involve apparent rates of reaction. Such data provides empirical description of chemical reactivity but cannot provide direct mechanistic evidence of reaction paths.

3.2 CHEMICAL EQUILIBRIUM AND KINETIC MODELS COMPARED

The concentration of a component in soil solution is governed by chemical reactions occurring in solution, fluxes through the solution, and phase transfer.[1] To compare idealized equilibrium and kinetic models for a stationary state system, consider the following simple chemical reaction,

$$A \rightleftharpoons B$$

occurring in soil solution in the absence of transfer between other soil phases (Fig. 3-1).

3.2.1 Stationary State Thermodynamic Equilibrium Model

For the assumption of a stationary state *thermodynamic equilibrium model,* the soil solution is viewed as a closed system; the fluxes of A and B through the system are zero ($dn_{A_{in}}$, $dn_{B_{in}}$, $dn_{A_{out}}$, $dn_{B_{out}}$ = 0, Fig. 3-1) temperature and pressure are fixed, and the system is homogeneous with respect to the distribution of A and B. The equilibrium constant (K_{AB}) and the initial concentration of one component (C_{Ai}) must be known to express the equilibrium distribution of A and B (C_A and

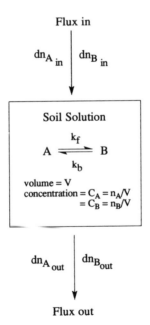

FIGURE 3.1 Simplified representation of soil solution as an open system receiving fluxes from its surroundings.

C_B). The effective equilibrium constant is

$$K_{AB} = \frac{C_B}{C_A} \qquad [3\text{-}1]$$

and the condition for mass balance is

$$C_{A_i} = C_A + C_B \qquad [3\text{-}2]$$

Thus, the equilibrium concentrations can be expressed as

$$C_A = \frac{C_{A_i}}{1 + K_{AB}} \qquad [3\text{-}3]$$

and

$$C_B = C_{A_i} - C_A \qquad [3\text{-}4]$$

3.2.2 Steady State Kinetic Model

A *steady state kinetic model* describes the stationary state for an open system; the fluxes of A and B through the system are constant, temperature and pressure are fixed, and the system is homogeneous with respect to A and B. Identification of transformations of A and B requires rate constants for the forward and backward reactions

$$A \underset{k_b}{\overset{k_f}{\rightleftharpoons}} B$$

The rate constants for this reaction take the form of simple velocity expressions,

$$v_f = k_f C_A \qquad [3\text{-}5a]$$

and

$$v_b = k_b C_B \qquad [3\text{-}5b]$$

where v_f and v_b are velocities in the forward and backward directions, respectively. The changes in concentration with time become,

$$\frac{dC_B}{dt} = k_f C_A - k_b C_B \qquad [3\text{-}6a]$$

and

$$\frac{dC_A}{dt} = -k_f C_A + k_b C_B \qquad\qquad [3\text{-}6b]$$

The necessary information for solution of the steady state model consists of the mole fluxes of A and B per unit volume,

$$dn_A = r\overline{C}_A \qquad\qquad [3\text{-}7a]$$

and

$$dn_B = r\overline{C}_B \qquad\qquad [3\text{-}7b]$$

where the rate of solution transfer (r) is the rate of flow divided by soil solution volume and $\overline{C}$ represents an inflow concentration. For the steady state situation,

$$\frac{dC_A}{dt} = \frac{dC_B}{dt} = 0 \qquad\qquad [3\text{-}8]$$

If rates of inflow and outflow are equal,

$$\frac{dC_A}{dt} = r\overline{C}_A - k_f C_A + k_b C_B - rC_A \qquad\qquad [3\text{-}9a]$$

$$\frac{dC_B}{dt} = r\overline{C}_B + k_f C_A - k_b C_B - rC_B \qquad\qquad [3\text{-}9b]$$

If equations 3-9a and b are applied to the soil solution composition illustrated in Figure 3-1, the concentration changes with time (dC_A/dt and dC_B/dt) are defined by the rates of flux into ($r\overline{C}_A$ and $r\overline{C}_B$) and out of ($-rC_A$ and $-rC_B$) the compartment, and by rates of transformation within the compartment ($k_f C_A$ and $k_b C_B$). Equating 3-9a with 3-9b for the steady state condition yields

$$\overline{C}_A - C_A = \overline{C}_B - C_B \qquad\qquad [3\text{-}10]$$

Substitution of equation [3-10] into equation 3-9a and b provides steady state expressions for C_A and C_B,

$$C_A = \frac{r\overline{C}_A - k_b(\overline{C}_A + \overline{C}_B)}{(k_f + k_b + r)} \qquad\qquad [3\text{-}11a]$$

and

$$C_B = \frac{r\overline{C}_B - k_f(\overline{C}_A + \overline{C}_B)}{(k_f + k_b + r)} \qquad [3\text{-}11b]$$

Therefore, for the steady state model, the reaction quotient (Q) is

$$Q = \frac{C_B}{C_A} = \frac{r\overline{C}_B - k_f(\overline{C}_A + \overline{C}_B)}{r\overline{C}_A - k_b(\overline{C}_A + \overline{C}_B)} \qquad [3\text{-}12]$$

As rates of flow become exceedingly small, $r\overline{C}_A$ and $r\overline{C}_B$ drop from equation 3-12 and a closed system is approached where equilibrium conditions are operative:

$$\text{as} \quad r \longrightarrow 0, \quad Q = \frac{r\overline{C}_B - k_f(\overline{C}_A + \overline{C}_B)}{r\overline{C}_A - k_b(\overline{C}_A + \overline{C}_B)} \longrightarrow \frac{k_f}{k_b} = \frac{C_B}{C_A} = K_{AB} \quad [3\text{-}13]$$

Soil solutions in situ are open systems with inflows and outflows of constituents. The applicability of either a thermodynamic equilibrium or kinetic model for these systems depends on the magnitude of flux for constituents of interest at the scale of resolution considered. Kinetic models, because they require time-dependent data, involve a greater degree of complexity than do thermodynamic equilibrium models. Static or dynamic models considering phase transfers and additional chemical reactions are more mathematically complex than the idealized model presented here (Fig. 3-1), but their development follows the same approach. A brief review of chemical thermodynamics and kinetics illustrates the fundamental tools by which static and dynamic models may be applied to the soil environment.

3.3 STATICS IN REVIEW

3.3.1 The Four Fundamental Equations of Thermodynamics

Usual thermodynamics involves only pressure–volume work and is resolved from the *four fundamental equations of thermodynamics*. The first equation arises from the combined first and second laws of thermodynamics, and the other three are convenient expressions of the first fundamental equation. For usual thermodynamic systems of constant composition, the fundamental equations of state are as follows.

Thermodynamic Potential	Thermodynamic Function	Fundamental Equation of State
Internal energy	E	$dE = TdS - PdV$
Enthalpy, H	$H \equiv E + PV$	$dH = TdS + VdP$
Helmholtz free energy, A	$A \equiv E - TS$	$dA = -SdT - PdV$
Gibbs free energy, F	$F \equiv H - TS$	$dF = -SdT + VdP$

3.3.2 Variables of State

The state of a thermodynamic system is described by *intensive* and *extensive* variables (Table 3.1). Intensive variables are properties of the system that are independent of the quantity of matter present; these variables cannot be summed over the parts of a system. Temperature and pressure are intensive variables in usual thermodynamics; chemical potential, μ_i, is an additional intensive variable for systems of variable composition. Extensive variables are properties of the system that are dependent on the quantity of matter present and, therefore, can be obtained as the sum over every part of the system. Entropy and volume are extensive variables in usual thermodynamics; the quantity of a component, n_i, is an extensive variable in systems of variable composition.

3.3.3 Useful Transformations of the Fundamental Equations

The fundamental equations as written are not of experimental usefulness because there is no direct measure of entropy, S, in the laboratory. The equations, however, can be recast in terms of experimental equivalencies. Maxwell relations, heat capacities, thermodynamic definitions, and combinations of these result in useful transformations of the fundamental equations that are of experimental utility.

Changes in Thermodynamic Properties. For a thermodynamic property X, $\Delta X = X_{final} - X_{initial}$ describes energy changes in a system that are useful in evaluating the tendency for a reaction to occur. Change in Gibbs free energy, $\Delta F = F_{final} - F_{initial}$, is particularly useful in this regard because ΔF varies as a function of temperature and pressure which are readily measured in experiments. Standard free energies of reaction (ΔF_r°) are frequently calculated from tabulated values of standard free energies of formation (ΔF_f°):

$$\Delta F_r^\circ = \Sigma F_{f_{products}}^\circ - \Sigma F_{f_{reactants}}^\circ \qquad [3\text{-}14]$$

Maxwell Relations. Cross differentiation of fundamental equations yields experimentally measurable relations (Maxwell relations, Table 3.2).

TABLE 3.1 Thermodynamic Variables for Soil Systems

Variable	Symbol	Property	Units
Temperature	T	intensive	deg K
Entropy	S	extensive	J deg^{-1}
Pressure	P	intensive	kPa
Volume	V	extensive	L
Chemical potential of ith component	μ_i	intensive	J mol^{-1}
Quantity of ith component	n_i	extensive	mol

TABLE 3.2 Maxwell Relations for Usual Thermodynamics[a]

Fundamental Equation of State	Cross Differential
$dE = TdS - PdV$	$\left[\dfrac{\partial S}{\partial V}\right]_T = \left[\dfrac{\partial P}{\partial T}\right]_V$
$dH = TdS + VdP$	$\left[\dfrac{\partial S}{\partial P}\right]_T = -\left[\dfrac{\partial V}{\partial T}\right]_P$
$dA = -SdT - PdV$	$\left[\dfrac{\partial T}{\partial V}\right]_S = -\left[\dfrac{\partial P}{\partial S}\right]_V$
$dF = -SdT + VdP$	$\left[\dfrac{\partial T}{\partial P}\right]_S = \left[\dfrac{\partial V}{\partial S}\right]_P$

[a]Additional Maxwell relationships can be obtained for systems of variable composition through cross differentiation of the appropriate forms of the fundamental equations of state with respect to n_i and μ_i.

Consider a change in enthalpy, ΔH, for a temperature–pressure process:

$$T_1P_1 \longrightarrow T_2P_2$$

This is important if ΔH at T_1P_1 is known and ΔH at T_2P_2 is desired; thus, $d(\Delta H)$ is needed for a change in pressure and temperature. Cross differentiation of the fundamental equation, $dH = TdS + VdP$ with respect to T and P yields[2]

$$\left(\frac{\delta S}{\delta P}\right)_T = -\left(\frac{\delta V}{\delta T}\right)_P \qquad [3\text{-}15]$$

This is a Maxwell relation; $(\delta S/\delta P)_T$ cannot be measured, but the experimental equivalent, $-(\delta V/\delta T)_P$, can.

Now consider that $d(\Delta H) = f(T,P)$; therefore,

$$d(\Delta H) = \left(\frac{\partial(\Delta H)}{\partial T}\right)_P dT + \left(\frac{\partial(\Delta H)}{\partial P}\right)_P dP \qquad [3\text{-}16]$$

For a constant P process,

$$\left(\frac{\partial(\Delta H)}{\partial T}\right)_P = \Delta c_P \qquad [3\text{-}17]$$

where Δc_P is the heat capacity at constant P. Differentiation of the fundamental equation for enthalpy with respect to P at constant T gives

$$\left(\frac{\partial(\Delta H)}{\partial P}\right)_T = \Delta\left[V + T\left(\frac{\partial S}{\partial P}\right)_T\right] \qquad [3\text{-}18]$$

Substituting the Maxwell relation (equation 3-15) into 3-18,

$$\left(\frac{\partial(\Delta H)}{\partial P}\right)_T = \Delta\left[V - T\left(\frac{\partial V}{\partial T}\right)_P\right] \qquad [3\text{-}19]$$

Thus, for $d(\Delta H) = f(T,P)$,

$$d(\Delta H) = \Delta c_p dT + \Delta\left[V - T\left(\frac{\delta V}{\delta T}\right)_P\right] dP \qquad [3\text{-}20]$$

The relation has been recast in terms of experimentally measurable quantities.

A simpler case involves a change in temperature at constant pressure, where standard state pressure is used ($P_{std} = 101.33$ kPa).

$$T_1 P_{std} \longrightarrow T_2 P_{std}$$

Now $d(\Delta H°) = f(T)$; and

$$d(\Delta H°) = \left(\frac{\partial(\Delta H°)}{\partial T}\right)_P dT = \Delta c_p° dT \qquad [3\text{-}21]$$

This is a Kirchoff relation which can be evaluated over T.

$$\int_{H_{T_1}°}^{H_{T_2}°} d(\Delta H°) = \int_{T_1}^{T_2} \Delta c_p° dT \qquad [3\text{-}22]$$

The Kirchoff relation allows for calculation of $d(\Delta H°)_r = f(T)$ for any change in T using literature values of $\Delta c_p°$ for products and reactants, as

$$\Delta c_{P_r} = \Sigma \Delta c_{p\ products}° - \Sigma \Delta c_{p\ reactants}° \qquad [3\text{-}23]$$

Interrelations Between Fundamental Equations. Thermodynamic data are usefully transformed through consideration of interrelations between thermodynamic functions.

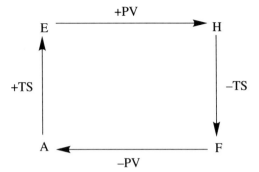

For example, $F = H - TS$; so at standard temperature and pressure (298.15°K and 101.33 kPa).

$$\Delta F_r^\circ = \Delta H_r^\circ - T\Delta S_r^\circ \qquad [3\text{-}24]$$

Reference data of ΔH_r° and partial molar entropy, S°, allow for calculation of values of ΔF_r°.

Consider the reaction at 298.15°K and 101.33 kPa.

$$Ca_{aq}^{2+} + SO_{4\ aq}^{2-} + 2H_2O_l \rightleftharpoons CaSO_4 \cdot 2H_2O_s$$

The following data are available from Stumm and Morgan (1981).

Species	H_f°, kJ mol^{-1}	S°, J mol^{-1}deg^{-1}
Ca_{aq}^{2+}	−542.83	−53.0
$SO_{4\ aq}^{2-}$	−909.20	20.1
H_2O_l	−285.83	69.91
$CaSO_4 \cdot 2H_2O_s$	−2022.60	194.1

$$\Delta H_r^\circ = \Sigma H_{f_{products}}^\circ - \Sigma H_{f_{reactants}}^\circ$$

$$= -2022.60 - [-542.83 - 909.20 + 2(-285.83)]$$

$$= 1.09 \text{ kJ mol}^{-1}$$

$$\Delta S_r^\circ = \Sigma S_{products}^\circ - \Sigma S_{reactants}^\circ$$

$$= 194.1 - [-53.0 + 20.1 + 2(69.91)]$$

$$= 87.18 \text{ J mol}^{-1}\text{deg}^{-1}$$

$$\Delta F_r^\circ = \Delta H_r^\circ - T\Delta S_r^\circ = 1.09 - (298.15)(87.18 \times 10^{-3}) = -24.90 \text{ kJ mol}^{-1}$$

3.3.4 Systems of Variable Composition

Transfers of mass occurring among soil solution and contacting phases and changes in mass due to chemical reactions in soil solution result in altered composition of the solution phase. Thus, thermodynamic functions of state must consider variation in T, P, and composition, n, where n is the mole number of components in the system. For a system of two components, n_i and n_j,

$$X = f(T, P, n_i, n_j) \tag{3-25}$$

$$dX = \left(\frac{\partial X}{\partial T}\right)_{P,n_i,n_j} dT + \left(\frac{\partial X}{\partial P}\right)_{T,n_i,n_j} dP + \left(\frac{\partial X}{\partial n_i}\right)_{T,P,n_j} dn_i + \left(\frac{\partial X}{\partial n_j}\right)_{T,P,n_i} dn_j \tag{3-26}$$

The partial derivative of a state function X with respect to the ith component of the system at constant T, P, and composition of the other components, n_j, is the *partial molar quantity;*

$$X_i \equiv \left(\frac{\partial X}{\partial n_i}\right)_{T,P,n_i} \tag{3-27}$$

Whereas X is an extensive property of the system, X_i is an intensive property of the system. Thus, when thermodynamic properties other than composition are fixed in a single-phase closed system,

$$dX = \sum_i X_i dn_i \tag{3-28}$$

The Phase Rule. A *phase* comprises a domain of uniform properties and homogeneous composition within a thermodynamic system. The uniformity of properties such as temperature, pressure, and electrolyte content is phase deliminating for soil solutions. The assumption of homogeneous composition for soils is dependent on experimental design; the translation of this assumption to field environments is dependent on the scale of resolution involved.

The *phase rule* describes variance or degrees of freedom, F, in a system as a function of the number of components (*C*), reactions (R), and phases (P) in the system:

$$F = C + 2 + R - P$$

The phase rule is usefully applied to complex systems because it defines the degrees of freedom—"the least number of intensive variables that must be specified to fix the values of all the remaining intensive variables"—necessary for describing the system by a thermodynamic model (Barrow, 1966). Conversely, it describes the maximum number of intensive variables which may be independently varied in experiments on the system.

For the reaction,

$$Ca^{2+}_{aq} + SO^{2-}_{4\ aq} + 2H_2O_l \rightleftharpoons CaSO_4 \cdot 2H_2O_s$$

the thermodynamic system is comprised of three components (Ca, SO_4, and H_2O) distributed between two phases; therefore, it possesses four degrees of freedom. If T and P are specified, the chemical potential of any two species can describe the composition of the system. If the system is broadened to encompass a binary exchange reaction as well,

$$2Ca^{2+}_{aq} + SO^{2-}_{4\ aq} + 2H_2O_l + 2NaX \rightleftharpoons CaSO_4 \cdot 2H_2O_s + CaX + 2Na^+$$

the system involves two reactions with five components (Ca, SO_4, H_2O, Na, and the exchange complex, X) occurring among three phases of the soil (aqueous solution, a precipitated phase of $CaSO_4 \cdot 2H_2O$, and an exchange complex. The system has six degrees of freedom; and, with T and P fixed, four components of the system could be independently varied to experimentally probe the system.

3.3.5 Chemical Potential

The free energy concept of chemical thermodynamics is a particularly powerful tool for predicting the *nature* and *extent* of reactions as they approach equilibrium in soil systems. The partial molar free energy or chemical potential, μ_i, describes the free energy contribution of the ith component driving the system toward equilibrium:

$$\mu_i \equiv F_i = \left(\frac{\partial F_i}{\partial n_i}\right)_{T,P,n_j} \tag{3-30}$$

The change in Gibbs free energy for a system of i components and α phases is obtained by summation over all components and all phases,

$$dF = \sum_\alpha S^\alpha dT^\alpha + \sum_\alpha V^\alpha dP^\alpha + \sum_\alpha \sum_i \mu_i dn_i \tag{3-31}$$

Consideration of a single-phase closed system at constant T and P reduces this expression to

$$dF = \sum_i \mu_i dn_i \tag{3-32}$$

The general conditions describing the nature or direction of a reaction occurring in this system are

$$\text{equilibrium:} \quad \sum_i \mu_i dn_i = 0$$

$$\text{spontaneous:} \quad \sum_i |\mu_i dn_i| > 0$$

Chemical potentials summed over a system describe the direction of reactions occurring in a single phase as well as the direction of transfers occurring between phases.

Free energy changes can further characterize the extent, ξ, of chemical reactions.

$$dF = \sum_i \nu_i \mu_i d\xi \qquad \text{[3-33]}$$

where $\nu_i d\xi \equiv dn_i$, ξ has units of moles, and ν_i (the stoichiometric coefficient for the ith component) is unitless. Extent of reaction is further described in section 3.4.7. Irreversible Thermodynamics.

Chemical Potential of Various Phases. The chemical potential expressions relevant to various phases and conditions are summarized in Table 3.3; the development of these various expressions follows.

Ideal Gas. Following the ideal gas law, the partial free energy with respect to P at constant T and n for a pure ideal gas is

$$\left(\frac{\partial F}{\partial P}\right)_{T,n} = V = \frac{nRT}{P} \qquad \text{[3-34]}$$

Rearrangement yields

$$dF = nRT \frac{dP}{P} = nRT \, d\ln P \qquad \text{[3-35]}$$

Integration over limits of pressure from a standard pressure, P°, to a pressure P yields

$$\int_{F^\circ}^{F} dF = nRT \int_{P^\circ}^{P} d\ln P \qquad \text{[3-36]}$$

$$F - F^\circ = nRT \ln \frac{P}{P^\circ} \qquad \text{[3-37]}$$

TABLE 3.3 Chemical Potential Expressions

		Standard State[a]	Reference State
Gases			
Ideal gas	$\mu_i = \mu_i^\circ + RT\ln P_i/P^\circ$	$T^\circ = 298.15$ K, $P^\circ = 1$ atm (101.33 kPa) pure ideal gas at $P_i = 1$ atm	$T = 298.15$ K
Real gas	$\mu_i = \mu_i^\circ + RT\ln f_i$	pure real gas at $f_i = 1$ and $\gamma_i = 1$	$\gamma_i \to 1$ as $P_i \to 0$
Condensed Phase Components			
Obey Raoult's law	$\mu_i = \mu_i^\circ + RT\ln\chi_i$	pure condensed phase component at $\chi_i = 1$	
Deviate from Raoult's law	$\mu_i = \mu_i^\circ + RT\ln A_i$	pure condensed phase component at $\chi_i = 1$ and $\gamma_i = 1$	$\gamma_i \to 1$ as $\chi_i \to 0$
Obey Henry's law	$\mu_i = \mu_i^\circ + RT\ln C_i$	pure condensed phase component at $C_i = 1$	
Deviate from Henry's law	$\mu_i = \mu_i^\circ + RT\ln A_i$	pure condensed phase component at $C_i = 1$ and $\gamma_i = 1$	$\gamma_i \to 1$ as $C_i \to 0$
Solutes in Ideal Dilute Solutions			
	$\mu_i = \mu_i^\circ + RT\ln A_i$	pure condensed phase component at $C_i = 1$ and $\gamma_i' = 1$	$\gamma_i \to 1$ as $C_i \to 0$
Solutions of Electrolytes			
	$\mu_S = \mu_S^\circ + \upsilon RT\ln A_\pm$	pure condensed phase component in the absence of ion interactions at $C_\pm = 1$ and $\gamma_\pm = 1$	$\gamma_\pm \to 1$ as $C_\pm \to 0$
	$\mu_\pm = \mu_\pm^\circ + RT\ln A_\pm$	at $C_\pm = 1$ and $\gamma_\pm = 1$	$\gamma_\pm \to 1$ as $C_\pm \to 0$
	$\mu_+ = \mu_+^\circ + RT\ln A_+$	at $C_+ = 1$ and $\gamma_+ = 1$	$\gamma_+ \to 1$ as $C_+ \to 0$
	$\mu_{+i} = \mu_{+i}^\circ + RT\ln f_i$	at $C_{+i} = 1$ and $f_i = 1$	$f_i \to 1$ as $C_{+i} \to 0$

Additional relations:

$f_i = P_i\gamma_i$

$\chi_i = P_i/P_i^\circ$

$A_i = \chi_i\gamma_i$

$C_i = P_i/K_{H_i}$

$A_i = C_i\gamma_i$

$A_i = C_i\gamma_i'$

$A_\pm = \gamma_\pm C_\pm$

$A_\pm = A_S^{1/\upsilon}$

$A_+ = \gamma_+ C_+$

$A_{+i} = f_i(Z_i, 1)C_{+i}$

[a]The standard state defines units of concentration (C). Depending on the rational state defined, C may carry units of molality or molarity; the activity coefficient (γ) will carry units of inverse concentration.

Differentiation with respect to *n* results in

$$\frac{\partial F}{\partial n} - \frac{\partial F^\circ}{\partial n} = RT \ln \frac{P}{P^\circ} = \mu - \mu^\circ \qquad [3\text{-}38]$$

For the *i*th ideal gas in a mixture of ideal gases,

$$\mu_i = \mu_i^\circ + RT \ln P_i \qquad [3\text{-}39]$$

where μ_i° is the chemical potential for the *standard state*[3] of the pure ideal gas at $P^\circ = 1$ atm (101.33 kPa) and $T^\circ = 298.15$ K (see the following section, Standard and Reference States).

Real Gas.   A correction factor, δ_i, accounting for deviations from ideality is introduced for the case of a real gas.

$$\mu_i = \mu_i^\circ + RT \ln P_i + \delta_i \qquad [3\text{-}40]$$

The correction factor $\delta_i \equiv RT\ln\gamma_i$, where γ_i is the fugacity coefficient.

$$\gamma_i = \frac{f_i}{P_i}$$

From equation 3-40,

$$\mu_i = \mu_i^\circ + RT\ln P_i + RT\ln\gamma_i$$
$$= \mu_i^\circ + RT\ln(P_i \, \gamma_i)$$
$$\mu_i = \mu_i^\circ + RT \ln f_i \qquad [3\text{-}41]$$

The fugacity, f_i, is an idealized pressure (effective pressure) representing the pressure of an ideal gas producing the same effect as observed for the real gas at P_i. As $P \to 0$, $f_i \to P_i$ and $\gamma_i \to 1$. For the real gas, μ_i° is the chemical potential for the standard state of a pure real gas at $P = 1$ atm, $T = 298.15$ K, and $f_i = 1$ atm; $f_i = P_i$ for most real gases at 1 atm (101.33 kPa). The reference state for a real gas is that of the pure substance at 298.15 K as $P_i \to 0$. Note that for the standard state of the real gas $P_i = P_i^\circ$, whereas for the reference state P_i approaches zero.

Standard and Reference States.   Thermodynamic description of chemical reactions requires that chemical potentials of substances are described in terms of standard and reference states. Table 3.3 summarizes common standard and references states. As this table indicates, the standard and reference states are seldom the same. The reference state describes an actual state of the component, whereas the standard state describes a hypothetical state.

The *standard state* of a substance is the state where concentration and activity coefficient both equal unity. The definition of standard state also specifies T and P as a convenience; most chemical reactions in soil systems occur under isothermal, isobaric conditions, or they are measured under specified conditions of constant T and P. The standard state for an element is selected for that form (solid, liquid, or gas) that is most stable under the standard state conditions of T and P.

The *reference state* of a substance is the state where the activity coefficient is equal to one. The reference state describes the state where the substance occurring in a specified phase approaches ideal behavior (the dilute solution limit).

Condensed Phase Components That Obey Raoult's Law. Raoult's law describes the observed behavior of binary liquid nonelectrolytic solutions,

$$P_i = \chi_i P_i^{\circ} \qquad [3-42]$$

where χ_i is the mole fraction of the ith component, P_i is the vapor pressure of i above the condensed phase, and P_i° is the vapor pressure above the pure condensed phase i. This relationship can be broadly applied to solvents occurring in multi-component condensed phase solutions of nonelectrolytes exhibiting ideal behavior.

For condensed phase components obeying Rauolt's law, chemical potential is expressed

$$\mu_i = \mu_i^{\circ} + RT \, \ln\chi_i \qquad [3-43]$$

where,

$$\mu_i^{\circ} = \mu_{i_{soln}}^{\circ} = \mu_{i_{gas}}^{\circ} + RT \, \ln P_i^{\circ} \qquad [3-44]$$

which is constant for constant T and P. The standard state for μ_i° is that of the pure condensed phase component at 1 atm (101.33 kPa) and 298.15 K. The reference state is that of the component at P = 1 atm and T = 298.15 K as $\chi_i \rightarrow 1$ in the physical state of the solution. Real solutions approach Raoult's law behavior as $\chi_i \rightarrow 1$.

Condensed Phase Components That Obey Henry's Law. Henry's law describes the relationship of the concentration of a volatile solute i in a condensed phase to its vapor pressure above the condensed phase. In a practical system,[4] Henry's law states

$$P_i = K_{H_i} C_i \qquad [3-45]$$

where K_{H_i} is the Henry's law constant and C_i is the concentration of solute i in moles L^{-1}.

For condensed phase components obeying Henry's law, chemical potential is expressed

$$\mu_i = \mu_i^\circ + RT \ln C_i \qquad [3\text{-}46]$$

The standard state for μ_i° is that of the condensed phase component where $C_i = 1$ in the physical state of the solution and where $f_i \rightarrow K_{H_i}$ at $P = 1$ atm and $T = 298.15$ K. The reference state is that of the pure substance at $P = 1$ atm and $T = 298.15$ K as $C_i \rightarrow 0$.

Solutes in Ideal Dilute Solutions. The chemical potential of condensed phase components of real solutions may be expressed in a practical system as

$$\mu_i = \mu_i^\circ + RT \ln C_i + RT \ln \gamma_i' \qquad [3\text{-}47]$$

This expression is useful for describing solutes in real solutions as they approach zero—the *ideal dilute solution*. In terms of the molar scale, C_i is the concentration of solute i in moles L^{-1}; γ_i' is the molar activity coefficient and carries units of inverse concentration. The *reference state* of the solute in a real solution as described by a practical system is the ideal dilute solution where the activity of i (A_i) is equal to its concentration,

$$A_i \longrightarrow C_i \longrightarrow 0 \quad \text{as} \quad \gamma_i' \longrightarrow 1$$

The activity is an idealized concentration; it represents the concentration of an ideal solute that would give the same effect as observed for the solute of a real solution at C_i. Thus, activity represents an "effective concentration," and γ_i' is the ratio of real to effective concentration.

$$\gamma_i' = \frac{A_i}{C_i} = \frac{\text{Effective concentration}}{\text{Real concentration}} \qquad [3\text{-}48]$$

The standard state for μ_i° is that of the condensed phase component ($C_i = 1$) in the physical state of the solution at $P = 1$ atm and $T = 298.15$ K.

Solutions of Electrolytes. Description of the chemical potential of electrolytes in dilute aqueous solution is of principal interest to the soil solution chemist. Unfortunately, it is not possible to describe the behavior of charged particles within the confines of classical thermodynamics; therefore, extrathermodynamic conventions are employed.

Consider a strong electrolyte,

$$M_mL_l \rightleftharpoons mM^{c+} + lL^{a-} \qquad [3\text{-}49]$$

For the individual ions,

$$\mu_M = \mu_M^\circ + RT \ln A_M^{c+}$$
$$\mu_L = \mu_L^\circ + RT \ln A_L^{a-} \qquad [3\text{-}50]$$

If $S \equiv M_mL_l$, then

$$\mu_S = m\mu_M + l\mu_L \qquad [3\text{-}51]$$

Thus, from 3-50

$$\mu_S = \mu_S^\circ + RT \ln A_S \qquad [3\text{-}52a]$$
$$\mu_S = m(\mu_M^\circ + RT \ln A_M^{c+}) + l(\mu_L^\circ + RT \ln A_L^{a-}) \qquad [3\text{-}52b]$$
$$\mu_S = m\mu_M^\circ + l\mu_L^\circ + RT \ln A_M^{c+} + RT \ln A_L^{a-} \qquad [3\text{-}52c]$$

The reference state for the electrolyte is an ideal dilute solution, and the standard state for μ_i° is that of the condensed phase component in the physical state of the solution. By definition,

$$\mu_S^\circ = m\mu_M^\circ + l\mu_L^\circ \qquad [3\text{-}53]$$

Substitution of 3-53 into 3-52c and simplification yields

$$\mu_S = \mu_S^\circ + RT \ln(A_M^{c+})^m (A_L^{a-})^l \qquad [3\text{-}54]$$

From 3-52a and 3-54,

$$\mu_S^\circ + RT \ln A_S = \mu_S^\circ + RT \ln(A_M^{c+})^m (A_L^{a-})^l \qquad [3\text{-}55]$$

Thus,

$$A_S = (A_M^{c+})^m (A_L^{a-})^l \equiv IAP \qquad [3\text{-}56]$$

The ion activity of the salt in solution is the product of the activities of the ions to which it dissociates in solution; this product is defined as the *ion activity product* (IAP).

MacInnes activity coefficients are defined as

$$\gamma_{S^+} \equiv \frac{A_M^{c+}}{mC_S}$$

$$\gamma_{S^-} \equiv \frac{A_L^{a+}}{lC_S} \qquad [3\text{-}57]$$

Substitution of 3-57 into 3-56 yields

$$A_S = (\gamma_+ C_S)^m (\gamma_- l C_S)^l \qquad [3\text{-}58a]$$

which can also be expressed

$$A_S = (\gamma_+^m \gamma_-^l)(m^m 1^l) C_S^{(m+1)} \qquad [3\text{-}58b]$$

The following additional definitions are made with respect to McInnes activity coefficients

$$\nu \equiv (m + 1)$$

$$\gamma_\pm \equiv \text{mean ionic activity coefficient} = (\gamma_+^m \gamma_-^l)^{1/\nu}$$

$$C_\pm \equiv \text{mean ionic concentration} = (m^m 1^l)^{1/\nu} C_S$$

Equation 3-58*b* can thus be expressed

$$A_S = [(\gamma_+^m \gamma_-^l)^{1/\nu}(m^m 1^l)^{1/\nu} C_S]^\nu \qquad [3\text{-}58c]$$

which reduces to

$$A_S = (\gamma_\pm C_\pm)^\nu \qquad [3\text{-}59]$$

From 3-59, the *mean ionic activity* is defined

$$A_\pm \equiv A_S^{1/\nu} = \gamma_\pm C_\pm \qquad [3\text{-}60]$$

The mean ionic activity coefficient and, therefore, the mean ionic activity are thermodynamically meaningful. McInnes activity coefficients, however, cannot be explained thermodynamically, as cannot the analogs for mixed electrolytic solutions—the *single ion activity coefficients.*

$$\gamma_+ \equiv \frac{A_+}{C_+}$$

$$\gamma_- \equiv \frac{A_-}{C_-} \tag{3-61}$$

For simple salt systems, $\gamma_\pm$ can be determined experimentally. This can be accomplished, for example, through measurements of electrical potential for systems at electrochemical equilibrium. Free energy relationships for such systems are described from

$$\Delta F = -n\mathfrak{F}E \tag{3-62}$$

where n is the moles of electric charge transferred, $\mathfrak{F}$ is the Faraday constant (96,490 A sec mol^{-1}), and E is the electric potential (J sec^{-1} A^{-1}). From equation 3-33,

$$\Delta F = \sum_i v_i \mu_i \tag{3-63}$$

since,

$$\Delta F = (\partial F/\partial \xi)_{T,P}$$

Thus,

$$-n\mathfrak{F}E = \sum_i v_i(\mu_i^\circ + RT \ln\gamma_i C_i) \tag{3-64}$$

or in terms of the mean ionic activity coefficient, $\gamma_\pm$, for a simple reaction such as that described in equation 3-49

$$-n\mathfrak{F}E = -n\mathfrak{F}E^\circ + vRT \ln\gamma_\pm C_\pm \tag{3-65}$$

which on rearrangement yields

$$\ln\gamma_\pm = \frac{n\mathfrak{F}(E^\circ - E)}{vRT} - \ln C_\pm \tag{3-66}$$

This expresses the mean ionic activity coefficient in terms that are measureable.

Using measured mean ionic activity coefficients, it is possible to calculate single ion activity coefficients for simple salt solutions. The relationship between mean and single ion activity coefficients for aqueous solutions serves as a starting point for this approach. For KCl,

$$\gamma_\pm = (\gamma_{K^+}\gamma_{Cl^-})^{1/2}$$

where it has been observed that $\gamma_{\pm} = \gamma_{K^+} = \gamma_{Cl^-}$ (apparently because of a close similarity in behavior of K^+ and Cl^- in solution). This allows for the calculation of single ion activities for simple electrolytic solutions of equivalent ionic strength.

Example. Consider solutions of constant ionic strength $(I = 1/2\Sigma C_i Z_i^2 = 0.3)$, experimentally determined mean ionic activity coefficients are as follows:

Solution	$\gamma_{\pm}$
0.3 M KCl	0.687
0.1 M CaCl$_2$	0.518
0.1 M Ca(NO$_3$)$_2$	0.485

Using these data, single ion activity coefficients for the various ions in aqueous solution can be calculated.

$$\gamma_{K^+} = \gamma_{Cl^-} = \gamma_{\pm} = 0.687$$

$$\gamma_{CaCl_2} = (\gamma_{Ca^{2+}}\gamma_{Cl^-}^2)^{1/3}$$

$$\gamma_{Ca^{2+}} = \frac{\gamma_{CaCl_2}^3}{\gamma_{Cl^-}^2} = \frac{(0.518)^3}{(0.687)^2} = 0.294$$

$$\gamma_{Ca(NO_3)_2} = (\gamma_{Ca^{2+}}\gamma_{NO_3^-}^2)^{1/3}$$

$$\gamma_{NO_3^-} = \frac{\gamma_{Ca(NO_3)_2}^3}{\gamma_{Ca^{2+}}} = \left(\frac{(0.485)^3}{(0.294)}\right)^{1/2} = 0.623$$

Unfortunately, this approach does not lend itself to considerations of complex multivalent electrolytic solutions, such as soil solutions.

Chemical potential expressions for electrolytic solutions can be summarized as follows

$$\mu_S = \mu_S^\circ + RT \ln A_S \qquad [3\text{-}52a]$$

$$\mu_{\pm} = \mu_{\pm}^\circ + RT \ln A_{\pm} = \mu_{\pm}^\circ + \nu^{-1}RT \ln A_S \qquad [3\text{-}67]$$

$$\mu_{+} = \mu_{+}^\circ + RT \ln A_{+} = \mu_{+}^\circ + RT \ln \gamma_{+} C_{+} \qquad [3\text{-}68]$$

In the case of equation 3-68, γ_{+} is not measurable but is calculable. For electrolytes in solution the reference state is that of the ideal dilute solution.

3.3.6 Calculation of Single Ion Activity Coefficients

It is not possible to resolve the chemical potential of an ionic species in dilute aqueous solution in an unambiguous manner. For an electrolyte S in solution,

$$\mu_S = \mu_S^{\circ} + RT \ln A_S \qquad [3\text{-}52a]$$

As previously derived,

$$A_S = (\gamma_{\pm} C_{\pm})^{\nu} \qquad [3\text{-}59]$$

Therefore,

$$\mu_S = \mu_S^{\circ} + \nu RT \ln\gamma_{\pm} + \nu RT \ln C_{\pm} \qquad [3\text{-}69]$$

The mean ionic activity coefficient for charged particles ($\gamma_{\pm}$) is primarily a consequence of the energy of interaction of the electrical charges on the particles. When modeling dilute electrolytic solutions, it is therefore necessary to account for the energy expended in charging particles in a solvent. The development of such a model is dependent on a value for the electrical potential on the surface of the charged particles. The Debye–Hückel theory provides a solution to this problem using laws of electrostatics and thermodynamics and the assumption that ions behave as point charges in a continuous medium with a dielectric constant equal to that of the solvent.

   The Debye–Hückel equation in its various forms as well as other empirically based equations provide a simple means for expression of single ion activity coefficients. A variety of these relationships are used by the soil solution chemist (Table 3.4).

The Limited Debye–Hückel Equation.

$$\log f_i = -A Z_i I^{1/2} \qquad [3\text{-}70]$$

where A is an empirical coefficient, Z_i is valence, and I is ionic strength (M); f_i denotes an empirically derived activity coefficient in contrast to γ_i which is a derived thermodynamic quantity. The coefficient A is equal to $1.82 \times 10^6 (\varepsilon T)^{-3/2}$, where ε is the dielectric constant (A $= 0.509$ for water at $298.15°K$ and 101.33 kPa). The limited Debye–Hückel equation shows close correspondence with experimental results for I < 0.005 M.

The Extended Debye–Hückel Equation.

$$\log f_i = \frac{-A Z_i^2 I^{1/2}}{1 + \beta a_i^{\circ} I^{1/2}} \qquad [3\text{-}71]$$

TABLE 3.4 Individual Ion Activity Coefficients

Approximation	Equation[a]	Approximate Applicability [ionic strength (M)]
Debye–Hückel	$\log f = -AZ^2 \sqrt{I}$	(1) $<10^{-2.3}$
Extended Debye–Hückel	$= -AZ^2 \dfrac{\sqrt{I}}{1 + Ba_i^{\circ}\sqrt{I}}$	(2) $<10^{-1}$
Güntelberg	$= -AZ^2 \dfrac{\sqrt{I}}{1 + \sqrt{I}}$	(3) $<10^{-1}$ useful in solutions of several electrolytes
Davies	$= -AZ^2 \left(\dfrac{\sqrt{I}}{1 + \sqrt{I}} - 0.2I \right)$	(4)[b] <0.5

[a] I (ionic strength) $= 1/2 \Sigma C_i Z_i^2$; $A = 1.82 \times 10^6 (\varepsilon T)^{-3/2}$ (where ε = dielectric constant); $A \approx 0.5$ for water at 25°C; Z = charge of ion; $B = 50.3(\varepsilon T)^{-1/2}$; $B \approx 0.33$ in water at 25°C; a = adjustable parameter (angstroms) corresponding to the size of the ion.
[b] Davies has proposed 0.3 (instead of 0.2) as a coefficient for the last term in parentheses.
From Stumm and Morgan, 1981.

includes an empirically derived constant β equal to $50.3(\varepsilon T)^{-1/2}$ ($\beta = 0.328$ at 298.15°K and 101.33 kPa) and an ion size parameter (a_i°) which accounts for the effective size of hydrated ions (Table 3.5). The extended Debye–Hückel equation holds for $I < 0.1$ $\underline{M}$. The *Guttelberg equation* is a simplified form of the extended Debye–Hückel equation where the product $\beta a_i^{\circ} = 1$.

The Davies Equation. Davies suggested an empirical correction of the Guttleberg equation that uses a correction factor containing an adjustable coefficient (0.2 or 0.3).

$$\log \mathbf{f}_i = -AZ_i^2 \left(\frac{I^{1/2}}{1 + I^{1/2}} - 0.2I^{1/2} \right) \tag{3-72}$$

This equation holds for $I < 0.5$ $\underline{M}$ and is often used in preference to the extended Debye–Hückel equation because (1) it is easily manipulated as a computer algorithm, (2) it reliably models mixed electrolytic solutions, and (3) it contains only a single adjustable coefficient that is independent of the charged species considered.

The robustness of the Davies equation has been described by Sposito (1984) for the case of predicting conditional stability constants for soluble complex formation. For a metal–ligand complex in an aqueous solution,

$$M^+ + L^- \rightleftharpoons ML$$

TABLE 3.5 **Values for the Parameter $a_i°$ Used in the Extended Debye–Hückel Equation**

$10^8 \, a_i°$	

Inorganic Ions: Charge 1

$10^8 \, a_i°$	Ion
9	H^+
6	Li^+
4–4.5	Na^+, $CdCl^+$, ClO_2^-, IO_3^-, HCO_3^-, $H_2PO_4^-$, HSO_3^-, $H_2AsO_4^-$
3.5	OH^-, F^-, NCO^-, HS^-, ClO_3^-, ClO_4^-, BrO_3^-, IO_4^-, MnO_4^-, NCS^-
3	K^+, Cl^-, Br^-, I^-, CN^-, NO_2^-, NO_3^-
2.5	Rb^+, Cs^+, NH_4^+, Tl^+, Ag^+

Inorganic Ions: Charge 2

8	Mg^{2+}, Be^{2+}
6	Ca^{2+}, Cu^{2+}, Zn^{2+}, Sn^{2+}, Mn^{2+}, Fe^{2+}, Ni^{2+}, Co^{2+}
5	Sr^{2+}, Ba^{2+}, Ra^{2+}, Cd^{2+}, Hg^{2+}, S^{2-}, $S_2O_4^{2-}$, WO_4^{2-}
4.5	Pb^{2+}, CO_3^{2-}, SO_3^{2-}, MoO_4^{2-}, $Co(NH_3)Cl_5^{2+}$, $Fe(CN)_5NO^{2-}$
4	Hg_2^{2+}, SO_4^{2-}, $S_2O_3^{2-}$, $S_2O_8^{2-}$, SeO_4^{2-}, CrO_4^{2-}, $S_2O_6^{2-}$, HPO_4^{2-}

Inorganic Ions: Charge 3

9	Al^{3+}, Fe^{3+}, Cr^{3+}, Se^{3+}, Y^{3+}, La^{3+}, In^{3+}, Ce^{3+}, Pr^{3+}, Nd^{3+}, Sm^{3+}
4	PO_4^{3-}, $Fe(CN)_6^{3-}$, $Cr(NH_3)_6^{3+}$, $Co(NH_3)_6^{3+}$, $Co(NH_3)_5 \cdot H_2O^{3+}$

Inorganic Ions: Charge 4

11	Th^{4+}, Zr^{4+}, Ce^{4+}, Sn^{4+}
6	$Co(S_2O_3)(CN)_5^{4-}$
5	$Fe(CN)_6^{4-}$

Inorganic Ions: Charge 5

9	$Co(SO_3)_2(CN)_4^{5-}$

Organic Ions: Charge 1

8	$(C_6H_5)_2CHCOO^-$, $(C_3H_7)_4N^+$
7	$[OC_6H_2(NO_2)_3]^-$, $(C_3H_7)_3NH^+$, $CH_3OC_6H_4COO^-$
6	$C_6H_5COO^-$, $C_6H_4OHCOO^-$, $C_6H_4ClCOO^-$, $C_6H_5CH_2COO^-$, $CH_2{=}CHCH_2COO^-$, $(CH_3)_2CHCH_2COO^-$, $(C_2H_5)_4N^+$, $(C_3H_7)_2NH_2^+$
5	$CHCl_2COO^-$, CCl_3COO^-, $(C_2H_5)_3NH^+$, $(C_3H_7)NH_3^+$
4.5	CH_3COO^-, CH_2ClCOO^-, $(CH_3)_4N^+$, $(C_2H_5)_4NH_2^+$, $NH_2CH_2COO^-$
4	$NH_3^+CH_2COOH$, $(CH_3)_3NH^+$, $C_2H_5NH_3^+$
3.5	$HCOO^-$, H_2-citrate, $CH_3NH_3^+$, $(CH_3)_2NH_2^+$

TABLE 3.5 *Continued*

$10^8\ a_i^{\,\circ}$

Organic Ions: Charge 2

7	$OOC(CH_2)_5COO^{2-}$, $OOC(CH_2)_6COO^{2-}$, Congo red anion^{2-}
6	$C_6H_4(COO)_2^{2-}$, $H_2C(CH_2COO)_2^{2-}$, $(CH_2CH_2COO)_2^{2-}$
5	$H_2C(COO)_2^{2-}$, $(CH_2, COO)_2^{2-}$, $(CHOHCOO)_2^{2-}$
4.5	$(COO_2)^{2-}$, H-citrate^{2-}

Organic Ions: Charge 3

5	Citrate^{3-}

Source: Kielland (1937).

The thermodynamic and conditional stability constants are

$$K_{ML} = \frac{A_{ML}}{A_M A_L}$$

$$K_{ML}^c = \frac{C_{ML}}{C_M C_L}$$

or, upon log transformation,

$$\log K_{ML} = \log A_{ML} - \log A_M - \log A_L$$

$$\log K_{ML}^c = \log C_{ML} - \log C_M - \log C_L$$

The difference in the thermodynamic and the conditional equilibrium stability constants ($\Delta \log K \equiv \log K_{ML} - \log K_{ML}^c$) resolves to

$$\Delta \log K = \log \gamma_{ML} - \log \gamma_M - \log \gamma_L \qquad \text{[3-73a]}$$

From the Davies equation 3-72,

$$\Delta \log K = A \left[\left(\frac{I^{1/2}}{1 + I^{1/2}} \right) - 0.2 I^{1/2} \right] (Z_{ML}^2 - Z_M^2 - Z_L^2) \qquad \text{[3-73b]}$$

If $\Delta Z^2 \equiv (Z_{ML}^2 - Z_M^2 - Z_L^2)$, then

$$\Delta \log K = A \left[\left(\frac{I^{1/2}}{1 + I^{1/2}} \right) - 0.2 I^{1/2} \right] \Delta Z^2 \qquad \text{[3-73c]}$$

If the Davies equation describes single ion activities, a plot of $\Delta \log K$ versus ΔZ^2 for various metal–ligand complexation reactions will yield a line of slope equal to the Davies equation. This proves to be the case as shown in Figure 3-2 for the data compiled by Sun et al. (1980).

3.4 DYNAMICS IN REVIEW

Chemical dynamics or kinetics entails motions and forces among microscopic particles involved in chemical reactions. Reaction rate, factors influencing reaction rate, and the mechanistic explanation of reaction rate are all considered by chemical dynamics. Chemical dynamics treats the transition process from one state of a system to another and the time-dependence of the transition (Moore and Pearson, 1981).

Because chemical dynamics details chemical reactions from the perspective of extent of reaction with time $(d\xi/dt)$, it is more fundamental than is chemical statics (thermodynamics). The complexities of kinetic theory, however, make accurate application of chemical dynamics uncertain for all but the simplest homogeneous reactions. Consequently, chemical statics can precisely describe extent of reaction,

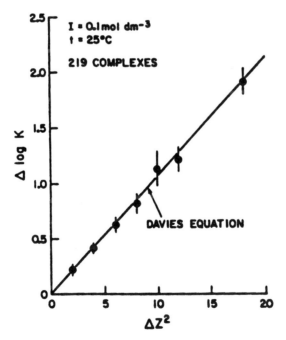

FIGURE 3.2 Graph of log K versus the parameter ΔZ^2 for 219 inorganic and organic metal–ligand complexes at $I_e = 0.1$ mol dm^{-3}. [Sposito, 1984. Published in *Soil Sci. Soc. Am. J.* 48:531–536. 1984. Soil Sci. Soc. Am.]

whereas chemical dynamics can only approximate rate of reaction in most cases. This is particularly true for the complexities of heterogeneous reactions occurring in soil systems where the application of chemical dynamics is mostly empirically based.

3.4.1 Empiricism of Soil Kinetics

Kinetically, *homogeneous reactions* occur in one phase, while *heterogeneous reactions* occur in more than one phase. An aqueous hydrolysis reaction occurring in soil solution exemplifies a homogeneous reaction, whereas an adsorption reaction at the solid–solution interface is a heterogeneous reaction. Clearly, the majority of reactions of interest in soil systems—adsorption/desorption, metabolism, precipitation/dissolution, exchange—occur at interfaces and therefore are dynamically defined by heterogeneous kinetics.

Rate theory allows for the calculation of innumerable reaction paths and reaction orders for processes occurring in a complex system such as soil. Assignment of reaction path, order, and mechanism to a process requires physical evidence for reaction products and intermediates. This evidence is mostly absent for processes of interest to the soil chemist, so chemical dynamics as applied to soil systems basically involves empirical description of rates—that is, apparent kinetics—and subjective assignment of pathways and mechanisms.

It is difficult to isolate chemical processes experimentally from physical processes occurring in soils. Therefore, apparent kinetics may describe rate-determining physical processes governing transport (diffusion) between chemically reactive compartments of the soil system—as between bulk soil solution and soil solution in micropores—rather than the chemical processes of interest (Skopp, 1986).

Consideration of chemical dynamics solely from the standpoint of apparent reaction order can offer significant intuitive insights as to reaction paths and mechanisms provided the physical limitations of drawing inferences exclusively from the fitting of rate data are borne in mind. As previously stated, these physical limitations may be due to (1) the absence of physical—that is, spectroscopic—evidence for reaction intermediates and products and/or (2) transport-limited processes driving the observed apparent kinetics in the experimental system. The utility of empirical description of rate-dependent data has been hindered by both the acceptance of apparent kinetics as indicative of mechanism and the use of linearization techniques in the data-fitting process. The advent of computer techniques utilizing nonlinear statistics and mechanistically based models that allow for the comparison of chemically versus physically restricted control of observed reaction rate has vastly improved the empirical utility of chemical dynamics for interpretation of soil processes (Alexander and Scow, 1989).

3.4.2 Irreversible Reactions

Rate Law. *Rate law* describes the dependence of reaction rate (velocity) on reactant concentrations and is based on the probability of reactants proceeding to products. The probability of reactants proceeding to products is decreased as the number of reactant molecules increases; thus, mechanism of elementary reactions at the molecular level can be described by *molecularity*—the number of reactant molecules involved in a single-step process. Unimolecular reactions involve a single reactant,

$$A \longrightarrow C$$

$$A \longrightarrow C + D$$

Bimolecular reactions involve two reactant molecules,

$$A + B \longrightarrow C + D$$

$$A + A \longrightarrow C + D$$

Simultaneous reaction of more than two molecules is very unlikely, so reactions occurring with molecularity of greater than bimolecular generally occur as a step-wise series of bimolecular reactions.

Description of reaction rate by rate law requires the concentration of reactants, a stoichiometric equation describing the reaction, and mechanisms of product formation. The rate equation for a reactant i in an elementary reaction of the type

$$aA + bB \longrightarrow products$$

can be described by empirical rate law as

$$rate = \frac{dC_i}{dt} = -kC_A^{n_a}C_B^{n_b} \qquad [3\text{-}74]$$

where the proportionality factor, k, is the rate constant; the exponents n_a and n_b are equal to the respective stoichiometric coefficients (v_i) a and b *for the case of an elementary reaction.*

Reaction Order. Reaction order is the sum of the exponents n_i in equation 3-74. Reaction order is defined by the sum of the stoichiometric coefficients for the case of an elementary reaction, but for complex reactions there is no necessary relationship between v_i and n_i, and reaction order must be experimentally determined.

Rate expressions for simple homogeneous reactions may be developed through consideration of the reaction (Table 3.6)

$$aA + other\ reactants \longrightarrow products$$

for which rate is defined as the loss of A with time of reaction,

TABLE 3.6   Rate Law for Simple Homogeneous Reactions

$$aA + bB \xrightarrow{k} products$$

Order with Respect to A	Differential Rate Expression	Linear Form[a]	Half-life, $t_{1/2}$
Zero Order	$dC_A/dt = -k$	$C_A = C_{A°} - kt$	$0.5C_{A°}k^{-1}$
First Order	$dC_A/dt = -kC_A$	$\ln C_A = \ln C_{A°} - kt$	$0.693k^{-1}$
Second Order	$dC_A/dt = -kC_A^2$ where $C_{A°} = C_{B°}$	$C_A^{-1} = C_{A°}^{-1} + kt$	$k^{-1}C_{A°}^{-1}$
Pseudo-First Order	$dC_A/dt = (kC_{B°})C_A = -k'C_A$ where $C_{A°} \ll C_{B°}$	$\ln C_A = \ln C_{A°} - k't$	$0.693k^{-1}$

[a]Integration between limits of $C_{A°} = 0$ and C_A and $t_0 = 0$ and t.

$$\frac{dC_A}{dt} = -kC_A^a \qquad [3\text{-}75]$$

Zero Order Reaction, a = 0. Reaction rate is independent of concentration of A.

$$\frac{dc_A}{dt} = -k \qquad [3\text{-}76]$$

This actually states a pseudo-zero-order situation where something other than concentration of A is driving the reaction. Integration yields, $C_A = C_{A_1} - kt$, where C_{A_0} is the concentration of A at t = 0.

First Order Reaction, a = 1. Rate of the reaction is dependent solely on C_A.

$$\frac{dC_A}{dt} = -kC_A \qquad [3\text{-}77a]$$

$$\frac{dC_A}{C_A} = -kdt \qquad [3\text{-}77b]$$

Integration yields

$$\ln C_A = -kt + constant \qquad [3\text{-}77c]$$

A plot of C_A as a function of t yields a straight line of slope $-k$, where k has units of time^{-1}. If C_{A_0} is the concentration of A at t = 0, integration of equation 3-77b between the limits of C_{A_0} and C_A results in

$$\ln c_A - \ln c_{A0} = -kt \qquad [3\text{-}78a]$$

The exponential form of the first order rate law is

$$c_A = c_{A0}e^{-kt} \qquad [3\text{-}78b]$$

The *half-life* or *half-time* of the reaction ($t_{1/2}$) is defined when $C_A = 1/2\ C_{A0}$. Substitution of $t_{1/2}$ and $1/2\ C_{A0}$ in equation 3-78b results in $t_{1/2} = 0.693k^{-1}$. The half-life is independent of concentration for the first order reaction.

Second Order Reaction, (a + b = 2). The bimolecular reaction,

$$aA + bB \longrightarrow \text{products}$$

may take several forms. The second order reaction exhibits a rate that is concentration-dependent. The rate law for two cases common to soil systems is developed here.

The first case considers initial conditions where $C_{A0} = C_{B0}$; this case also applies to reactions of the type

$$A + A \longrightarrow \text{products}$$

In this instance, reaction rate is proportional to C_A^2.

$$\frac{dC_A}{dt} = -kC_A^2 \qquad [3\text{-}80a]$$

$$\frac{dC_A}{C_A^2} = -kdt \qquad [3\text{-}80b]$$

Integration between the limits of C_{A0} and C_A resolves to the linearized form of the rate law for the second order reaction

$$\frac{1}{C_A} = \frac{1}{C_{A0}} + kt \qquad [3\text{-}81]$$

A plot of C_A^{-1} as a function of t gives a line of slope k and intercept C_{A0}^{-1}. The half-life for the second order reaction is dependent on initial concentration, $t_{1/2} = k^{-1}\ C_{A0}^{-1}$.

The second case considers initial conditions where $C_{A0} \ll C_{B0}$; this describes the *pseudo-first order reaction*. The pseudo-first-order reaction empirically describes many soil reactions, for example, where the concentration of adsorbate $\ll$ the concentration of adsorbent.

$$\frac{dC_A}{dt} = -(kC_{B_0})C_A = -k'C_A \qquad [3\text{-}82]$$

In equation 3-82, C_{B_0} is incorporated into the pseudo-first-order rate constant, k'. Changes in C_{B_0} are assumed to be negligible in comparison with C_{A_0} because of the much greater concentration of component B relative to component A.

3.4.3   Complex Reaction Mechanisms

Many processes of interest to soil chemists and biochemists involve reaction sequences considerably more involved than those described by elementary reactions. These complex reactions take the form of opposing, parallel, and concurrent reaction mechanisms as well as mixed heterogeneous reactions.

Opposing Reactions. Opposing or reversible reactions proceed in both the forward and backward direction:

$$A \rightleftharpoons C$$

$$A + B \rightleftharpoons C$$

$$A \rightleftharpoons C + D$$

If the concentrations of other reactants and products are constant, a reaction of the type

$$A + \text{other reactants} \rightleftharpoons C + \text{other products}$$

simplifies to

$$A \rightleftharpoons C$$

If the forward and backward reaction rates are apparent first order with respect to A and C, the reaction can be described empirically as two opposing first-order reactions:

$$\frac{dC_A}{dt} = -k_1 C_A + k_{-1} C_C \qquad [3\text{-}83a]$$

where k_1 and k_{-1} represent the reaction rates for the forward and backward reaction, respectively. If the initial concentration of A is C_{A_0}, then $C_C = (C_{A_0} - C_A)$,

$$\frac{dC_A}{dt} = -k_1 C_A + k_{-1}(C_{A_0} - C_A) \qquad [3\text{-}83b]$$

Integration between the limits of C_{A_0} and C_A for t = 0 and t, respectively, results in

$$c_A = \frac{c_{A_0}}{(k_1 + k_{-1})} (k_{-1} + k_1 \, e^{-(k_1 + k_{-1})t}) \qquad [3\text{-}84]$$

Notice that in the absence of the reverse reaction (when $k_{-1} = 0$), equation 3-84 reduces to the first order rate law described in equation 3-78b. An alternative form of equation 3-84 can be developed for the equilibrium condition of equation 3-83a,

$$\frac{dC_A}{dt} = 0 = -k_1 C_{A_{eq}} + k_{-1} C_{C_{eq}} \qquad [3\text{-}85]$$

which results in

$$\frac{k_1}{k_{-1}} = \frac{C_{C_{eq}}}{C_{A_{eq}}} \qquad [3\text{-}86]$$

Since $C_{C_{eq}} = C_{A_0} - C_{A_{eq}}$,

$$\frac{k_1}{k_{-1}} = \frac{C_{A_0} - C_{A_{eq}}}{C_{A_{eq}}} \qquad [3\text{-}87a]$$

Solving for C_{A_0},

$$C_{A_0} = C_{A_{eq}} \frac{(k_1 + k_{-1})}{k_{-1}} \qquad [3\text{-}87b]$$

Substitution of equation 3-87b into equation 3-84 and rearrangement results in

$$\frac{C_A}{C_{A_{eq}}} = 1 + \frac{k_1}{k_{-1}} e^{-(k_1 + k_{-1})t} \qquad [3\text{-}88a]$$

The ratio k_1/k_{-1} defines a conditional equilibrium constant K and $k \equiv (k_1 + k_{-1})$; substitution into equation 3-88a yields

$$\frac{C_A}{C_{A_{eq}}} = 1 + Ke^{-kt} \qquad [3\text{-}88b]$$

Equation 3-88b can be rearranged to produce the following linear form.

$$\ln \left(\frac{C_A - C_{A_{eq}}}{C_{A_{eq}}} \right) = \ln(K) - kt \qquad [3\text{-}88c]$$

If a reaction of interest obeys rate law for opposing first order reactions, a plot of experimental measurements of $\ln[(C_A - C_{Aeq})/C_{Aeq}]$ versus t will give a straight line of slope $-k$ and intercept $\ln K$.

Competing Reactions. Competing, parallel, concurrent, or simultaneous reactions are of the general type,

If k_1 and k_2 are both first order, the rate law becomes

$$\frac{dC_A}{dt} = -(k_1 + k_2)C_A \qquad [3\text{-}89a]$$

$$\frac{dC_A}{C_A} = -(k_1 + k_2)dt \qquad [3\text{-}89b]$$

Integrating equation 3-89b between the limits of C_{A_0} and C_A for $t = 0$ and t gives

$$C_A = C_{A_0}e^{-(k_1+k_2)t} \qquad [3\text{-}89c]$$

If $k \equiv (k_1 + k_2)$, equation 3-89c takes the same form as the simple first order expression (equation 3-78b) and $t_{1/2} = 0.693k^{-1}$ for the *net* reaction. The linear form of equation 3-89c is

$$\ln C_A - \ln C_{A_0} = -(k_1 + k_2)t \qquad [3\text{-}89d]$$

A plot of $\ln C_A - \ln C_{A_0}$ against t results in a straight line of slope $(k_1 + k_2)$. To develop the individual rate constants, the rates of formation of products must be considered.

$$\frac{dC_C}{dt} = k_1 C_A \qquad [3\text{-}90]$$

Substitution of equation 3-89c into equation 3-90 and rearrangement gives

$$dC_C = k_1 C_{A_0}e^{-(k_1+k_2)t}dt \qquad [3\text{-}91]$$

Integrating equation 3-91 between the limits of $C_{C_0} = 0$ and C_C results in

$$C_C = \frac{k_1 C_{A0}}{(k_1 + k_2)} [1 - e^{-(k_1+k_2)t}] \qquad [3\text{-}92a]$$

A similar result is obtained for D.

$$C_D = \frac{k_2 C_{A0}}{(k_1 + k_2)} [1 - e^{-(k_1+k_2)t}] \qquad [3\text{-}92b]$$

From equations [3-92a] and [3-92b],

$$\frac{k_1}{k_2} = \frac{C_C}{C_D} \qquad [3\text{-}93]$$

According to equation 3-93, the relative yield of products of competing reactions at some finite period of time is determined only by the rate constants. The analytical information necessary to determine values of k_1 and k_2 is C_A at various times and C_C and C_D at *one* time.

Consecutive Reactions. Consider the consecutive reactions.

$$A \xrightarrow{k_1} C \xrightarrow{k_2} E$$

where k_1 and k_2 are both first order. If $C_{A0} = C_A + C_C + C_E$, solution of the appropriate rate law expressions results in the following for the concentrations of products and reactants with time.

$$C_A = C_{A0} = C_{A0} e^{-k_1 t} \qquad [3\text{-}94]$$

$$C_C = C_{A0} \frac{k_1}{k_1 - k_2} (e^{-k_1 t} + e^{-k_2 t}) \qquad [3\text{-}95]$$

$$C_E = C_{A0} \left(1 + \frac{k_2}{k_1 - k_2} e^{-k_1 t} - \frac{k_1}{k_1 - k_2} e^{-k_2 t}\right) \qquad [3\text{-}96]$$

Mixed Reactions. The principles demonstrated for these various types of complex reactions can be extended to mixed reactions of multiple steps involving differing orders of reaction; solution, however, becomes mathematically cumbersome and is probably unwarranted as the mechanisms evoked cannot be conclusively demonstrated. Solution of rate law for mixed heterogeneous reactions is more usefully developed through model optimization and approximation techniques. An example of this approach for empirical description of dynamic data is developed in section 3.4.5.

3.4.4   Common Empirical Models

The complexity of kinetic rate law for mixed heterogeneous systems such as soils, coupled with uncertainties as to the mechanisms involved have lead to the use of numerous empirical kinetic models for describing dynamic processes. These models are used because they provide useful empirical descriptions of observations; they cannot be used to infer processes without auxiliary data to support proposed mechanisms. Common models considered here for description of dynamic soil processes are fractional order kinetics, the Elovich equation, the Michaelis–Menten equation, Monod kinetics, and multicompartment models.

Fractional Order Kinetics. The power function, $f(x) = ax^k$, has been used by soil chemists since the late nineteenth century (Cameron, 1911), mainly to describe concentration dependence of sorptivity (the Freundlich equation), but also to describe time-dependent data (Bloom and Erich, 1987). Fractional order kinetics for a reactant A going to products is expressed

$$\frac{dc_A}{dt} = -k_1 c_A^n \qquad\qquad [3\text{-}97a]$$

$$\frac{dc_A}{c_A^n} = -k_1 t \qquad\qquad [3\text{-}97b]$$

where n is an empirical fitting constant (n $\neq$ an integer). The integrand form of the fractional order rate law is

$$C_A^{(1-n)} - C_{A_0}^{(1-n)} = (n + 1)kt \qquad\qquad [3\text{-}98]$$

Elovich Equation. The Elovich equation is commonly evoked to describe sorptive processes and has experienced recent popularity in soils literature (Chein and Clayton, 1980; Hodges and Johnson, 1987; Sikora et al., 1991). The model, however, is not indicative of any one mechanism and in fact can describe diffusion-control versus chemisorptive control of sorption equally well (see, for example, the discussion by Sparks, 1986). The Elovich equation takes the general form

$$\frac{dq}{dt} = \alpha e^{-\beta q} \qquad\qquad [3\text{-}99]$$

where the appearance of a product q with t is a function of the empirical fitting constants α and β. In terms of a product C, with initial concentration C_{C_0} at t = 0,

$$\int_{q=C_0}^{q=C} e^{\beta q} dq = \alpha \int_{t=0}^{t} dt \qquad\qquad [3\text{-}100a]$$

which simplifies to the linear form

$$(C_{C_0} - C_C) = \beta^{-1}\ln\alpha\ \beta + \beta^{-1}\ln t \qquad [3\text{-}100b]$$

The Elovich equation is empirically useful in that irregularities in kinetic data within and among experiments are evident, and these irregularities may lend interpretive insight regarding the nature of the dynamic processes involved.

Michaelis–Menten Equation. The kinetics of biodegradative processes are frequently described by a mixed model involving an enzyme-catalyzed step.

$$E + S \underset{k_{-1}}{\overset{k_1}{\rightleftharpoons}} ES \overset{k_2}{\longrightarrow} P + E$$

where E, S, ES, and P represent the enzyme, substrate, enzyme-substrate complex, and product. The rate of formation of P relative to concentrations of E and S is[5]

$$\frac{dC_P}{dt} = \frac{k_1 k_2}{(k_{-1} + k_2)}\ C_E C_S \qquad [3\text{-}101]$$

or in terms of the total enzyme E_t, $C_{E_t} = C_E + C_{ES}$,

$$\frac{dC_P}{dt} = \frac{k_2 C_{E_t} C_S}{K_M + C_S} \qquad [3\text{-}102]$$

where K_M is the Michaelis–Menten constant ($K_M \equiv [k_{-1} + k_2]/k_1$). From equation 3-102, the maximum rate of this reaction (V_{max}) occurs when $C_S \gg K_M$,

$$\left(\frac{dC_P}{dt}\right)_{max} \equiv V_{max} = k_2 C_{E_t} \qquad [3\text{-}103]$$

Substitution of equation 3-103 into equation 3-102 gives the Michaelis–Menten equation

$$\frac{dC_P}{dt} = \frac{V_{max} C_S}{K_M + C_S} \qquad [3\text{-}104]$$

The Michaelis–Menten equation assumes a constant amount of reactive material (enzyme), so it best describes biodegradation occurring in nongrowing biotic systems (Alexander and Scow, 1989).

Several linear forms of the Michaelis–Menten equation are used to develop K_M and V_{max} from experimental data for the initial rate of reaction (v_0). The most useful takes the form

$$v_0 = -K_M \frac{v_0}{C_S} + V_{max} \qquad [3\text{-}105]$$

A plot of v_0 as a function of v_0/C_S has a slope $= -K_M$ and an intercept $= V_{max}$.

Monod Kinetics. Monod (1949) mathematically described the biodegradative relationship existing when a population of cells in a favorable environment utilizes an organic molecule as a carbon source. The solution, describing depletion of a substrate S with time, takes the same form as the Michaelis–Menten equation (equation 3-104).[6]

$$\frac{dC_S}{dt} = \frac{\mu_{max}C_S}{K_S + C_S} \qquad [3\text{-}106]$$

where μ_{max} is the maximum specific growth rate of cells (occurring at higher substrate concentrations) and the constant K_S represents the substrate concentration, C_S, when $-dC_S/dt$ is one-half the maximal rate (Alexander and Scow, 1989). Monod kinetics appear to describe biodegradation when C_S is approximately the same order of magnitude as K_S.

Equation 3-106 can be adapted to include growth of cells as they utilize S (*Monod-with-growth-kinetics*; Alexander and Scow, 1989) through consideration of the initial substrate concentration C_{S_0} and the substrate concentration necessary to produce the initial population of biodegraders (C_{X_0}).

$$-\frac{dC_S}{dt} = \frac{\mu_{max}C_S(C_{S_0} + C_{X_0} - C_S)}{K_S + C_S} \qquad [3\text{-}107]$$

The integrated form of equation [3-107] is

$$K_S\ln\frac{C_S}{C_{S_0}} = (C_{S_0} + C_{X_0} + K_S)\ln\left(\frac{C_X}{C_{X_0}}\right) - (C_{S_0} + C_{X_0})\,\mu_{max}t \qquad [3\text{-}108]$$

where C_X is the substrate concentration producing the population density. The Monod equation with simplification will yield other common degradation equations (Simkins and Alexander, 1984).

Multicompartment Models. Degradation and sorption/desorption are concurrent processes for molecules undergoing chemical and/or biological transformation in soils; therefore, the kinetic expression of the fate of such molecules in complex environments requires mixed heterogeneous models. Hamaker and Goring (1976) described degradation using a *two-compartment model* where a substrate is distributed between an available and an unavailable compartment—for example, between soil solution and a sorbed phase.

$$S_{\text{sorbed}} \underset{k_1}{\overset{k_{-1}}{\rightleftharpoons}} S_{\text{solution}} \xrightarrow{k_2} \text{Products}$$

Multicompartment models can take numerous forms. A common approach is to allow degradation to occur in both compartments, a fast-reacting compartment and a slow-reacting compartment—for example, soil solution where biodegradative processes occur, and a solid surface where surface hydrolysis may occur.

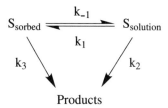

A simplified solution for this reaction sequence is (Scow et al., 1986).

$$\frac{dC_{S_{\text{solution}}}}{dt} = k_{-1} C_{S_{\text{sorbed}}} - (k_1 + k_2) C_{S_{\text{solution}}} \qquad [3\text{-}109]$$

$$\frac{dC_{S_{\text{sorbed}}}}{dt} = k_1 C_{S_{\text{solution}}} - (k_{-1} + k_3) C_{S_{\text{sorbed}}} \qquad [3\text{-}110]$$

Gustafson and Holden (1990) extended the two-compartment model to a general consideration of i compartments degrading S by unique first-order rate constants k_i,

$$C_S = \sum_i C_{S_i} e^{-k_i t} \qquad [3\text{-}111]$$

If p_i is the fraction of C_{S_0} occurring in the ith compartment, equation 3-111 becomes

$$C_S = \sum_i p_i C_{S_0} e^{-k_i t} \qquad [3\text{-}112]$$

Extension of the multicompartment model over a continuum of compartments yields

$$C_S = \int p(k) C_{S_0} e^{-k_i t} dk \qquad [3\text{-}113]$$

This allows the relative distribution of S to be described by a probability density function $p(k)$. The following form of equation 3-113 is obtained using the gamma distribution to describe $p(k)$ (Johnson and Kotz, 1970).

$$C_S = C_{S_0}(1 + \beta t)^{-\alpha} \qquad\qquad [3\text{-}114]$$

where degradation of S is modeled with the use of the empirical fitting parameters α and β.

Table 3.7 summarizes the various kinetic models commonly used to empirically describe time-dependent data describing soil processes.

3.4.5 Nonlinear Model Optimization and Approximation of Time-Dependent Data.

The foregoing examples for treating time-dependence of chemical reactivity demonstrate that, because of complexities associated with kinetic treatment of data, simplifying assumptions and linearization techniques are frequently employed. Nonlinear model optimization and approximation is an alternative approach for treating time-dependent data that has been described by Reilly and Blau (1974): "In the past much effort and mathematical ingenuity has been expended on getting models into convenient forms for . . . the resulting linear model. . . . In most cases this convenience is achieved at the cost of distorting the error structure. When curves are fitted by eye or simple least squares constant variance is almost invariably assumed. . . . It is generally much better to write the model in such a way that the errors are independent and of constant variance."[7] The following example outlines this approach for the case of a two-compartment model applied to concurrent sorption/desorption and biodegradation.

McCall and Agin (1985) investigated picloram fate in seven soils by measuring the loss in applied mass with time after application to soils in closed systems, concurrent with measurement of the solid–liquid partitioning, again, as a function of time. Their data can be recast to provide picloram phase distribution and degradation with time.[8] Therefore, substrate (picloram) fate in their test systems takes the form of the two-compartment model of Hamaker and Goring (1976).

$$S_{\text{sorbed}} \underset{k_1}{\overset{k_{-1}}{\rightleftharpoons}} S_{\text{solution}} \xrightarrow{k_2} \text{Products}$$

Solution of models such as these can be accomplished through *compartment models,* a special class of nonlinear models used in chemical kinetics and pharmacokinetics where responses are defined by a linear system of ordinary differential equations (Bates and Watts, 1988). The fate of substrate is described by two rate expressions,

$$\frac{dC_{S_{\text{solution}}}}{dt} = k_{-1}C_{S_{\text{sorbed}}} - (k_1 + k_2)C_{S_{\text{solution}}} \qquad\qquad [3\text{-}109]$$

$$\frac{dC_{S_{\text{sorbed}}}}{dt} = k_1 C_{S_{\text{solution}}} - k_{-1}C_{S_{\text{sorbed}}} \qquad\qquad [3\text{-}115]$$

TABLE 3.7 Empirical Rate Laws Frequently Applied to Soil Systems

Forward Reaction Rate Laws (dC_A/dt)	Description
	$aA + \text{other reactants} \xrightarrow{k} \text{products}$
$-k$	Zero order reaction ($a = 0$)
$-kC_A$	First order reaction ($a = 1$)
$-kC_A^n$	Fractional order reaction ($a = n \neq$ an integer)
$\alpha e^{-\beta C_P}$	Elovich kinetics (for appearance of a product P, dC_P/dt)
	$aA + bB + \text{other reactants} \xrightarrow{k} \text{products}$
$-kC_A^2$	Second order reaction ($a = 2$ or $C_{A_0} = C_{B_0}$)
$-kC_A C_B$	Overall second order reaction ($a = b = 1$)
$-k'C_A = -(kC_B)C_A$	Pseudo-first order reaction ($C_{A_0} \ll C_{B_0}$)
	$aA + \text{other reactants} \underset{k_{-1}}{\overset{k_1}{\rightleftharpoons}} cC + \text{other products}$
$-k_1 C_A + k_{-1}(C_{A_0} - C_A)$	First order opposed by first order ($a = c = 1$)

$$aA + \text{other reactants} \quad \begin{array}{c} \xrightarrow{k_1} cC + \text{other products} \\ \xrightarrow{k_2} bB + \text{other products} \end{array}$$

$-(k_1 + k_2)C_A$	Competing first order ($a = 1$)

$$aA + E \underset{k_{-1}}{\overset{k_1}{\rightleftharpoons}} aEA \xrightarrow{k_2} E + \text{products}$$

$V_{max}C_A/(K_M + C_A)$	Michaelis–Menten kinetics [V_{max} = maximum reaction rate; $K_M = (k_{-1} + k_2)/k_1$] ($a = 1$, for appearance of a product P, dC_P/dt)
$-\mu_{max}C_A/(K_S + C_A)$	Monod kinetics [μ_{max} = maximum specific growth rate; $K_S = C_A$ at 1/2 maximum growth rate]
$-\mu_{max}C_A(C_{A_0} + C_{X_0} + C_A)/(K_S + C_A)$	Monod-with-growth kinetics [C_{X_0} = concentration required to produce the initial population of degraders]

$$aA \xrightarrow{k_2} \text{products}, \quad k_{-1} \updownarrow k_1, \quad cC \xrightarrow{k_3}$$

$-(k_1 + k_2)C_A + k_{-1}C_C$	Two-compartment model ($a = c = 1$)

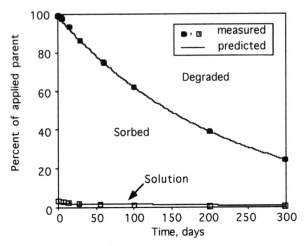

FIGURE 3.3 Predicted fate of picloram in soil as determined by optimized nonlinear fitting of a two-compartment kinetic model. [After Sims et al., 1992.]

These ordinary differential equations can be solved computationally by various integration algorithms and when coupled with a general nonlinear optimization process (Reilly and Blau, 1974) yield results such as those of Fig. 3-3 and Table 3.8, where parameter estimation is based on model-fitting by the likelihood function.

This approach offers the following advantages over more conventional means of modeling time-dependent data:

- The model simultaneously treats the processes of sorption/desorption and biodegradation to allow for a clearer understanding of their interrelationship.

TABLE 3.8 Optimized Rate Constants and Overall Model Fit for a Two-Compartment Kinetic Model of Simultaneous Sorption/Desorption and Degradation of Picloram in Soil

Soil	Rate Constant, Day^{-1}			Model Fit, Percent of Variation Explained
	Sorption k_1	Desorption k_{-1}	Degradation k_2	
Catlin	0.2230	24.6	0.532	98
Commerce	0.0148	17.0	1.65	51
Fargo	0.0537	27.1	0.926	89
Holridge	0.1410	21.5	0.610	99
Kawkawlin	0.0905	23.3	0.402	85
Norfolk	0.1120	22.9	0.360	82
Walla-Walla	0.0717	19.0	0.865	98

Modeled from the data of McCall and Agin, 1985

The modeled results, for the data for the present case, indicate that the relative rate of degradation (k_2) compared to rates for sorption (k_1) and desorption (k_{-1}) is the factor controlling loss of mass from the system (Table 3.8). Meanwhile, conventional data on whole soil half-life of the substrate is generated and is in good agreement with that obtained by conventional apparent first-order fits using linearization techniques (Table 3.9).

- Estimates are generated for the error bounds around individual kinetic parameters as well as for the model as a whole. This allows for model discrimination in the instance where two equally plausible mechanisms for reaction are being considered and additionally allows for determination of the effect of added parameters on the modeled fit of the data (Wolt et al., 1992).

3.4.6 Activation Energy

The commonly observed increase in reaction rate with increase in temperature was stated by Arrhenius as

$$k = Ae^{-E^*/RT} \qquad [3\text{-}116a]$$

where the pre-exponential factor A—the *frequency factor*—is an empirical constant for the reaction and E^* is the *activation energy*. Activation energy represents the forces overcome during a chemical reaction—the *activation barrier*. From equation 3-116a, E^* is inversely proportional to k. Activation energy is used to distinguish chemical versus physical control of reactions; in soils, $E^* < 42$ kJ mol^{-1} typically identifies diffusion-limited control over processes (Sparks, 1986).

The logarithmic form of equation 3-116a is

TABLE 3.9 Comparison of Picloram Half-Life in Soil as Determined by Apparent First-Order Kinetics and a Two-Compartment Model

	Substrate Half-Life, Days	
Soil	Apparent First Order[a]	Two-Compartment Model
Catlin	148	144
Commerce	440	522
Fargo	372	385
Holridge	172	175
Kawkawlin	429	445
Norfolk	361	393
Walla-Walla	229	219

[a]McCall and Agin, 1985.

$$\ln k = \ln A - \frac{E^*}{RT} \qquad \text{[3-116}b\text{]}$$

which on rearrangement gives

$$E^* = -RT\ln k + RT\ln A \qquad \text{[3-116}c\text{]}$$

The empirical constants E^* and A can be obtained from equation 3-116b through a linear plot of ln k against T^{-1}.

The Arrhenius equation in the form of equation 3-116c is reminiscent of the free energy relationship

$$\Delta F^\circ = -RT\ln K_{eq} \qquad \text{[3-117}a\text{]}$$

Evaluation of change in ΔF° for a change in T first proceeds with rearrangement of equation 3-117a and obtaining the partial differential with respect to T at constant P.

$$\ln K_{eq} = -\frac{1}{R}\frac{\Delta F^\circ}{T} \qquad \text{[3-117}b\text{]}$$

$$\left[\frac{\partial \ln K_{eq}}{\partial T}\right]_P = -\frac{1}{R}\left(\frac{T[\partial \Delta F^\circ/\partial T]_P - \Delta F^\circ}{T^2}\right) \qquad \text{[3-118]}$$

Since, $\Delta F^\circ = \Delta H^\circ - T\Delta S^\circ$, equation 3-118 becomes

$$\left[\frac{\partial \ln K_{eq}}{\partial T}\right]_P = -\frac{1}{R}\left(\frac{T[-\Delta S^\circ] - \Delta H^\circ + T\Delta S^\circ}{T^2}\right) \qquad \text{[3-119}a\text{]}$$

$$\left[\frac{\partial \ln K_{eq}}{\partial T}\right]_P = \frac{\Delta H^\circ}{RT^2} \qquad \text{[3-119}b\text{]}$$

Equation 3-119b is the van't Hoff equation and can also be expressed

$$d\ln K_{eq} = \frac{\Delta H^\circ}{RT^2}\,dT \qquad \text{[3-119}c\text{]}$$

For a reaction,

$$A + B \underset{k_{-1}}{\overset{k_1}{\rightleftharpoons}} C$$

$K_{eq} = k_1/k_{-1}$, from equation 3-13. Substitution into 3-119c yields

$$\frac{d(\ln k_1)}{dT} - \frac{d(\ln k_{-1})}{dT} = \frac{\Delta H°}{RT^2}$$ [3-120]

at constant P, where

$$\frac{d(\ln k_1)}{dT} = \frac{E_1^*}{RT^2} \quad \text{and} \quad \frac{d(\ln k_{-1})}{dT} = \frac{E_{-1}^*}{RT^2}$$ [3-121]

Reaction rate (first order with respect to A) can be expressed as

$$\text{rate} = -\frac{dC_A}{dT} = k_1 C_A = Z e^{-E_1^*/RT}$$ [3-122]

where Z is the number of collisions per second (Castellan, 1971). Substitution of the Arrhenius equation (equation 3-116a) into equation 3-122 results in, $AC_A = Z$. Thus, the frequency factor appears representative of the fraction of collisions occurring for reacting molecules with energies $> E^*$.

3.4.7 Irreversible Thermodynamics

A general theory of reaction rates, linking static and dynamic consideration of chemical reactivity, may be developed from quasi-thermodynamic principles through the assumption that systems slightly removed from equilibrium can be treated as systems at equilibrium.[8] *Irreversible thermodynamics* is a method for treating systems near equilibrium and is built on three basic assumptions:

1. Entropy is created for every irreversible process and for every finite rate process.
2. Entropy of an isolated system is maximum at equilibrium.
3. Any molecular process at equilibrium occurs at the same rate as the reverse of that process—the *principle of microscopic reversibility*.

The change in entropy for a process (dS) can be related to the heat (dq) absorbed by the process in a system through the inequality

$$dS \geq \frac{dq}{T}$$ [3-123a]

that can also be stated

$$dS - \frac{dq}{T} \geq 0 \qquad [3\text{-}123b]$$

The entropy increase in the system plus its surroundings—the *entropy production,* $d\sigma$—is defined such that $d\sigma \geq 0$; equation 3-123b becomes

$$dS - \frac{dq}{T} = d\sigma \qquad [3\text{-}124a]$$

which on rearrangement is

$$Td\sigma = TdS - dq \qquad [3\text{-}124b]$$

Entropy production is positive for an irreversible process and zero for a reversible process.

For a constant T, P process, $TdS = d(TS)$ and $dq = dH$; equation 3-124b becomes

$$Td\sigma = -d(H - TS) \qquad [3\text{-}125a]$$

or,

$$Td\sigma = -dF \qquad [3\text{-}125b]$$

From equation 3-33

$$dF = \left[\frac{\partial F}{\partial \xi}\right]_{T,P} d\xi \qquad [3\text{-}126]$$

so

$$Td\sigma = -\left[\frac{\partial F}{\partial \xi}\right]_{T,P} d\xi \qquad [3\text{-}127]$$

The *reaction affinity,* **A,** is defined

$$A \equiv -\left[\frac{\partial F}{\partial \xi}\right]_{T,P} \qquad [3\text{-}128]$$

(Reaction affinity is positive for a spontaneous reaction in the forward direction.) Equation 3-127 becomes

$$Td\sigma = Ad\xi \qquad [3\text{-}129]$$

The rate of entropy increase, $d\sigma/dt$, from equation 3-129 is

$$\frac{d\sigma}{dt} = \frac{A}{T}\frac{d\xi}{dt} \qquad [3\text{-}130]$$

Since $d\sigma \geq 0$, $d\sigma/dt \geq 0$; it follows that

$$A\frac{d\xi}{dt} \geq 0 \qquad [3\text{-}131]$$

Equation 3-131 is DeDonder's inequality and can be used to develop a useful linear law by combining rate law with thermodynamic law.

Consider the reaction

$$A + B \underset{k_{-1}}{\overset{k_1}{\rightleftharpoons}} C$$

The rate equation for this reaction is

$$\frac{dC_A}{dt} = -\frac{1}{V}\frac{d\xi}{dt} = -k_1 C_A C_B + k_{-1} C_B \qquad [3\text{-}132a]$$

This equation can be recast as

$$\frac{1}{V}\frac{d\xi}{dt} = k_1 C_A C_B \left[1 - \frac{k_{-1} C_B}{k_1 C_A C_B} \right] \qquad [3\text{-}132b]$$

Since, $k_1/k_{-1} = K$ and $C_C/C_A C_B = Q$,

$$\frac{1}{V}\frac{d\xi}{dt} = k_1 C_A C_B \left[1 - \frac{Q}{K} \right] \qquad [3\text{-}133]$$

As the reaction approaches equilibrium, $k_1 C_A C_B \rightarrow r$ (the exchange rate for the reaction). Substituting r and rearranging results in

$$\frac{d\xi}{dt} = Vr \left[1 - \frac{Q}{K} \right] \qquad [3\text{-}134]$$

Two additional free energy relationships need be considered,

$$\frac{dF}{d\xi} = \Delta F^\circ + RT \ln Q \qquad [3\text{-}135]$$

$$\Delta F^\circ = RT\ln K \qquad [3\text{-}136]$$

Combining equations 3-128, 3-135, and 3-136 gives

$$\left[\frac{dF}{d\xi}\right]_{T,P} = RT\ln \frac{Q}{K} = -\mathbf{A} \qquad [3\text{-}137a]$$

or,

$$\frac{Q}{K} = e^{-\mathbf{A}/RT} \qquad [3\text{-}137b]$$

Combination of equation 3-134 with the expansion of equation 3-137b yields the following approximation near equilibrium.

$$\frac{d\xi}{dt} = Vr \frac{\mathbf{A}}{RT} \qquad [3\text{-}138]$$

The reaction affinity **A,** a thermodynamically-derived parameter, is proportional to the exchange rate of reaction r, a kinetic parameter.

NOTES

1. The development here follows that found in W. Stumm and J. J. Morgan, 1981, pp. 8–11.
2. The cross differentiation proceeds as follows:

$$dH = TdS + VdP$$

Differentiate with respect to T at constant P.

$$\left(\frac{\partial H}{\partial T}\right)_P = T\left(\frac{\partial S}{\partial T}\right)_P + V\left(\frac{\partial P}{\partial T}\right)_P$$

$$\left(\frac{\partial H}{\partial T}\right)_P = T\left(\frac{\partial S}{\partial T}\right)_P + 0$$

Differentiate with respect to P at constant T.

$$\left(\frac{\partial H}{\partial P}\right)_T = T\left(\frac{\partial S}{\partial P}\right)_T + V\left(\frac{\partial P}{\partial P}\right)_T$$

$$\left(\frac{\partial H}{\partial P}\right)_T = T\left(\frac{\partial S}{\partial P}\right)_T + V$$

Differentiate with respect to P at constant T.

$$\left[\frac{\partial}{\partial P}\left(\frac{\partial H}{\partial T}\right)_P\right]_T = T\left[\frac{\partial}{\partial P}\left(\frac{\partial S}{\partial T}\right)_P\right]_T$$

Differentiate with respect to T at constant P.

$$\left[\frac{\partial}{\partial T}\left(\frac{\partial H}{\partial P}\right)_T\right]_P = T\left[\frac{\partial}{\partial T}\left(\frac{\partial S}{\partial P}\right)_T\right]_P$$
$$+ \left(\frac{\partial S}{\partial P}\right)_T + \left(\frac{\partial V}{\partial T}\right)_P$$

The order of differentiation is immaterial, so

$$\left[\frac{\partial}{\partial P}\left(\frac{\partial H}{\partial T}\right)_P\right]_T = \left[\frac{\partial}{\partial T}\left(\frac{\partial H}{\partial P}\right)_T\right]_P$$
$$\left[\frac{\partial}{\partial P}\left(\frac{\partial S}{\partial T}\right)_P\right]_T = \left[\frac{\partial}{\partial T}\left(\frac{\partial S}{\partial P}\right)_T\right]_P$$

thus,

$$\left(\frac{\partial S}{\partial P}\right)_T = -\left(\frac{\partial V}{\partial T}\right)_P$$

3. The standard state for any chemical element is defined as the most stable phase of the element at $298.15°K$ and 101.33 kPa. In its standard state, chemical potential of the element equals zero.

4. Selection of concentration scales for expression of chemical potential of condensed phase components is largely a matter of convenience. The *rational scale,* where concentrations are expressed in terms of mole fractions (χ_i) applies well to solvents in solution where $\chi_i \rightarrow 1$. The *practical scale* of concentration is useful for solutes occurring in small amounts. A practical scale based on molarity is useful for analytical reasons, whereas a practical scale based on molality is sometimes preferred because it is independent of temperature. For ideal dilute solutions both molarity and molality are equivalent to mole fraction.

5. The mixed model for enzyme catalysis can be broken into three component reactions:

$$\text{Reaction}_1: E + S \xrightarrow{k_1} ES$$
$$\text{Reaction}_{-1}: ES \xrightarrow{k_{-1}} E + S$$
$$\text{Reaction}_2: ES \xrightarrow{k_2} P + E$$

Rate of the net reaction is described in terms of the appearance of product (P):

$$\text{rate} = \frac{dC_P}{dt} = k_2 C_{ES}$$

The concentration of the enzyme-substrate complex, C_{ES}, can be expressed in terms of

E and S.

$$\text{Reaction}_1: \quad \frac{dC_{ES}}{dt} = k_1 C_E C_S$$

$$\text{Reaction}_{-1}: \quad \frac{dC_{ES}}{dt} = -k_{-1} C_{ES}$$

$$\text{Reaction}_2: \quad \frac{dC_{ES}}{dt} = -k_2 C_{ES}$$

The net rate of change in C_{ES} for the mixed reaction is

$$\left(\frac{dC_{ES}}{dt}\right)_{net} = k_1 C_E C_S - k_{-1} C_{ES} - k_2 C_{ES}$$

Under steady state conditions, $(dC_{ES})/dt)_{net} = 0$. Solving for C_{ES},

$$C_{ES} = \frac{k_1 C_E C_S}{k_{-1} + k_2}$$

The net rate for the overall reaction becomes

$$\text{rate} = \frac{dC_P}{dt} = k_2 C_{ES} = \frac{k_1 k_2}{k_{-1} + k_2} C_E C_S \qquad [3\text{-}101]$$

Typical assay methods determine total enzyme concentration, C_{E_t}; therefore, equation 3-101 must be recast in terms of C_{E_t}.

$$C_{E_t} = C_E + C_{ES}$$

$$C_{E_t} = C_E + \frac{k_1}{k_{-1} + k_2} C_E C_S$$

$$C_{E_t} = C_E \left(1 + \frac{k_1}{k_{-1} + k_2} C_S\right)$$

Solving for C_E in terms of C_{E_t},

$$C_E = \frac{C_{E_t}(k_{-1} + k_2)}{k_{-1} + k_2 + k_1 C_S}$$

This expression describing C_E in terms of C_{E_t} can now be substituted back into the rate expression (equation 3-101).

$$\text{rate} = \frac{k_1 k_2 C_{E_t} C_S}{k_{-1} + k_2 + k_1 C_S} = \frac{k_2 C_{E_t} C_S}{\left(\dfrac{k_{-1} + k_2}{k_1}\right) + c_S}$$

Define the Michaelis–Menten constant, $K_M = (k_{-1} + k_2)/k_1$, and V_{max}, the maximum rate of the reaction when $C_S \gg K_M$,

$$\text{maximum rate} = \left(\frac{dC_P}{dt}\right)_{max} = V_{max} = k_2 C_{E_t} \qquad [3\text{-}103]$$

Substitution of K_M and V_M into the rate expression results in the Michaelis–Menten equation,

$$\frac{dC_P}{dt} = \frac{V_{max} C_S}{K_M + C_S} \qquad [3\text{-}104]$$

where $dC_P/dt \equiv v_0$ when initial rates are described.

The Michaelis–Menten constant, K_M, is characteristic of an enzyme system at a given temperature. Two common cases of the Michaelis–Menten equation are when $C_S \ll K_M$ and when $C_S \gg K_M$. When $C_S \ll K_M$,

$$v_0 = \frac{k_2 C_{E_t} C_S}{K_M} = k' C_{E_t} C_S$$

The overall reaction is second order; it is first order with respect to E_t and first order with respect to S. When $C_S \gg K_M$, $v_0 = k_2 C_{E_t}$. The overall reaction is first order; it is first order with respect to E_t, and zero order with respect to S.

6. The specific growth rate, μ, describes change in microbial biomass with time,

$$\frac{dC_B}{dt} = \mu C_B$$

where μ is constant for a particular micro-organism or microbial consortia, B, acting on a specific substrate, S. Rearrangement of this rate expression gives

$$\mu = \frac{dC_B}{dt}\frac{1}{C_B} = \frac{d(\ln C_B)}{dt}$$

The observed relationship between growth rate and substrate concentration is expressed by Monod kinetics as

$$\mu = \frac{\mu_{max} C_S}{K_S + C_S}$$

In a system closed to influx and outflow of S and B, and where S is the sole carbon source for B, the following equality will hold

$$C_{S_0} + \frac{dC_S}{dC_B} C_{B_0} = C_S + \frac{dC_S}{dC_B} C_B$$

where C_{S_0} and C_{B_0} represent initial concentrations of S and B, respectively, and dC_S/dC_B

$= Y^{-1}$ when Y is the yield in biomass from conversion of S. This equation can be rewritten

$$C_{S_0} + C_{X_0} = C_S + C_X$$

where $C_X = C_B Y^{-1}$; C_X represents the substrate concentration necessary to produce biomass concentration C_B. Similarly, $C_{X_0} = C_{B_0} Y^{-1}$. Thus, the expression for μ can be rewritten as

$$\mu = \frac{dC_B}{dt} \frac{1}{C_B} = \frac{Y^{-1}dC_X}{dt} \frac{1}{Y^{-1}C_X} = \frac{dC_X}{dt} \frac{1}{C_X} = \frac{d(\ln C_X)}{dt}$$

Rearrangement and substitution of the equation for Monod kinetics results in

$$\frac{dC_X}{dt} = \frac{\mu_{max} C_S C_X}{K_S + C_S}$$

Substitution of

$$C_X = C_{S_0} + C_{X_0} - C_S$$

results in

$$\frac{dC_X}{dt} = \frac{\mu_{max} C_S (C_{S_0} + C_{X_0} - C_S)}{K_S + C_S}$$

Also, since $C_{S_0} - C_S = C_X - C_{X_0}$, it follows that $-dC_S/dt - dC_X/dt$; thus,

$$-\frac{dC_S}{dt} = \frac{\mu_{max} C_S (C_{S_0} + C_{X_0} - C_S)}{K_S + C_S} \qquad \text{[3-107]}$$

Equation 3-107 is the expression of Monod-with-growth kinetics as a function of rate of substrate transformation $(-dC_S/dt)$. Equation 3-107 may be rewritten

$$-\frac{dC_S}{dt} = \frac{\mu_{max} C_S C_X}{K_S + C_S}$$

since, $C_{S_0} + C_{X_0} = C_S + C_X$. For the steady state condition, $d(\ln C_X)/dt = 0$ and $C_X = 1$. The steady state is one of no biomass accumulation; therefore, equation 3-107 reduces to an expression for Monod kinetics with no growth,

$$-\frac{dC_S}{dt} = \frac{\mu_{max} C_S}{K_S + C_S} \qquad \text{[3-106]}$$

7. Courtesy *Canadian Journal of Chemical Engineering.*

8. McCall and Agin (1985) express sorptive distribution as a function of time conventionally as a K_d, where the K_d in this instance is determined by rapid desorptive extraction of the substrate from soil after various times of incubation. The K_d expresses phase distribution

of substrate as a ratio of concentrations (See Section 8.3). Expressing phase distribution for the purpose of dynamic modeling requires conversion from a ratio of disparate concentration units (mg kg^{-1} and mg L^{-1} for the sorbed and solution phases, respectively) to a mass basis applied over the entire system (% of applied mass),

$$S_{solution} = R/[1 + (K_d/f]$$

$$S_{sorbed} = R - S_{solution}$$

where,

R = applied substrate recovered at time t (as a normalized percent of applied at time 0, $R_0 = 100$ at t_0)

$S_{solution}$ = percent of applied substrate in solution

S_{sorbed} = percent of applied substrate in the sorbed phase

K_d = desorptive K_d (given)

f = moisture fraction of soil as incubated (given)

9. The development here follows that of Castellan, G. W., 1971, *Physical Chemistry*, 2nd ed., Addison–Wesley, Reading, MA, pp. 777–780.

CHAPTER 4

MASTER VARIABLES

Master variables are those soil solution parameters that have an overt controlling influence over chemical concentration, speciation, and activity in the soil solution. Soil pH, ionic strength (I), equilibrium redox potential (E_H), and suspended colloidal materials—in their respective order of general influence—have comprehensive effects on soil solution composition. The effects of these master variables as impacted by temperature, pressure, and moisture should be considered in the application and interpretation of soil solution chemistry. The extent that a method for obtaining soil influences the values of solution master variables greatly affects its reliability as a measure of "unaltered" soil solution.

4.1 SOIL pH

4.1.1 Theoretical Basis of pH

Sørenson originally defined pH as $p_sH = -\log C_{H^+}$, but conventional definitions and electrochemical measurement techniques express pH as

$$pH = -\log A_{H^+} \qquad [4\text{-}1]$$

The correspondence of pH expressed as a concentration term (p_sH) relative to its expression as an activity term (pH) is governed by the operational definition of activity. Using the infinite dilution scale described in section 3.4.5,

$$-\log A_{H^+} = -\log C_{H^+} + \log \mathbf{f}_{H^+} \qquad [4\text{-}2]$$

As I $\rightarrow$ 0, $f_{H^+} \rightarrow$ 1, and $C_{H^+} \rightarrow A_{H^+}$. For practical application, this resolves to the approximation pH $\approx$ p_sH, since for I $<$ 0.1, p_sH and pH agree to within $\pm$ 0.02 units. (Because of the various problems inherent with measurement of soil solution pH, it is unlikely soil pH measurements would mandate greater resolution than that dictated by the forgoing approximation.)

4.1.2 Measurement of pH

Hydrogen-ion activities are not measurable in a strict thermodynamic sense because single-ion activities and junction potentials are not thermodynamically resolvable (Sparks, 1984). Measurement of pH by electromotive force (emf), however, provides a reasonable approximation of A_{H^+}.

Electrometric measurement of pH utilizes a voltmeter sensitive to extremely high resistances in series with reference and indicating membrane electrodes—often configured as combination electrodes. The potential of the hydrogen-indicating electrode is measured relative to a reference electrode. The reaction

$$H^+ + Ag_s + Cl^- \rightleftharpoons 0.5H_{2g} + AgCl_s$$

considered as an idealized measurement of A_{H^+} is illustrated by the following phase boundary program.

$$Pt_s, H_{2g} \quad | \quad H^+, Cl^- \quad | \quad AgCl_s, Ag_s$$

| H^+-indicating electrode | test solution | reference electrode |

where the single vertical line represents a phase boundary. From equation 3-64,

$$-n\mathcal{F}E = -n\mathcal{F}E° + RT\ln A_{H^+} + RT\ln A_{Cl^-} \qquad [4\text{-}3a]$$

or, on rearrangement,

$$pH = (E - E°)\frac{\mathcal{F}}{2.303RT} + \log A_{Cl^-} \qquad [4\text{-}3b]$$

where, n = 1. If $E°' = E° - E'$, where $E' = (2.303\ RT/\mathcal{F}) \log A_{Cl^-}$,

$$pH = (E - E°')\frac{\mathcal{F}}{2.303RT} \qquad [4\text{-}4]$$

Typically electrometric pH measurement utilizes a glass H^+-indicating electrode and a calomel electrode as the reference. The calomel electrode additionally acts as a salt bridge, forming a liquid junction across saturated KCl within the electrode and the test system. The glass-calomel electrode assembly can be diagrammatically illustrated through a phase boundary diagram.

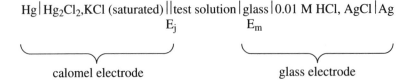

$$Hg \mid Hg_2Cl_2, KCl \text{ (saturated)} \| \text{test solution} \mid \text{glass} \mid 0.01 \text{ M HCl, AgCl} \mid Ag$$

$$\underbrace{\phantom{Hg \mid Hg_2Cl_2, KCl \text{ (saturated)} \| E_j}}_{\text{calomel electrode}} \quad \underbrace{\phantom{\text{glass} \mid 0.01 \text{ M HCl, AgCl} \mid Ag}}_{\text{glass electrode}}$$

where the double vertical line represents the liquid junction; E_j and E_m represent the junction and membrane potentials, respectively. From equation 4-4,

$$pH = (E_{mea} - E') \frac{\mathcal{F}}{2.303RT} \qquad [4\text{-}5]$$

where E_{mea} is the observed emf and $E' = E_j + E_m + E_c + E_{Ag|AgCl}$ (potentials due to the liquid junction, glass membrane, calomel, and Ag-AgCl internal electrodes, respectively). Measurements of pH are made relative to standards of known pH. Thus, $\Delta pH = (pH_{unk} - pH_{std})$ or,

$$pH_{unk} = (E_{unk} - E_{std}) \frac{\mathcal{F}}{2.303RT} + pH_{std} \qquad [4\text{-}6]$$

when E', T, P, and I are constant—or are corrected—for the standard and unknowns. Equation 4-6 constitutes an operational definition of pH.

Effect of Temperature. The relationship of pH to the dissociation constant of water (K_W) is expressed

$$pH = pK_W - pOH \qquad [4\text{-}7]$$

At $298.15°K$ and 101.33 kPa, $pK_W = 14.00$. Since $\Delta F = \Delta F° + RT\ln Q$ and from equation 3-64, $\Delta F = -n\mathcal{F}E$,

$$E = E° + \frac{2.303RT}{\mathcal{F}} pK_W \qquad [4\text{-}8]$$

where E is the acidity potential—that is, the measured emf for a glass-calomel electrode combination. Equation 4-8 demonstrates the thermodynamic temperature dependence of E, and therefore pH, through its relationship to pK_W. The pH of neutrality (pH = pOH = 7 at $298.15°K$ and 101.33 kPa) shifts with a change in T (Table 4.1).

TABLE 4.1 Temperature Effect on pK$_W$^{a} and pH

T, °C	pK$_W$	pH of Neutrality
0	14.93	7.47
5	14.73	7.37
10	14.53	7.27
15	14.35	7.18
20	14.17	7.09
25	14.00	7.00
30	13.83	6.92
50	13.26	6.63

^{a}pK$_W$ = 4470.99 T^{-1} − 6.0875 + 0.01706T (T in °K) from Harned and Owen (1958).

Effect of Pressure. The fundamental equation describing ΔF (from equation 3-34) for a system of constant T and composition under standard state conditions becomes

$$\left[\frac{\partial \Delta F^\circ}{\partial P}\right]_{T,n_i} = \Delta \overline{V}^\circ \qquad [4\text{-}9]$$

Since ΔF° = −RTlnK, equation 4-9 yields

$$d \ln K = -\frac{\Delta \overline{V}^\circ}{RT} dP \qquad [4\text{-}10]$$

If partial molar volume, $\Delta \overline{V}^\circ$, is independent of pressure, equation 4-10 can be integrated between limits of P = 101.33 kPa and P.

$$\ln \frac{K_P}{K_{\cdot P=101.33}} = -\frac{\Delta \overline{V}^\circ}{RT}(P - 101.33) \qquad [4\text{-}11]$$

This effect of P on K$_W$ is illustrated in Table. 4.2.

Practical effects of pressure on measured pH are of a lesser magnitude than those of Table 4.2 and arise through the buildup of a pressure differential between

TABLE 4.2 Effect of Pressure on K$_W$

P, MPa	K$_P$/K$_{P=0.101}$			
	5°C	15°C	25°C	35°C
0.101	1	1	1	1
20.26	1.24	1.225	1.202	1.18
40.52	1.54	1.49	1.435	1.384

Owen and Brinkley, 1941.

the test solution and the interior of the glass electrode. Such pressure effects are typically avoided through venting of the glass electrode to the ambient atmosphere.

Effect of Ionic Strength. For the idealized pH measurement of a dilute HCl solution, Equation 4-3 can be recast as

$$pH = (E - E°') \frac{\mathcal{F}}{2.303RT} + \log C_{Cl^-} + \log f_{Cl^-} \qquad [4\text{-}12]$$

The single ion activity coefficient, f_{Cl^-}, can be calculated by the Davies equation (equation 3-72, where A = 0.5 and $Z^2 = 1$):

$$\log f_{Cl^-} = -0.5 \left[\frac{I^{1/2}}{1 + I^{1/2}} - 0.2I \right] \qquad [4\text{-}13]$$

Considering solutions of equivalent pH ($pH_A = pH_B$) and C_{Cl^-} ($C_{Cl_A^-} = C_{Cl_B^-}$) where I is varied—for example, through varying the concentration of a background electrolyte such as $NaClO_4$—the change in observed emf, ΔE, is

$$\Delta E = \frac{2.303RT}{\mathcal{F}} (\log f_{Cl_A^-} - \log f_{Cl_B^-}) \qquad [4\text{-}14]$$

The *apparent* ΔpH associated with uncorrected comparisons between solutions of differing I is

$$\Delta pH_{apparent} = \Delta \log f_{Cl^-} \qquad [4\text{-}15]$$

Thus, significant decreases in apparent pH can occur with slight increases in I (Table 4.3). The effect is most pronounced for dilute unbuffered systems, such as rainwater or low ionic strength displaced soil solutions, where it may be necessary

TABLE 4.3 Effect of Ionic Strength on Apparent pH of a Simple Electrolytic Solution[a]

ΔI	$-\log f$	f	ΔE	Apparent Change in pH
			J sec^{-1} A^{-1}	
0.000	0.000	1.00	0.000000	0.00
0.001	0.016	1.04	0.000917	-0.02
0.005	0.033	1.08	0.001958	-0.03
0.010	0.045	1.11	0.002677	-0.05
0.050	0.088	1.22	0.005202	-0.09
0.100	0.112	1.29	0.006633	-0.11

[a]Dilute HCl solutions where pH and C_{Cl^-} are constant; I is varied by varying background electrolyte.

to utilize an ionic strength adjusting solution to mitigate the potential effect of differences in ionic strength between test solutions and standard pH buffer solutions.

4.1.3 Measurement of Soil Solution pH

Common systems for measurement of soil pH are saturated pastes and $1:1$ or $1:5$ soil–water slurries. The intent of such measurements is to represent the "active acidity" of the soil, that is, the hydrogen ion activity of the soil solution. The utility of any one of these common measurements is limited to the extent to which they represent the pH of ambient soil solution. There is little unanimity as to the proper system for assessment of soil pH although the effect of dilution to increase soil pH has been widely recognized (Peech, 1965). At one time there was a tendency toward measurement of soil pH in saturated pastes as a preferred approach (Peech, 1965), but this may be problematic due to extraneous potentials that may be generated by the influence of colloids on electrodes—the *suspension effect* (Sparks, 1984).

Since in situ measurements of soil solution pH can lead to erroneous results, measurement of displaced soil solution pH is preferred for unambiguous measurement of "active acidity." In fact, the impetus for development of certain soil solution displacement techniques has been the desire to represent pH of ambient soil solution more realistically (Mubarak and Olsen, 1976b). Methodology for displacement of soil solution and subsequent sample handling prior to and during pH measurement, however, introduces biases that must be considered for reliable measurement of soil solution pH. The buffering effect of "reserve acidity" associated with the soil solid phase is lost as soil solution is displaced from the soil solid matrix; thus, shifts in pH of weakly buffered displaced soil solutions may significantly alter chemical speciation in the aqueous phase. Principal factors affecting pH of displaced soil solutions are temperature, pressure, ionic strength, and carbonate equilibria.

Pressure and Temperature Effects During Displacement. Pressure or centrifugal displacement techniques can affect pH both directly and indirectly through changes in temperature and pressure. Effects of pressure are negligible when driving pressures are < 4.5 MPa. Effects of temperature, however, may be evident even with low-pressure centrifugal displacement (< 500 kPa). Low-speed centrifugal displacement of soil solution without temperature control can increase soil solution temperatures by $\approx 5°C$ depending on speed and duration of centrifugation.

Consider the case where high-speed centrifugal displacement results in an increase in midpoint driving pressure from 101.3 kPa to 4.5 MPa and where temperature increases from 25 to 30°C. First, from Table 4.2, K_w will increase 1.033 times when pressure is increased (pK_w decreases from 14.00 to 13.98 when temperature is held constant at 298.15°K). If the ambient soil solution is at neutrality, soil solution initially at pH 7.00 will decrease to pH 6.99 when acted on by a pressure of 4.5 MPa. If temperature additionally increases by 5° (to 303.15°K) during displacement, from Table 4.1, an additional 0.08 unit decrease in soil solution pH will occur. Thus, there is a net decline in soil solution pH of ≈ 0.1 unit due to the combined effects of temperature and pressure *during the time course of the displacement*. In the absence of additional confounding effects, such as CO_2-degassing (see the next section, Effects of Carbonate Equilibria), pH would be expected to return to its original value when the original temperature and pressure are restored.

Actually, speciation of all soil solution components are affected directly by pressure and temperature during displacement and indirectly by pressure- and temperature-induced shifts in solution pH. Composition of the soil solution obtained may be incontrovertibly altered to the extent that mass is transferred between soil solution and other soil phases during the course of displacement (in other words, failure to achieve a closed system model for soil solution during the period of displacement may invalidate the analytical results arising from the displacement).

Effects of Carbonate Equilibria. The most common and consequential effects on soil solution pH arise from the aqueous phase distribution of carbonate species and the gas-phase/liquid-phase partitioning of CO_2. The equilibria presented in Table 4.4 describe a model for a simple natural water carbonate system open to the atmosphere. Shifts in carbonate equilibria described in Table 4.4 may result from pressure and temperature changes during certain types of soil solution displacement or from CO_2-degassing that may occur when soil solutions are separated from the buffering influences of soil solid and biotic components.

The total aqueous phase carbonate concentration ($C_T = [H_2CO_3{}^*] + [HCO_3-] + [CO_3^{2-}]$) can be described as a function of the solution partial pressure of CO_2 ($P_{CO_{2g}}$) and pH. From Table 4.4,

$$C_T = K_H RT[CO_{2g}] \left(1 + \frac{K_1}{[H^+]} + \frac{K_1 K_2}{[H^+]^2} \right) \qquad [4\text{-}16]$$

TABLE 4.4   Equilibrium Distribution of Species in a Dilute Aqueous Carbonate System[a]

$$CO_{2g} + H_2O \rightleftharpoons H_2CO_3^* \qquad\qquad K_H' = K_H RT = \frac{[H_2CO_3^*]}{[CO_{2g}]}$$

where CO_{2g} behaves ideally, $[H_2CO_3^*] = [CO_{2aq}] + [H_2CO_3]$; K_H' is the dimensionless Henry's law constant, and K_H is the Henry's law constant.

$$H_2CO_3^* \rightleftharpoons H^+ + HCO_3^- \qquad\qquad K_1 = \frac{[H^+] [HCO_3^-]}{[H_2CO_3^*]}$$

$$HCO_3^- \rightleftharpoons H^+ + CO_3^{2-} \qquad\qquad K_2 = \frac{[H^+] [CO_3^{2-}]}{[HCO_3^-]}$$

[a]Conditional equilibrium constants for constant ionic strength systems.

where $[CO_{2g}]$ is a concentration term ($= mol\ m^{-3}$, when $R = 8.143\ J\ mol^{-1}\ deg^{-1}$). Dalton's law of partial pressure gives

$$P_{CO_2} = RT[CO_{2g}] \qquad [4\text{-}17]$$

where P_{CO_2} is in Pa. Substitution of equation 4-17 into equation 4-16 gives

$$C_T = K_H P_{CO_2} \left(1 + \frac{K_1}{[H^+]} + \frac{K_1 K_2}{[H^+]^2}\right) \qquad [4\text{-}18]$$

The Interaction of Soil Solution pH and Carbonate Equilibria. If soil solution carbonate equilibria are modeled by the simple system described in Table 4.4, varied displacement conditions as described previously (T increases by 5°C and P increases to 4.5 MPa) result in changes in the equilibria constants for carbonate. Table 4.5 summarizes several displacement scenarios where soil solution carbonate equilibria is initially fixed at a solution CO_2 partial pressure of 304 Pa (approximately ten times atmospheric CO_2) for a closed system at 297.15°K and 101.3 kPa total pressure. Carbonate is assumed to be controlled by respiration of soil biota for this natural water model; the influence of a carbonate-controlling solid phase component—such as calcium carbonate—is not considered.

When soil solution is displaced under conditions of a closed system, $H_2CO_3{}^*$ is treated as an nonvolatile acid and C_T is constant as conditions of P and T vary; equation 4-18 simplifies to

$$C_T = [H_2CO_3{}^*] \left(1 + \frac{K_1}{[H^+]} + \frac{K_1 K_2}{[H^+]^2}\right) \qquad [4\text{-}19]$$

Displaced soil solution pH varies according to the relationship described in equation 4-19 where the influence of T and P is on the magnitude of the equilibrium constants K_1 and K_2. Table 4.6 shows the decreases in K_1 and K_2 accompanying an increase in T. The effect of P is analogous to that on K_W as previously described (equations 4-9 to 4-11),

$$\left[\frac{\partial \ln K}{\partial P}\right]_{T,n_i} = -\frac{\Delta \overline{V}{}^\circ}{RT} \qquad [4\text{-}20]$$

$$\log \frac{K_{P_2}}{K_{P_1}} = -\frac{\Delta \overline{V}{}^\circ}{2.303RT}(P_2 - P_1) \qquad [4\text{-}21]$$

When P increases from 101.3 kPa to 4.5 MPa and T is constant ($= 298.15°K$), the value of the equilibrium constants K_1 and K_2 increase by a factor of 1.049 and 1.050, respectively (interpolation from the data of Owen and Brinkley, 1941). The effect of closed system displacement is an approximate 0.05 unit decrease in soil solution pH *during the time course of the displacement* (Table 4.5).

TABLE 4.5 Effect of Displacement Conditions on Total Carbonate ($C_T = [H_2CO_3^*] + [HCO_3^-] + [CO_3^{2-}]$) Concentration in Soil Solution as Modeled for a Simple Natural Water Carbonate System[a]

System	Temperature K	Midpoint Driving Pressure kPa	P_{CO2g} Pa	pH	$-\log K'_H$	$-\log K_1$	$-\log K_2$	C_T mol m^{-3}
				Initial Condition; In Situ Soil Solution				
Open[b]	298.15	101.3	304	7.00	0.081	6.352	10.329	0.55
				5 C Rise in T during Displacement at 101.3 kPa				
Closed	303.15	101.3		6.96		6.327	10.290	0.55
				T is Constant during Displacement at 4.5 MPa				
Closed	298.15	4500		6.98		6.331	10.308	0.55
				Displacement at 303.15 C and 4.5 MPa				
Closed	303.15	4500		6.94		6.306	10.269	0.55
				5 C Rise in T during Displacement at 101.3 kPa				
Open[b]	303.15	101.3	304	7.03	0.134	6.327	10.290	0.54
				T is 298.15 and P is 101.3 kPa following Displacement				
Open[c]	298.15	101.3	30	7.99	0.081	6.352	10.329	0.45

[a]The effect of ionic strength is not considered, equilibrium constants are extrapolated to zero ionic strength (Stumm and Morgan, 1981) for the reactions described in Table 4.4.
[b]Gas exchange occurs between soil solution and soil atmosphere.
[c]Gas exchange occurs between soil solution and ambient atmosphere.

TABLE 4.6 Temperature Dependence of Constants Describing Carbonate Equilibria[a]

Temperature °C	pK_H[b]	pK_1	pK_2
20	6.416	6.381	10.377
25	6.476	6.352	10.329
30	6.536	6.327	10.290
35	—	6.309	10.250

[a]Equilibrium constants extrapolated to zero ionic strength, after Stumm and Morgan (1981).
[b]$K_H = [H_2CO_3^*]/P_{CO_2}$, mol L^{-1} Pa^{-1}.

Soil solution displacement more frequently occurs under conditions where solution is removed from the soil matrix in a system open to exchange with atmospheric CO_2 (C_T is no longer constant). In open system scenarios (Table 4.5), variation in T and P (causing slight depressions in soil solution pH and decreases in carbonate equilibrium constants) is secondary to the effect of CO_2-degassing in affecting carbonate status of displaced soil solution. The net effect is increased soil solution pH and decreased C_T over that of in situ soil solution. Here, change in P_{CO_2} from that of in situ soil atmosphere to that of ambient atmosphere has the controlling influence on soil solution pH–a tenfold decrease in P_{CO_2} as a consequence of soil solution degassing results in a full unit increase in pH of the displaced solution.

Considerable research attention has been given to CO_2 effects on soil solution pH and to a means for avoiding or correcting for these effects. Kittrick (1983) employed special precautions to avoid exposure of solutions displaced from calcite suspensions to elevated P_{CO_2} that occurred from analyst breath contamination during pH measurement. More commonly researchers are concerned with degassing effects as soil solution displacement techniques cause CO_2-degassing and elevated solution pH. Suarez (1986) developed a multichamber tension lysimeter to avoid degassing effects when sampling soil water in field situations. Alternatively, Suarez (1987) developed a predictive model to allow for correcting lysimeter water pH back to ambient soil water pH, provided precipitation did not occur as CO_2 degassed from the lysimeter water. The increased pH expected from CO_2 degassing as a conventional single-chamber tension lysimeter fills with soil solution is illustrated in Figure 4-1; for air-flushed lysimeters with liquid-to-air volume ratios of < 0.2, the increase in pH approaches 1 unit, consistent with the open system degassing described earlier (Table 4.5).

The influence of CO_2-degassing on acid soil solutions is related to C_T; increasing C_T resulting from respiratory enrichment of CO_2 in the soil atmosphere increases the potential for solution pH to rise as displaced soil solutions are degassed (Table 4.7). The magnitude of respiratory enrichment is significant in natural environments ($P_{CO_2} > 800$ Pa are frequently reported for forested soils and can exceed 1 kPa in cultivated soils), but factors other than CO_2 may buffer pH of soil solution once displaced from the soil matrix. Soil solutions displaced from acid soils may be

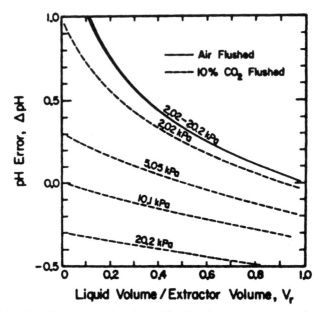

FIGURE 4.1   The pH error as a function of liquid volume to extractor volume for single-chamber extractors. The uppermost solid line represents predicted values for an air filled extractor evacuated to 2.02, 5.05, 10.1, and 20.2 kPa and then sealed. The dashed lines represent predictions for an extractor filled with 10% CO_2-air mixture, evacuated to total pressures of 2.02, 5.05, 10.1, and 20.2 kPa and then sealed. [Suarez, 1987. Published in *Soil Sci. Soc. Am. J.* 51:64–67. 1987. Soil Sci. Soc. Am.]

buffered by Al speciation or organic acids so that the effect of CO_2-degassing is less strongly evidenced. CO_2-degassing has an insignificant effect on soil solution pH when pCO_2 exceeds 600 Pa in forested soil solutions where dissolved organic acids maintain solutions at $<$ pH 4.5 (insignificant proton contributions from carbonic acid; Zabowski and Sletten, 1991). Soil solutions of calcareous soils will have higher alkalinity and therefore are buffered against significant shifts in pH from CO_2-degassing.

TABLE 4.7   Effect of Total Dissolved Carbonate (C_T) on Change in pH with Degassing of Spodosol B Horizon Soil Solutions

C_T mmol L^{-1}	Initial pH	Degassed pH[a]	ΔpH
0.21	4.9	5.1	0.2
1.1	5.1	5.5	0.4
4.3	4.7	6.8	2.1
13.7	4.7	8.5	3.8

[a]Degassed with N_2 followed by requilibration with atmospheric CO_2.
After David and Vance, 1989.

4.2 IONIC STRENGTH

The term *ionic strength* was introduced by Lewis and Randall (1961) to express the nonideality imposed by an electrolytic solution on any dissociated electrolyte in solution. Ionic strength (I) is characteristic of the solution and is defined as

$$I = 1/2 \Sigma C_i Z_i^2 \tag{4-22}$$

where C_i and Z_i are the concentration and valence, respectively, of the *i*th component. The definition of ionic strength arose from evaluation of the mean activity coefficients of the ions of electrolytes in various electrolytic solutions (see Solutions of Electrolytes under section 3.3.5). Generalization led to the empirical observation that "in dilute aqueous solutions the activity coefficient of a given strong electrolyte is the same in all solutions of the same ionic strength." (Barrow, 1966). This generalization is reflected in the Debye–Hückel theory and in empirical relationships expressing ion activity coefficients (Table 3.4); for example, for the Limited Debye–Hückel Equation (equation 3-70),

$$\log f_i = -A Z_i^2 I^{1/2}$$

Ionic strength, therefore, represents a master variable of the soil solution that must be accounted for when expressing composition of soil solution in terms of ion activities.

4.2.1 Electrical Conductivity

Electrical conductivity (EC) is the inverse of electrical resistivity and expresses the ability of a solution to conduct electricity. Electrical conductivity is proportional to the concentration of electrolytes in solution; therefore, measurement of electrical conductivity affords a simple means whereby the ionic strength of dilute aqueous solutions may be estimated.

Electrical conductivity expresses the *specific conductance* of a solution at 25°C between electrodes of 1 cm^2 cross section and placed 1 cm apart (EC_{25}). Measurement of the resistance (R_T) imparted within a cell containing platinized electrodes is accomplished using a conductivity meter and a conductivity cell with a specified cell constant ($k \approx 1$ cm^{-1} is appropriate for soil solutions). Thus,

$$EC_{25} = \frac{k f_T}{R_T} \tag{4-23}$$

where f_T corrects the measured resistance to 25°C. The traditional units for expressing of EC are mmhos cm^{-1}; in SI units these are equivalent to dS m^{-1} (dS is decisiemens where 1 siemen = amp volt^{-1}).

4.2.2   Electrical Conductivity as an Estimate of Ionic Strength

Ionic strength of soil solutions and natural waters demonstrates a strong linear correlation with EC. Empirical estimates of I (mol L^{-1}) range from 0.011 to 0.016 times EC (dS m^{-1}) (Table 4.8). Average results for the measurements reported in Table 4.8 (exclusive of those for soil–water extracts) indicate a relative range in estimates of I from EC of 36%. This variance may relate to the differing nature of solutions and ionic strengths considered, the methods by which solutions were obtained and prepared for measurement, and the degree of ion-pair correction made for I when computed from solution electrolyte composition. Electrical conductivity is frequently measured for soil–water extracts, but this measurement is not directly translatable to EC of soil solutions at field moisture contents because of solubilization of salts in more highly buffered soils or because of a dilution effect in more highly weathered, weakly buffered soils (Table 4.9).

4.3   ELECTRICAL POTENTIAL

4.3.1   Theoretical Basis of E_H

Oxidation-reduction or *redox equilibria* describe reactions where electron transfer results in changes in the oxidation state of products and reactants. *Oxidation state* "represents a hypothetical charge that an atom would have if the ion or molecules were to dissociate" (Stumm and Morgan, 1981).

Redox equilibria may be interpreted as comprising two steps (half-reactions) describing an oxidation and a corresponding reduction. Redox equilibria are described by the equilibrium redox potential (E_H). The E_H is an electrochemically measurable quantity that is thermodynamically defined by the Nernst equation:

$$E_H = E_H^0 - \frac{2.303RT}{n\mathscr{F}} \log \frac{\prod\limits_j A_{Red}^{n_i}}{\prod\limits_i A_{Ox}^{n_i}} \qquad [4\text{-}24]$$

The subscript H in E_H reflects that E_H is expressed on a hydrogen scale where E_H (volts) equals zero for the half-reaction

$$2H^+ + 2e^- \rightleftharpoons H_{2(g)}$$

(By IUPAC convention, all half-reactions are written as reductions with the same sign as the equilibrium constant for the reduction reaction. Sign conventions, however, may vary among authors.)

The standard state E_H (E_H°, where products and reactants are in the standard state of unit activity) is related to the equilibrium constant for the reduction reac-

TABLE 4.8 Empirical Relationships for Ionic Strength of Electrolyte Solutions as a Function of Solution Electrical Conductivity

Type of Electrolytic Solution	Range in Measured EC dS m^{-1}	Empirical Fit for Ionic Strength	Correlation Coefficient	Source
Flooded soil extracts and electrolyte solutions[a]	<6.25	I = 0.016EC		Ponnamperuma et al., 1966
Displaced soil solution[b]	<0.02 to 1.4	I = 0.015EC − 0.0006	0.994	Alva et al., 1991
Saturation paste extracts and river waters[d]	0.64 to 32.4	I = 0.0127EC − 0.002[c]	0.996	Griffin and Jurinak, 1973
Displaced soil solution[e]	0.034 to 1.0	I = 0.0120EC − 0.0004[c]	0.993	Gillman and Bell, 1978
Flooded soil leachates[f]	1.67 to 33.2	I = 0.0116EC − 0.0001[c]	0.982	Pasricha, 1987
Flooded soil extracts[g]	9.65 to 47.6	I = 0.0114EC[c]		Leffelaar et al., 1983
Displaced soil solution and soil-water extracts[e]	0.034 to 1.0	I = 0.0109EC − 0.0002[c]	0.984	Gillman and Bell, 1978

[a]221 dilute ionic strength soils.
[b]Six weathered soils from the southeastern USA.
[c]Corrected for ion-pair formation
[d]27 alkaline soils and 124 river waters.
[e]Six weathered tropical soils.
[f]Three salt-impacted soils cultivated to rice.
[g]Salt-impacted soils.

TABLE 4.9   Relationship Between Electrical Conductivity (dS m^{-1}) of a Soil Extract and Electrical Conductivity of the Soil Solution for Six Highly Weathered Soils at 10 kPa Moisture Potential

Soil–Water Extract	Regression Equation	r
1:1	$C(1:1) = 0 \cdot 721\, C_{10kPa} - 0 \cdot 188$	$0 \cdot 982$
1:2.5	$C(1:2.5) = 0 \cdot 492\, C_{10kPa} - 0 \cdot 168$	$0 \cdot 966$
1:5	$C(1:5) = 0 \cdot 258\, C_{10kPa} - 0 \cdot 021$	$0 \cdot 971$
1:10	$C(1:10) = 0 \cdot 139\, C_{10kPa} - 0 \cdot 014$	$0 \cdot 984$

G. P. Gillman and L. C. Bell, 1978, *Australian Journal of Soil Research* 16:67–77.

tion; following from equation 4-24,

$$E_H^0 = \frac{2.303RT}{n\mathcal{F}} \log K^\circ \qquad [4\text{-}25]$$

Consideration of the general redox reaction

$$m/4\ O_{2(g)} + mH^+ + ne^- + Ox \rightleftharpoons Red + m/2\ H_2O$$

results in the following relationship for E_H

$$E_H = E_H^0 - \frac{2.303RT}{n\mathcal{F}} \log \frac{A_{Red}}{A_{Ox}} \qquad [4\text{-}26]$$

4.3.2   Measurement of E_H

Reliable measurement of E_H is achievable for well-defined systems at equilibrium. The equilibrium condition for the redox couple of interest is that of no net current—that is, the rate of reduction is offset by the opposing rate of oxidation, both of which can be evaluated by the rate of electron transfer in the forward and back directions. Electrometric measurements of E_H evaluate the magnitude of the *exchange current* (i_0) in a single direction. Since measurements assume a system at equilibrium, activities of oxidized and reduced species at the surface of the Pt electrode are equal to those in the bulk solution, from equation 4-26.

$$\frac{A_{Red}}{A_{Ox}} = e^{n\mathcal{F}/RT(E_H^0 - E)} \qquad [4\text{-}27]$$

The sensitivity of measured E_H depends on (1) the magnitude of the extraneous potential applied to null the i_0 of the test solution relative to the magnitude of i_0 and (2) the rapidity for which a response to applied current is measured. Test systems comprising concentrated solutions of a couple with relatively large E_H—

that is, strongly poised systems—will provide more reliable measurements of E_H than will weakly poised systems. Systems of mixed couples or systems containing O_2 generally exhibit slower response times and thus afford less reliable measurements of E_H. Strongly reducing systems, however, may be problematic from the standpoint of extraneous potentials generated by reactions at the surface of the Pt electrode.

4.3.3 Interpretation of Measured E_H

Measured redox potentials of soil solutions and other natural waters reflect a mixed potential measurement of an electrically dynamic system. This arises because soils are typically not well-poised with respect to redox—that is, the redox behavior of soil systems is not fixed by any one redox couple. Thus, measured electrode potential reflects a nonequilibrium condition along a continuum of redox that approaches a steady state only under strongly reducing conditions. These static anaerobic conditions are infrequently manifested in soil, surface, or ground waters since these systems are never totally isolated from the atmosphere. A further complication arises from the low concentrations of redox species existing in most natural aqueous solutions, as this limits the ability to reliably measure i_0. Thus, there is little expectation that measured E_H of dilute aqueous solutions of mixed redox couples will reflect the theoretical E_H of the system (Lindberg and Runnells, 1984).

Table 4.10 summarizes the principal redox reactions in soils and the range of measured E_H where they may be operative. Oxidizing and reducing agents present in soils will be bracketed between the H_2O—O_2 couple and the H^+—H_2 couple (Fig. 4-2). In the face of limited O_2 in the soil environment, microbes will utilize successively less effective electron donors as sources of reducing power. In aerobic soils, oxygen is the major electron acceptor and acts to buffer electron activity; thus, for aerobic soils measured E_H is not particularly informative because it largely reflects the hydrogen ion potential of the system. When making comparisons of E_H between systems, this limitation may be counteracted in part through correction of E_H to neutrality (the E_{H_7}). As soils become progressively anaerobic, pH tends to stabilize due to the buffering effect of accumulated CO_3^{2-} and HCO_3-. Under these conditions, redox processes are better reflected in measured E_H. In flooded soils, the redox response attributable to C, N, S, Fe, and Mn is more readily reflected in measured E_H.

4.3.4 Electron Activity

The redox status of soil solutions can also be usefully described as a function of the electron activity (pe $= -\log A_{e-}$). Considering the general redox reaction described earlier (4.3.1),

$$m/4\ O_{2(g)} + mH^+ + ne^- + Ox \rightleftharpoons Red + m/2\ H_2O$$

TABLE 4.10 Order of Utilization of Principal Electron Acceptors in Soils, Equilibrium Potentials of These Half-Reactions at pH 7, and Measured Potentials of These Reactions in Soils

Reaction	E_{H7} (V)	Measured Redox Potential in Soils (V)
O_2 Disappearance	0.82	0.6 to 0.4
$1/2\ O_2 + 2e^- + 2H^+ = H_2O$		
NO_3^- Disappearance	0.54	0.5 to 0.2
$NO_3^- + 2e^- + 2H^+ = NO_2^- + H_2O$		
Mn^{2+} Formation	0.4	0.4 to 0.2
$MnO_2 + 2e^- + 4H^+ = Mn^{2+} + 2H_2O$		
Fe^{2+} Formation	0.17	0.3 to 0.1
$FeOOH + e^- + 3H^+ = Fe^{2+} + 2H_2O$		
HS^- Formation	-0.16	0 to -0.15
$SO_4^- + 9H^+ + 6e^- = HS^- + 4H_2O$		
H_2 Formation	-0.41	-0.15 to -0.22
$H^+ + e^- = 1/2\ H_2$		
CH_4 Formation (example of fermentation)	—	-0.15 to -0.22
$(CH_2O)_n = n/2\ CO_2 + n/2\ CH_4$		

Bohn et al., 1985.

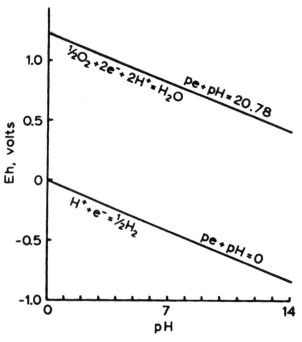

FIGURE 4.2 Redox limits in soil systems.

the equilibrium constant can be expressed as

$$K = \frac{A_{Red}}{A_{Ox}A_{H^+}^m A_{e^-}^n} \qquad [4\text{-}28a]$$

or

$$\log K = -\log A_{e^-}^n + \log \frac{A_{Red}}{A_{Ox}A_{H^+}^m} \qquad [4\text{-}28b]$$

Rearrangement of equation 4-28b provides an expression for pe,

$$pe = \frac{1}{n} \log K - \frac{1}{n} \log \frac{A_{Red}}{A_{Ox}A_{H^+}^m} \qquad [4\text{-}28c]$$

For the condition where all species other than the electron have unit activity, equation 4-28c reduces to

$$pe^0 = \frac{1}{n} \log K^0 \qquad [4\text{-}29]$$

The general case for equation 4-28c is

$$pe = pe^0 - \frac{1}{n} \log \frac{\prod_j A_{Red}^{n_j}}{\prod_i A_{Ox}^{n_i}} \qquad [4\text{-}30]$$

From equations 4-24 and 4-30,

$$pe = \frac{E_H \mathfrak{F}}{2.303RT} \qquad [4\text{-}31]$$

Substitution of appropriate values of $\mathfrak{F}$, R, and T yields E_H (mVolts) = 59.2pe at 298.15°K.

pe + pH. The relationship pe + pH provides a useful expression of the two master variables most responsible for description of chemical and microbiological status of soil solutions. The general relationship for pe + pH follows from equation 4-28c

$$npe + mpH = \log K - \log \frac{A_{Red}}{A_{Ox}} \qquad [4\text{-}33]$$

As reflected by Figure 4-2, dissociation of water to $H_{2(g)}$ or $O_{2(g)}$ imposes redox limits on soil solutions and natural waters. Using the expression pe + pH results in the following expressions for the limits to oxidation and reduction, respectively (Lindsay, 1979).

$$\text{limit to oxidation:}\quad \text{pe} + \text{pH} = 20.78 + 1/4 \log O_{2(g)}$$

$$\text{limit to reduction:}\quad \text{pe} + \text{pH} = -1/2 \log O_{2(g)}$$

Construction of similar expressions for other redox couples of interest in soil solutions and their graphical presentation affords an effective means for interpretation of the importance of redox reactions to speciation in the soil solution.

4.4 SUSPENDED COLLOIDAL MATERIALS

Consideration of aqueous solution in the strict sense of "a single-phase liquid system" is not entirely applicable to soil solution and other natural waters. This is because in natural waters, "a variety of organic and inorganic materials exist as colloids . . . including macromolecular components of 'dissolved' organic carbon (DOC) such as humic substances, 'biocolloids' such as microorganisms, microemulsions of nonaqueous phase liquids, mineral precipitates and weathering products, precipitates of transuranic elements such as plutonium, and rock and mineral fragments."[1] The concept of "soil solution" has occasionally been broadened to encompass suspended colloids (Wolt, 1993) in recognition that these colloids may frequently be inextricable components of what is described as soil solution. Suspended colloidal material needs be considered as a delimitating variable of soil solution since it may affect transport, speciation, bioavailability, and total aqueous-phase concentrations of metals and ligands with which it interacts.

4.4.1 Implications of Colloid Presence in Soil Solution

Aqueous-phase colloids are particles of low solubility that do not dissolve in water but remain as an identifiable solid phase in suspension. The distinct behavior of colloids relates to the high surface area of interaction between the suspended colloidal solids and the suspending aqueous solution. The high surface-to-mass ratio of colloids imparts high chemical reactivity to these particles that is largely independent of the nature of the colloid. A typical clay colloid increases in specific surface area from 2.5 to 2500 m^2 g^{-1} as equivalent spherical diameter decreases from 1 to 0.001 μm (Bohn et al., 1985). In addition, the concentrations of colloids in natural water may be substantial (Fig. 4-3). Humic materials in groundwaters commonly range from 30 to 100 μg C L^{-1} but may approach 10^3 μg C L^{-1} in colored waters (Thurman, 1985).

Interpretations of both mobility and bioavailability of chemicals in soils require consideration of the effects of suspended colloidal materials. Facilitated transport

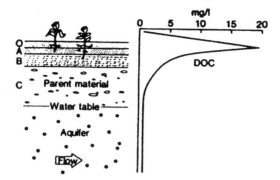

FIGURE 4.3 Changes in aqueous concentration of organic carbon in the vadose and saturated zones. [Thurman, 1985]

of chemicals in association with suspended colloidal material may confound conventional efforts to model chemical mobility in the vadose zone solely as a consequence of water solubility. Similarly, soil solution colloids may alter bioavailability of chemicals so as either to restrict or facilitate chemical exposure in the environment. Suspended colloids in solution are anticipated to act as sites for the adsorption, exchange, binding, and lipophilic dissolution of chemicals, and may, thus, significantly affect availability of chemicals in the aqueous phase. The significance of soil solution colloids in affecting chemical transport or bioavailability in soils depends on the degree to which colloidal particles remain stable in suspension; electrolyte concentration, solution pH, ion adsorption, and molecular retention will all influence colloidal stability.

4.4.2 Characterization of Suspended Colloidal Material in Soil Solution

The inability to eliminate or characterize conclusively the colloidal components of displaced soil solution is a factor contributing to the conditional nature of soil solution definitions. Many researchers employ the cutoff limits afforded by filtration of soil solutions through filters with 0.45 or 0.2 μm pore-size openings to obtain "colloid-free" soil solution for analytical characterization. This is not entirely appropriate; for example, determination of total Al in water has been compromised by the ability of fine-grained Al-bearing solids to pass through 0.45 or 0.2 μm pore-size openings (Kennedy et al., 1974). Turbidimetric inspection of displaced soil solution for the scattering of light by suspended colloids—the Tyndall effect—may serve to screen solutions for the presence of colloidal materials.

Colloidal components are infrequently quantified in soil solution and, if quantified, nonspecific measurements are employed. Colloidal organic matter in solution, for example, is frequently inferred from measurements of color or DOC. Refined measurements are needed to distinguish the partitioning of DOC between suspended colloidal organic matter and low molecular weight soluble organic acids.

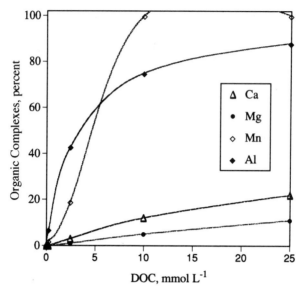

FIGURE 4.4 Effect of dissolved organic carbon (DOC) on complexation of soil solution cations with colloidal organic materials in the aqueous phase. Modeled from data of van Pragg and Weissen [1984] for soil solution displaced from a brown acid Of2 horizon (pH 3.85 and total ion concentrations of 1.0, 0.315, 0.145, and 0.211 mmol L^{-1} for Ca, Mg, Mn, and Al, respectively).

Quantitation of inorganic colloids is even more complex and may require kinetic studies to discriminate chemicals interacting with colloidal inorganics in soil solution.

Quantitation of total suspended colloidal materials leaves unresolved the characterization of the colloidal pool. Colloidal size and, secondarily, nature and density of reactive functionalities as well as colloidal stability will influence the degree of chemical reactivity associated with suspended colloidal materials. Because of the difficulties associated with characterization of soil solution colloids, the preferred approach at present is to apply a typical case model of colloidal reactivity to a measurement such as DOC (Plankey and Patterson, 1987; Sposito et al., 1981). Figure 4.4 presents an example of this approach, where metal complexation by model organic colloids is used to predict ion complexation by DOC in soil solution.

NOTE

1. Reprinted with permission from McCarthy and Zachara, "Subsurface transport of contaminants," *Environ. Sci. Technol.* 23:496–502. Copyright 1989 American Chemical Society.

CHAPTER 5

OBTAINING SOIL SOLUTION: LABORATORY METHODS

Methodologies for obtaining "unaltered" soil solution in a laboratory setting may be broadly defined as displacement techniques and encompass column displacement (including pressure or tension displacement, with or without a displacing head solution), centrifugation (including centrifugation with immiscible liquid), and saturation extracts (including saturation pastes). Additionally, a miscellaneous grouping of methods may be described that—although not intended to obtain "unaltered" soil solution per se—are frequently employed as models of soil solution; these methods include water extracts of soils and a variety of complexation and exchange techniques. Although various lysimetric methods have been used to obtain soil solution in laboratory settings, these methods are better addressed in a subsequent chapter on field sampling methodologies (Chapter 6).

5.1 OPERATIONAL NEEDS AND PRACTICAL CONSTRAINTS

Methods for obtaining soil solution in the laboratory share several common aspects that influence the degree to which they represent true soil solution. Composition of displaced soil solution is affected by soil moisture content, sample pretreatment, and duration of displacement. These operational constraints to the composition of displaced soil solution must be considered and described in order that the result of soil solution displacement and analysis be properly interpreted.

5.1.1 Soil Moisture Content

The intent of soil solution displacement methods is to obtain a solution representing ambient soil solution as it occurs in situ. The displacement of soil solution at

moistures representative of "field moisture contents"—typically unsaturated conditions—is a key to the concept of soil solution reflected in the definition presented
in Chapter 1. Consideration of soil at field moisture contents is necessitated by the
inability to predict consistently the effects of variation in soil to water ratios across
broad ranges of soil solution composition; neither variation in total electrolyte concentrations or the activity ratios of specific ion components of the soil solution can
be adequately resolved when water to soil ratios vary from field moisture contents
to ratios > 1 (United States Salinity Laboratory Staff, 1954; Khasawneh and Adams, 1967). This is the main limitation to the use of water extracts as models of
soil solution (see section 5.6, Water Extracts of Soil).

Soil Solution Recovery. Obtaining soil solution at low field moisture contents
is operationally constrained by the difficulty in freeing water from the soil matrix.
Percent moisture recovered from soils displaced under equivalent conditions generally reflects the effect of soil texture on the soil moisture release curve; increased
clay content generally results in increased energy of water retention by soil. Therefore, the threshold soil moisture content necessary for recovery of soil solution by
displacement increases with increasing clay content of soil. Threshold moisture
content for soil solution recovery depends on the displacement method used. For
the high-pressure immiscible displacement method of Whelan and Barrow (1980),
threshold moisture contents occurred at moisture tensions of 10 to 70 kPa; above
threshold moisture to saturation, there was a linear increase in fractional moisture
recovery from soils of varied texture (Fig. 5.1). Mubarak and Olsen (1976a), using
a similar displacement technique, showed a more curvilinear response in moisture
recovery with moisture content; fractional recoveries increased more rapidly with
changes in soil moisture > 60%, probably because of low energies of soil moisture
retention as soils approached saturation.

Low-pressure displacement techniques exhibit higher threshold moisture contents for solution recovery. Column displacement techniques utilizing a displacing
head, for example, require soils to be near field capacity ($\approx$33 kPa moisture tension) for effective displacement. This may be a disadvantage if in situ soil moisture
contents fall below threshold moisture recoveries for solution displacement, or if
volumes of soil solution displaced are insufficient for the analytical techniques used.
This limitation must be balanced against the greater potential for high-pressure
displacement techniques to modify composition of displaced soil solution.

Soil Solution Composition. Variation in soil moisture within the range of field
moisture contents results in substantially different effects on soil solution composition. This depends largely on the particular soil and chemical investigated and,
perhaps secondarily, on the choice of displacement or extraction technique. Chemical concentrations in soil solution are buffered to varying degrees against shifts in
soil moisture by exchangeable, precipitated, or adsorbed pools in the soil solid
phase. Changes in effective water content—that is, "plant-available" water—may
have a lesser effect on chemical concentration in soil solution than on the transport
of chemicals from the sorbed phase to bulk soil solution (Green and Obien, 1969).

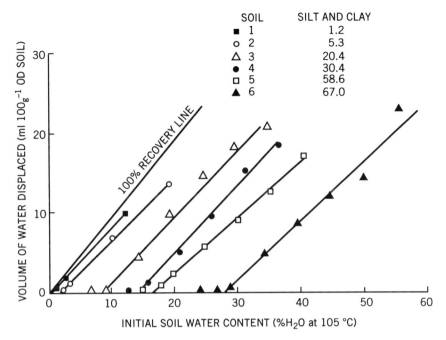

SOIL	SILT AND CLAY
■ 1	1.2
○ 2	5.3
△ 3	20.4
● 4	30.4
□ 5	58.6
▲ 6	67.0

FIGURE 5.1   Recovery of water from a range of soils at different initial water tensions. A 100% recovery line is shown for comparison. (Displacement: carbon tetrachloride; centrifuge speed: 265,000 m s^{-2}; centrifuge time: two hours; silt+clay % as indicated). [Whelan and Barrow, 1980. Published in *J. Environ. Qual.* 9:315–319. 1980. Am. Soc. Agron., Crop Sci. Soc. Am., Soil Sci. Soc. Am.]

Short contact times and low ratios of soil to water may allow for water extracts to model soil solution composition effectively for soils with relatively high buffer capacities (Hoagland et al., 1920).

Mubarak and Olsen (1976b) demonstrated dramatically different effects of dilution on soil pH measured for a clay-textured soil as compared to a fine sandy loam soil (Fig. 5.2). Even though both soils exhibited equivalent soil pH when soil moisture exceeded 100%, the soil pH measured at 30% moisture (that is, field moisture contents) for the clay was approximately 0.5 pH units higher and that of the fine sandy loam was approximately 0.5 pH units lower than for the saturated soils. Such effects of dilution are largely unpredictable, thus there is a likelihood of drawing differing interpretations of chemical reactivities in soils when interpretations are drawn from techniques that inadequately model in situ soil solution.

Table 5.1 represents the data of Moss (1963) for cation composition of three tropical soils as influenced by variation in soil moisture contents. Soil solutions at field moisture contents were obtained by column displacement while soil-water slurries were employed for obtaining "soil solution" outside the range of field moisture contents. Significant declines in soil solution Ca, Mg, and K concentrations are apparent for all soils with increasing moisture content; this effect was

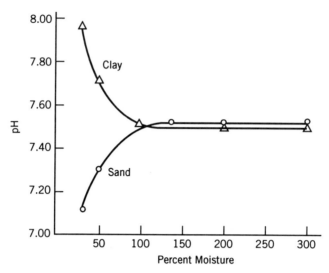

FIGURE 5.2   Variation in displaced soil solution pH as a function of initial soil moisture content for a clay and fine sandy loam soil. [Mubarak and Olsen, 1976b. Published in *Soil Sci. Soc. Am. J.* 40:880–882. 1976. Soil Sci. Soc. of Am.]

most dramatic over the range of field moisture contents. The activity ratio, pK − $1/2p(Ca+Mg)$, was constant over the entire range of moisture content for the high-CEC Montserrat soil (CEC=53.3 cmol(+)kg^{-1}), but not for the high-CEC Princes-Town marl (CEC=59.5 cmol(+)kg^{-1})—the latter contained appreciable Ca in the form of CaCO$_3$—or for the low-CEC River Estate soil (CEC=10.2 cmol(+)kg^{-1}).

5.1.2   Effects of Soil Pretreatment

Soil moisture content in the interval between sampling and soil solution displacement additionally affects soil solution composition. Soil moisture content when stored, soil drying and rewetting, as well as time and temperature of storage, may significantly alter composition of displaced soil solution.

Treatment of field-sampled soil prior to displacement and analysis introduces considerable uncertainty as to the analytical result obtained. Displacement and analysis of soil solutions immediately following sampling is preferred for unambiguous characterization of soil solution composition, but practical considerations frequently necessitate a period of storage following sampling. When circumstances require storage prior to displacement and analysis, maintenance at field moisture content, aerobically and at ambient conditions, is preferred. However, this approach may be inconvenient and unnecessary in some instances.

Soil Drying and Rewetting. Storage of field-sampled soil in an air-dry condition following sieving is a widespread soil storage technique, but it represents the most problematic form of storage for obtaining soil solution representative of in

TABLE 5.1 Change in Cation Composition and pK−1/2p(Ca+Mg) Values of Soil Solutions with Change in Moisture Content[a]

Moisture Content, %	Ca + Mg mmol L^{-1}	Ca mmol L^{-1}	K mmol L^{-1}	Activity Ratio	
				pK − 1/2pCa	pK − 1/2p(Ca + Mg)
			Montserrat clay[b]		
40	5.70	3.60	0.68	1.89	1.98
45	5.55	3.45	0.67	1.88	1.98
50	5.35	3.35	0.67	1.87	1.98
100	3.95	2.70	0.61	1.87	1.96
200	3.65	2.50	0.57	1.89	1.97
300	3.40	2.40	0.55	1.90	1.97
400	3.20	2.25	0.53	1.91	1.99
500	2.90	2.10	0.50	1.91	1.98
750	2.60	1.90	0.48	1.91	1.98
1000	2.20	1.55	0.45	1.89	1.97
2500	0.91	0.65	0.30	1.90	1.97
5000	0.50	0.36	0.23	1.90	1.97
10000	0.27	0.20	0.17	1.90	1.96
			River Estate fine sandy loam[c]		
20	6.40	4.35	0.90	1.80	1.89
25	5.37	3.70	0.80	1.83	1.91
30	5.00	3.38	0.77	1.82	1.90
35	4.75	3.25	0.75	1.82	1.91
40	4.50	3.00	0.73	1.82	1.90
45	4.33	2.93	0.72	1.83	1.91
70	3.90	2.50	0.65	1.84	1.93
80	3.70	2.30	0.64	1.82	1.93
100	3.60	2.30	0.56	1.88	1.98
300	2.90	1.76	0.34	2.04	2.15
500	2.37	1.42	0.28	2.07	2.20
750	1.80	1.20	0.24	2.12	2.20
1000	1.10	0.80	0.18	2.16	2.23
2500	0.80	0.60	0.15	2.18	2.24
5000	0.42	0.24	0.12	2.09	2.22
7500	0.30	0.20	0.10	2.14	2.23
10000	0.29	0.20	0.09	2.19	2.27
			Princes-Town marl[d]		
30	17.70		1.36		1.90
35	17.25		1.36		1.90
40	16.10		1.33		1.89
45	14.60		1.19		1.91
50	11.75		1.09		1.91
100	7.08		0.81		1.95
300	4.75		0.59		2.02

TABLE 5.1 *Continued*

Moisture Content, %	Ca + Mg mmol L^{-1}	Ca mmol L^{-1}	K mmol L^{-1}	Activity Ratio pK − 1/2pCa	pK − 1/2p(Ca + Mg)
500	4.28		0.49		2.07
750	3.45		0.36		2.16
1000	3.03		0.33		2.18
2500	2.23		0.30		2.15
5000	1.68		0.28		2.12
7500	1.22		0.27		2.09
10000	0.80		0.25		2.02

[a]P. Moss (1963).
[b]pH 6.3, CEC 43.3 cmol(+) kg-1, 4.8% OM.
[c]pH 6.0, CEC 10.2 cmol(+) kg-1, 3.1% OM.
[d]pH 7.3, CEC 55.2 cmol(+) kg-1, 4.0% OM.

situ conditions. Drying and rewetting affects both chemical and biological transformations in soils; any changes will be reflected in the subsequent analysis of soil
solution composition (Bartlett and James, 1980; Qian and Wolt, 1990; Walworth,
1992). Figure 5.3 illustrates the variable effect of equilibration period following
rewetting of air-dry soil on cation concentration of soil solution. Differential effects
of equilibration time are observed for the soil solution concentrations of total Ca,
Mg, and K.

Increased proton donation from highly polarized water associated with the surfaces of air-dried clays is hypothesized to contribute to increased surface acidity

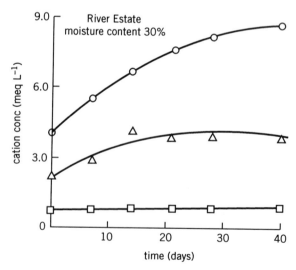

FIGURE 5.3   Effect of time of contact between soil and its solution phase on total ion
concentrations in soil solution (Ca: ○; Mg: △; K: ▢). [Moss, 1963. Reprinted by permission
of Kluwer Academic Publishers.]

and increased water soluble organic acids in dried soils. Further increases in water soluble organic acids occur as soil microbiota succumb to desiccation. Organic matter desiccation and oxidation in combination with increased acidity may shift available pools of specific cations and anions either through release from disrupted microbial biomass, changes in cation and anion exchange capacities, or reduction reactions.

The effects of drying on chemical transformations in soil involve interactions with other environmental conditions during drying such as light, humidity, and temperature that all may influence soil biota. These effects appear to be enhanced with time of storage in a dry state; increased times of air-dry soil storage result in greater stimulation of microbial activity upon rewetting. The well-recognized microbial burst that occurs when dry soils are rewetted may result in soil solution composition that is uncharacteristic of ambient soil solution in soils continuously maintained at field moisture contents.

The variable effects of sample storage have been demonstrated by comparing composition of soil solutions displaced from Ultisols that were incubated continuously field-moist at ambient temperature ($\approx 25°C$) with soils that were air-dried, remoistened, and then incubated (Qian and Wolt, 1990). Soil solution composition was frequently influenced by the interaction of sample treatment (field-moist versus air-dried and rewet) and time of incubation (up to 32 days) such that the differences observed were either unpatterned in time or were inconsistent across various soils and horizons sampled. Storage effects were more pronounced in A horizons than in B horizons, probably because of higher organic carbon content and greater microbial activity in A horizons. Short-term incubation (up to five days), however, had a negligible effect on both cation concentrations and ion activity ratios of soil solution displaced from low ionic strength Oxisol subsoils (Le Roux and Sumner, 1967).

Storage Temperature. Refrigeration or freezing of field-moist soils is frequently employed to retard changes in soil solution composition (Edmeades et al., 1985; Ross and Bartlett, 1990; Whelan and Barrow, 1980; Walworth, 1992). Table 5.2 summarizes the effect of freezing (96 hours) as well as thawing and refreezing on ion composition of displaced soil solutions. These treatments minimally affected NH_4^+, NO_3^-, Cl^-, and total P concentration. In contrast, frozen storage ($-17°C$) of Spodosol soil samples for longer periods (36 days) significantly increased concentrations of SO_4^{2-}, Cl^-, F^-, and total Al in displaced soil solutions (Ross and Bartlett, 1990). These effects were attributed to physical disruption of the soil and solubilization of organic matter. Desiccation of soil subjected to prolonged frozen storage is an additional confounding influence on the integrity of displaced soil solution composition. Frozen storage of soil may also disrupt soil structure and affect handling properties sufficiently to limit the success of soil solution displacement by most column displacement techniques.

Refrigeration can bring about consequential changes in concentration of biologically mediated components in soil solution. Refrigerated storage (5°C) resulted in significant changes in macronutrient ions displaced from a New Zealand soil within

TABLE 5.2 **Comparison of Ion Concentrations in Displaced Soil Solution as Affected by Freezing as Well as by Thawing and Refreezing**[a]

	Before	Frozen	Bulk sample thawed at		
Soil	Freezing	for .96 hours	24 hours	48 hours	96 hours
			mg L^{-1}		
			NH_4^+-N		
3	64	63	53	59	6
6	25	23	25	26	24
4	37	40	31	42	41
			P		
3	2.0	1.7	1.8	1.5	1.8
6	1.6	1.3	1.7	1.5	1.6
4	2.0	2.2	2.1	2.0	2.4
			NO_3^--N		
3	76	69	81	75	72
6	28	30	32	27	26
4	52	41	52	40	42
			Cl^-		
3	1261	1256	1298	1248	1277
6	559	459	644	501	523
4	853	834	804	767	807

[a]B. R. Whelan and N. J. Barrow (1980).

1 day of storage (Edmeades et al., 1985; Fig. 5-4). Refrigerated storage appears to result in a rapid shift to NO_3^- as the dominant anion in soil solution perhaps as a consequence of bacterial nitrification.

Effects of Added Moisture. Moisture status of field-moist soils is frequently supplemented with distilled water addition prior to displacement (Wolt and Graveel, 1986; Gillman and Bell, 1978; Le Roux and Sumner, 1967). This approach is desirable in order to achieve uniform moisture content thereby improving consistency of displacement. Adjustment of moisture status requires an equilibration period prior to displacement; thus, the suitability of this approach depends on the ability to maintain integrity of the ambient soil solution. In view of the substantial effects of storage conditions on soil solution composition, it is best to limit equilibration following added moisture to short periods (< 1 day) at ambient temperatures or moderate periods (1 to 5 days) for frozen samples. Refrigerated storage during equilibration results in rapid shifts in anion composition of soil solution and should be avoided.

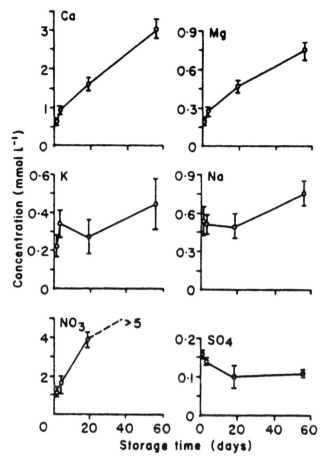

FIGURE 5.4 The effect of time of storage of soil samples on soil solution concentrations (mean and standard error of six soils). [Edmeades et al., 1985]

5.1.3 Displacement Duration

The duration of displacement can influence soil solution composition through effects on the portion of soil water sampled (Chapter 2) as well as biologically mediated alteration of soil solution. Variation in the rate of displacement by high-pressure filtration (1.7 MPa) from 2 seconds to 6 minutes per mL filtrate did not influence cation concentrations or activity ratios for soil solution displaced from an Oxisol subsoil (Le Roux and Sumner, 1967). Conversely, variation in centrifugation time for high-speed immiscible displacement of soil solution from 5 to 180 minutes resulted in slight but consistently significant changes in the concentrations of N forms, total P, and Cl in soil solution (Whelan and Barrow, 1980; Table 5.3). Gillman (1976) noted insignificant effects of low-pressure centrifugal displacement on EC of soil solution but slight variation in soil solution cation concentrations when times of displacement were varied from 15 to 60 minutes. Low-pressure

TABLE 5.3 Ion Concentration in Displaced Soil Solution as Affected by Duration of Centrifugation[a]

Centrifugation time minutes	NH_4^+-N		NO_3^--N		P		Cl	
	Soil 2	Soil 4	Soil 2	Soil 4	Soil 2	Soil 4	Soil 2	Soil 4
				mg L^{-1}				
5	263	34	52	2	5.0	11.2	984	966
10	260	37	148	1	5.4	12.2	964	986
15	257	34	149	2	5.9	12.2	988	1009
20	267	30	148	2	5.2	11.6	1005	978
30	263	30	150	2	5.1	11.4	1034	1004
60	257	30	148	3	4.7	11.6	977	1004
120	259	30	158	5	4.7	11.3	958	982
180	260	30	165	4	4.6	10.3	986	975
SE^b	6	2	9	1	0.5	0.7	21	32

[a]B. R. Whelan and N. J. Barrow (1980).
[b]Standard error of the mean.

centrifugation (73.5 kPa) at 20°C resulted in altered anion and cation composition in soil solution of a freshly sampled Ultisol when centrifugation time was varied from 20 to 240 minutes (Walworth, 1992). Column displacement techniques generally demonstrate constancy of composition of successive increments of displaced soil solution when soil solution breakthrough times occur within two to four hours following initiation of the displacement (Table 2.2).

5.2   CENTRIFUGATION

5.2.1   Low-Pressure Centrifugal Displacement

Centrifugal displacement at low pressures ($<$ 500 kPa) represents the most widely employed approach to obtaining soil solution (Adams et al., 1980; Davies and Davies, 1963; Gillman, 1976; Aitken and Outhwaite, 1987; Edmunds and Bath, 1976). Cameron and coworkers developed methods for centrifugal displacement of soil solution early in the twentieth century (Cameron, 1911), but the approach was soon largely supplanted by column displacement procedures. Reintroduction of the technique is generally attributed to Davies and Davies (1963). Modifications of their basic technique typically entail adaptations to accommodate differing sample sizes. Gillman (1976) originally adapted this procedure to large-sized samples and low-pressure centrifugation. Figure 5.5 illustrates some types of centrifugation apparatus which have been used for low-pressure centrifugal displacement.

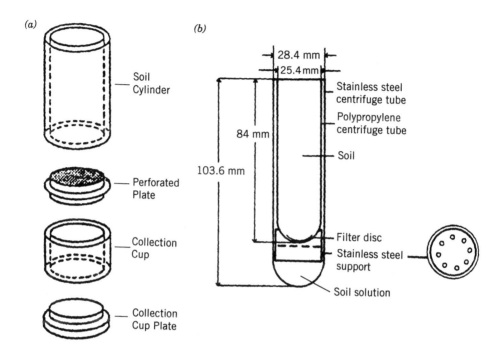

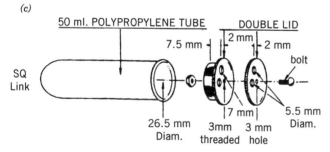

FIGURE 5.5 Assemblies for centrifugal displacement of soil solution. (*a*) Plexiglass apparatus for low-speed centrifugal displacement. The perforated plate is welded to the bottom of the soil cylinder, and the collection cup is welded to the bottom of the collection cup. [Adams et al., 1980. Published in *Soil Sci. Soc. Am. J.* 44:733–735. 1980. Soil Sci. Soc. Am.] (*b*) Modification of centrifuge tubes for high-speed centrifugal displacement. [Elkhatib et al., 1987. Published in *Soil Sci. Soc. Am. J.* 51:578–583. 1987. Soil Sci. Soc. Am.] (*c*) Sealed centrifuge assembly for immiscible displacement. [Mubarak and Olsen, 1976b. Published in *Soil Sci. Soc. Am. J.* 40:880–882. 1976. Soil Sci. Soc. of Am.]

Recommended Procedure. The method described here is that of Gillman (1976) as described by Adams et al. (1980) with some additional modifications. This method accommodates large sample sizes (up to 1 kg dry soil); it results in volumes of recovery in the general range 5 to 50 mL dependent on soil texture, field moisture content, and time and speed of centrifugation.

Apparatus. The centrifuge apparatus is constructed from Plexiglass tubing and sheets to form a two-tier apparatus (Figure 5.5*a*). The apparatus must be milled to exact tolerances to accommodate the centrifuge bucket employed in order to avoid shifting of the sample load during centrifugation. The upper tier of the centrifuge apparatus is fitted with a 7-mm thick milled Plexiglass base plate containing an array of 1-mm diameter holes. The lower tier serves as the cup for collection of displaced soil solution; this cup is milled to nest tightly with the upper tier and has cemented to it a plate milled from 7-mm thick Plexiglass.

Several modifications of the basic assembly described by Adams et al. (1980) have been successfully employed. Adams and coworkers describe an apparatus designed to fit 0.6 L centrifuge buckets, but the same design can be easily modified to accommodate centrifuge buckets as small as 0.25 L using readily available Plexiglass stock. In place of Plexiglass sheet, the base plate may be constructed from milled perforated stainless steel stock with a nominal thickness of 1 mm. If the base plate for the upper tier is constructed of stainless steel, a center support of solid Plexiglass rod ($\approx$15-mm diameter and of a height equal to the inner depth of the inner-tier assembly) is used to provide support against collapse during centrifugation. The welded construction of the lower-tier collection cup frequently results in some sample loss due to leakage at the weld joint; fabrication of the lower tier in one piece by milling solid Plexiglass stock may be preferred.

Aspects of Method. A sheet of filter paper is cut to fit over the perforated plate in the upper tier of the centrifugation apparatus. Selection of the filter paper is an important consideration, especially when trace constituents in soil solution are to be determined (see, for example, the discussion of Al determination in Chapter 14). A hardened ash-free filter paper, such as Whatman No. 42, appears appropriate for most routine applications.

Field-moist soil screened to pass 2-mm diameter mesh openings is uniformly packed into the upper tier. Uniform packing is an important qualitative aspect of the displacement methodology that influences the ability to obtain representative soil solution in a consistent manner. Uniform packing is achieved by progressive, gentle tamping of the upper tier against a tabletop as the centrifugation apparatus is filled with soil. Once the desired quantity of soil is delivered to the apparatus, the soil is packed with firm pressure applied by hand to the soil surface using a rubber stopper or other device milled to fit the interior of the upper tier of the apparatus. Following packing, immediately proceed with displacement or, if warranted, add additional moisture to the surface of the packed soil and allow an appropriate interval for infiltration and equilibration.

Subsamples of soil solution should be removed and analyzed for master components (most importantly for pH as well as E_H, if it is integral to interpretation of the analytical result; see Chapter 4) and volatile components immediately following displacement. A primary consideration in this respect is to minimize sample disruption so as to maintain the integrity of solution components that may shift following separation from the contacting soil solid phase.

Comments. The principal advantages of this methodology are the common availability of the needed equipment and the relative ease of the displacement procedure. Most centrifuges can be utilized for centrifugal displacement with the construction of the appropriate-sized vessels for displacement. The type of compositional analysis desired will determine the necessary volume of recovered soil solution. Thus, it must be considered when selecting the size of displacement appoaratus and centrifuge head necessary to accommodate the appropriate volume of moist soil. The methodology and apparatus described for high-pressure centrifugal displacement (section 5.2.2) is amenable to low-speed centrifugation as well, except that the size of displacement apparatus described may severely limit the amount of soil solution recovered with low pressure centrifugation.

Low-pressure centrifugal displacement frequently employs nonrefrigerated centrifuges. Changes in solution temperature with use of nonrefrigerated centrifuges may be substantial and should be accounted for when expressing compositional analysis of soil solutions (Chapter 4).

Wendt (1992) describes a modification of the low-speed centrifugal displacement apparatus that obtains clarified soil solution without using filter paper. This apparatus is intended to avoid potential contamination from filter paper and to ease the clean up of the centrifugation apparatus following displacement. In the presence or absence of filter paper, it is advisable to ascertain whether interfering colloids are present. This is because colloid interactions are frequently important to the interpretation of soil solution compositional analysis (see section 4.4). Qualitative screening for the presence of colloids can be accommodated through spectroscopic or turbidimetric scattering of light passed through the displaced soil solution.

5.2.2 High-Pressure Centrifugal Displacement

High-pressure centrifugal displacement of soil solution as described by Davies and Davies (1963) has been widely adapted by numerous authors (Edmunds and Bath, 1976; Kinniburgh and Miles, 1983; Reynolds, 1984; Elkhatib et al., 1987; Ross and Bartlett, 1990). These various methodologies vary only slightly with regard to design and materials employed in constructing the centrifugation apparatus, and in the speed and duration of centrifugation employed.

Recommended Procedure. The methods of Davies and Davies (1963) and Elkhatib et al. (1987) are essentially identical and are described here.

Apparatus. Figure 5.5*b* illustrates a typical apparatus for high-pressure centrifugal displacement. The apparatus employs a standard 50-mL high-speed stainless steel centrifuge tube, a 38-mL high-speed polypropylene centrifuge tube with holes drilled in the bottom, and a milled stainless steel support with holes bored in it. The polypropylene centrifuge tube serves to contain the soil being displaced and nests within the stainless steel centrifuge tube; it is supported by the milled stainless steel insert. Soil solution is collected in the bottom of the stainless steel centrifuge tube.

Aspects of Method. The methodology is the same as described for low-pressure centrifugal displacement. A sheet of filter paper (commonly Whatman No. 42) is cut to fit over the perforations in the base of the internal centrifugation tube. Pre-screened field-moist soil is uniformly delivered to the internal tube and packed by tamping the tube. Centrifugation immediately follows, unless additional moisture is added after which an appropriate period for infiltration and equilibration is employed. Post-centrifugation sampling and analytical details are the same as described for low-pressure displacement (section 5.2.1).

Comments. Procedurally, high- and low-pressure centrifugal displacement differ only in those design criteria needed to accommodate displacement at higher pressures—that is, construction of the displacement apparatus to withstand higher centrifugal forces. But, as described in Chapter 4, high-pressure displacement techniques need to give greater consideration to the effects of pressure and temperature on soil solution composition. Zabowski (1989) found filter paper rupture with high-speed centrifugation to complicate the analysis of displaced soil solution. Therefore, Spodosol soil solutions were displaced in the absence of filter paper and then 0.22 μm cellulose acetate syringe filters were used to remove particulates following centrifugation.

5.2.3 Centrifugation with Immiscible Liquid

Immiscible displacement uses a centrifugation technique where soil solution is displaced by high-pressure centrifugation of a moist soil in the presence of a dense, water-immiscible liquid. An excess of the immiscible liquid is added to moist soil, and during centrifugation this liquid penetrates downward through the soil and soil solution is displaced upward. Following centrifugation, displaced soil solution is recovered as an aqueous phase overlying the displacing liquid and the soil pellet. This technique was originally devised by Mubarak and Olsen (1976a) and has been widely adapted (Whelan and Barrow, 1980; Kittrick, 1980; Elkhatib et al., 1986; Phillips and Bond, 1989).

Recommended Procedure. The method of Whelan and Barrow (1980) is described here with the modification that ethyl benzoylacetate (Elkhathib et al., 1986) is used in place of carbon tetrachloride as the displacing liquid. Twenty-five grams of field-moist soil are placed in a 50-mL capped nonreactive centrifuge tube, and 30 mL of displacing liquid (ethyl benzoylacetate) are added. Following centrifugation (typically at >1MPa for from 30 minutes to 2 hours), the surface liquid (soil solution) is decanted to a funnel containing phase-separating filter paper (Whatman silicone-treated 1PS). Soil solution in the funnel is recovered after particle-settling, and the volume recovered is measured. This solution is next passed through an 0.2 μm membrane filter (preceded with a prefilter if necessary). The membrane filter is flushed with several rinses of deionized water, the filtrate combined with the recovered soil solution, and the combined filtrate taken to volume prior to analysis.

Comments. A wide variety of water-immiscible displacing liquids have been investigated: carbon tetrachloride, ethyl benzoylacetate, and a variety of chloro- and fluorocarbons. Ethyl benzoylacetate is recommended here as it appears to have the most favorable combination of properties for use as a displacing liquid; high density (1.12 Mg m^{-3}), low toxicity, water insolubility, low volatility, and limited environmental risk. Other aspects that should be considered in selecting a displacing liquid are the partitioning of analytes into the displacing liquid phase and the potential interfering effects on compositional analysis of soil solutions that may occur if traces of the displacing liquid contaminate the displaced soil solution. Because of these uncertainties, immiscible displacement would not be the method of choice when trace organic analysis in soil solution is desired.

The method of Whelan and Barrow (1980) requires extensive post-displacement cleanup of soil solution prior to analysis, perhaps because decantation causes appreciable contamination of the soil solution obtained with displacing liquid and soil particulates. This can be circumvented by centrifuging decanted soil solution and checking for the Tyndall effect in place of filtration (Kittrick, 1980). Other researchers have pipetted rather than decanted the aqueous phase; this may reduce contamination but published methods seldom describe precautions to assure the integrity of the solutions that are obtained.

Regardless of cleanup method (filtration or centrifugation) the integrity of soil solution pH measurement may be altered by sample-handling following centrifugation. This can be avoided by using the specially modified centrifuge cap described by Mubarak and Olsen (1976b). The apparatus (Fig. 5.5c) consists of a two-part cap that restricts gas exchange between headspace above the displaced soil solution and the ambient atmosphere. Rotation of the outer portion of the cap allows access to a pH-sensing combination microelectrode for measurement of soil solution pH prior to removal of the soil solution. However, one must account for the contamination of the soil solution from the pH-electrode filling solution. Thus, separate samples may be necessary for pH characterization and ion compositional analysis, respectively.

5.3 COLUMN DISPLACEMENT

5.3.1 Traditional Column Displacement

The historical development of column displacement techniques is well described by Parker (1921) and is discussed in Chapter 2. The technique consists of packing field-moist soil into a glass cylinder followed by addition of a displacing liquid which moves soil solution downward through the column preceding the advancing front of displacing liquid. Successive increments of displaced soil solution are checked for compositional integrity and composited for subsequent analysis. The apparatus for column displacement is illustrated in Figure 5.6.

Recommended Procedure. Techniques for traditional column displacement have been described in length by Adams (1974); the preferred method described here is quoted from Adams et al. (1980). The displacement apparatus consists of a 3-cm diameter glass column, 60-cm long, drawn and fitted at its bottom with a small-bore drain tube (Fig. 5.6a). The drain tube is packed with glass wool and the column is filled with field-moist soil that has been sieved through 10-mm mesh screen. During filling the column is gently jarred by rocking and light tapping against a rubber stopper—a rubber stopper bored to fit the drain tube is used to protect the drain tube during filling and subsequent packing.

Following filling with the desired quantity of soil (typically 1 to 1.5 kg field-moist soil), the soil is compressed by tapping the packed column, with the attached bored rubber stopper, against a larger stopper placed on a bench top. The degree of compression necessary is dependent on soil moisture content and texture; generally only light tapping is needed for soils with appreciable clay content and the force of tapping must increase substantially as the sand content increases. The packed soil is next firmed by applying pressure to the surface with a rubber stopper attached to a rod. A clay seal is formed at the soil surface through application of a few mL of displacing liquid followed by gently rotating the rubber stopper as it is held just at the soil surface.

After packing, firming, and sealing the soil, 100 mL of displacing solution is added to the top of the soil by decanting the solution down the rod with attached stopper. The displacing solution consists of saturated calcium sulfate containing 4% w/v potassium thiocyanate (KCNS). The column is placed on a support rack and the drain tube is attached to a capped, 15-mL conical centrifuge tube that is vented to the atmosphere via capillary tubing (Fig. 5.6c). Successive increments of displaced soil solution ($\approx$5mL each) are recovered from the centrifuge tube and aliquots are removed for measurement of pH and EC. An aliquot ($\approx$50 μL) is delivered to a spot plate and checked for intrusion of displacing liquid through addition of a drop of 5% w/v ferric chloride solution in 0.1 M HCl; contamination is indicated by a tinge of red color occurring when Fe is oxidized in the presence of CNS^-. Increments of displaced soil solution that are not contaminated with displacing solution and that have a similar pH and EC are composited for analysis. Displacement is discontinued when displacing solution intrusion becomes evident.

Comments. Experience indicates that a properly prepared soil yields soil solution within two to four hours following the onset of displacement and that the period of recovery of "unaltered" soil solution will be from one to four hours. Yields by the procedure described range from 5 to 50 mL of displaced soil solution.

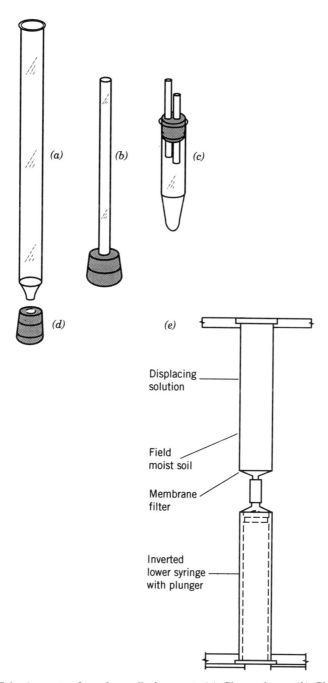

FIGURE 5.6 Apparatus for column displacement. (*a*) Glass column. (*b*) Glass rod and stopper. (*c*) Collection tube. (*d*) Rubber stopper assembly for tamping soil. [Adams et al., 1980. Published in *Soil Sci. Soc. Am. J.* 44:733–735. 1980. Soil Sci. Soc. Am.] (*e*) Apparatus for vacuum displacement of soil solution.

The proper degree of packing, firming, and sealing of soil is very much dependent on soil moisture and texture and can only be ascertained by trial packing and displacement of the soils of interest. Success of the technique, therefore, is dependent on operator skill and experience. Because of the labor-intensive and time-consuming nature of this procedure, it is difficult for a single operator to accomplish more than a dozen displacements in an eight-hour period. For these reasons, this method is not widely utilized, even though when properly executed it is probably the most reliable means to obtain "unaltered" soil solution.

Saturated gypsum is favored as the displacing solution because it mimics "typical" soil solution in ionic strength and dominant cation; additionally the gypsum solution will not adversely effect soil structure as it infiltrates the packed soil column, thus consistency of displacement is improved. Parker (1921) favored ethyl alcohol as a displacing liquid, but this may unfavorably alter soil structure and, thus, consistency of displacement.

Use of the capillary-vented centrifuge tube for collection of displaced soil solution restricts gas-mixing between soil atmosphere, reflected in the collection tube headspace, and ambient atmosphere. The collection tube headspace is assumed to represent soil atmosphere; this can be syringe-sampled via the vent tube in the event that analysis for volatile components is desired. Soil solution can also be syringe-sampled through this vent to afford measurement of solution pH with minimum CO_2 degassing effect.

Morgan (1916) described a high-pressure column displacement technique whereby displacement was afforded by use of viscous petroleum-based liquids as the displacing liquid. The technique was complicated and messy and has never received attention in modern research applications.

5.3.2 Vacuum Displacement

Vacuum displacement is a modified column displacement procedure devised to achieve displacement of "unaltered" soil solution in a manner consistent with traditional column displacement but without the operational restrictions that impede the utilization of traditional column displacement. The principal attributes differentiating vacuum displacement from traditional column displacement are a lesser degree of operator experience is needed, smaller sample sizes (<100 g) are used, and sample turnaround time is reduced. These attributes are facilitated through the use of a mechanical vacuum extractor.

Recommended Procedure. The method of vacuum displacement has been described by Wolt and Graveel (1986) and modifications have been employed to better facilitate routine analysis of xenobiotics in soil solution (Wolt et al., 1989; Wolt, 1993).

Apparatus. Vacuum displacement utilizes a mechanical vacuum extractor designed for leachate extraction of soil and vacuum extraction of saturated pastes (Holmgren et al., 1977). The mechanical extractor accommodates simultaneous displacement of up to 24 soil solutions. The mechanical extractor utilizes 60-mL polyethylene syringe barrels and plungers.

Aspects of Method. Field-moist soil is screened to pass 2-mm mesh and approximately 75 g is uniformly delivered to a syringe barrel containing a 25-mm diameter 0.45 μm Gelman polysulfone filter. Uniformity of packing is achieved through gentle side-to-side rocking of the syringe barrel and tamping against a bench top. After the desired mass of soil is delivered to the syringe barrel, the soil is packed to a volume of 55 mL by firming the soil with a 25-mm diameter rod (modified syringe plunger).

Soil-packed syringe barrels are mounted to the upper stage of the mechanical extractor where they are connected with tubing to an inverted 60-mL volume syringe with plunger (Fig. 5.6e). Displacing solution (9 mL saturated gypsum solution containing 4% KCNS w/v) is applied to the surface of packed soils and displacement is facilitated by operating the mechanical extractor for 2 hours at a displacement volume of 5 mL hr^{-1}.

The volume of solution recovered is determined by weighing. A 500-μL aliquot of displaced soil solution is removed to a polypropylene culture tube for pH measurement immediately following displacement. A 50-μL aliquot of displaced soil solution is delivered to a spot plate where it is reacted with 50 μL of 5% w/v FeCl$_3$ in 0.1 M HCl. Intrusion of displacing liquid into the displaced soil solution is indicated by a tinge of red indicating oxidation of Fe by CNS$^-$. The remaining displaced soil solution is used for EC measurement and then transferred to a polypropylene culture tube until compositional analysis is completed. Replicate soils are displaced and the results compared for constancy of the displacement procedure.

Comments. The analysis of trace inorganics and xenobiotics necessitates modification of the apparatus to accommodate 50-mL volume glass syringe barrels with Teflon plungers. This is achieved by milling the head and base of the mechanical extractor to fit syringes and plungers of varied diameters.

Improved consistency of displacement is achieved if soils are moistened to slightly above field capacity (typically 125% of 33 kPa) following packing of soils to syringe barrels. In this case, a two-hour period is used to allow moisture to infiltrate and equilibrate.

As with traditional column displacement, some degree of experience is necessary in order to properly pack soils so that unadulterated soil solution is obtained in a consistent manner. Typical volumes of recovery are from 2 to 5 mL.

5.3.3 Syringe-Pressure Displacement

Ross and Bartlett (1990) have designed a syringe-pressure method for obtaining soil solution that is intended to afford immediate displacement of soil solution from soils as they are sampled in the field. The method thus avoids some of the pitfalls associated with soil storage effects on displaced soil solution composition.

Recommended Procedure. Field-moist soil is packed into 60-mL polyethylene syringes fitted with 2.4-cm diameter Whatman No. 540 filter paper. If NO_3- determination is desired, filters are prerinsed free of anions with distilled, deionized H_2O. The soil-filled device is placed in an apparatus that allows compression of the sample by hand-turning a threaded plunger. The first 5 to 10 (usually cloudy) drops are discarded and the remaining solution is collected over a period of about 15 minutes with continued compression.

Comments. This procedure has been used to displace soil solution from Haplorthod Oa and Bh horizon samples (Ross and Bartlett, 1990). Displaced soil solution represented about 20% of field-moisture content. Volume yields were appreciably greater for high organic matter content Oa horizon samples that contained > 200% water compared to mineral soil samples that contained 58 to 72% water. Therefore, yields of soil solution may be somewhat limited in mineral soils containing lower moisture at field moisture contents than soils investigated by these authors.

5.4 PRESSURE MEMBRANE EXTRACTION

Pressure membrane extraction is a technique that uses pressurization of a specially designed pressure filtration apparatus where a column of moist soil is supported over a pressure filtration membrane that may be composed of ceramic, cellulose, or glass fibers. The method was originally described by Richards (1941) and was subsequently modified by Reitemeier and Richards (1944).

Recommended Procedure. The procedure described here is essentially that of Reitemeier and Richards (1944) utilizing an updated apparatus constructed of modern materials.

Apparatus. A soil column apparatus amendable to pressure membrane extraction of soil solution is illustrated in Figure 5.7. A principal design consideration is construction of the apparatus such that all surfaces contacting the soil or displaced soil solution are nonreactive and noncontaminating—in this case, contacting surfaces are stainless steel, Teflon, and glass. Dimensions of the glass cylinder supporting the soil sample can be varied to meet the needs of individual applications. Both the base and top plate are grooved to accommodate the soil-filled glass column and are fitted with inlet and outlet tubes. An airtight seal is achieved using O-rings placed in the grooves. The assembly is secured by bolts extending from the base to top plate.

Aspects of Method. Field-moist soil is sieved and delivered to the glass cylinder that rests upon a base plate containing a glass microfiber filter pad supported on stainless steel screen. Soil is packed to uniform bulk density using a hand-tamp designed to fit the internal diameter of the glass cylinder. A filter pad, stainless

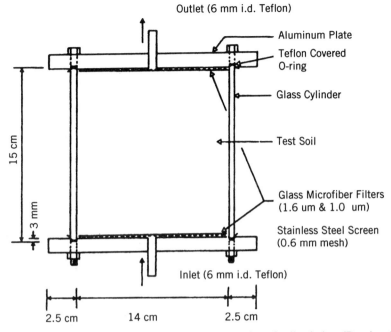

FIGURE 5.7   Apparatus for pressure membrane extraction of soil solution. [Reprinted with permission from Seigrist and Jenssen, 1990. Copyright 1990 American Chemical Society.]

steel screen, and top plate are placed on the surface of the packed soil and the assembly is secured and placed on a support rack. The inlet tube is connected to a source of nonreactive gas (commonly nitrogen for most displacements), and the outlet is connected to a vented collection source. Pressures of approximately 1.7 to 1.8 MPa are applied to displace soil solutions from soils initially at field moisture contents.

Comments. The procedure requires a considerable mass of soil; a device of the dimensions shown in Figure 5.7 (a column of 14 cm i.d. × 15 cm long) requires approximately 3.6 kg moist soil. Seigrist and Jenssen (1990) achieved uniform distribution of a sand in this device by packing soil sieved to pass a 4-mm screen to a bulk density of 1.64 g cm^3.

Time to achieve displacement varies according to soil texture and initial soil moisture content. Solution recovery with time from a sandy loam soil extracted at 1.52 MPa is shown in Figure 5.8. More than 16 hours extraction time were required to recover ≈ 50 mL soil solution from 3.5 kg of moist soil initially at 33 kPa (10.4%) moisture. Typical displacements by this method may require up to 4 days to complete (United States Salinity Laboratory Staff, 1954).

Large sample size requirements, relatively low fractional moisture recoveries, and prolonged extraction periods appear characteristic of this procedure (Reitemeier

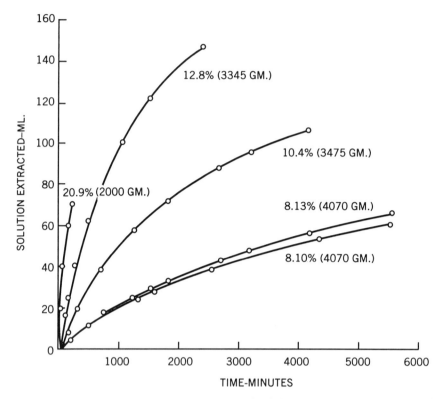

FIGURE 5.8 Effect of moisture content on rate of solution extraction for Hesperia sandy loam. [R.F. Reitemeier and L.A. Richards, 1944, "Reliability of the pressure-membrane method for extraction of soil solution," *Soil Science* 57:119–135.]

and Richards, 1944). Thus, it is not surprising that the procedure is of limited use for obtaining soil solution.

5.5 SATURATION EXTRACTS

Extracts of soils saturated and equilibrated with water have been widely used to represent soil solution. The technique is especially suitable where soil solution composition is buffered by high contents of soluble salts. Methods whereby saturation extracts are obtained are infrequently characterized in published research. The preferred approach is to obtain a saturation paste extract.

5.5.1 Saturation Paste Extracts

Recommended Procedure. The procedure for obtaining saturation paste extracts described herein was developed for the analysis of soil solutions from saline and alkali soils (United States Salinity Laboratory Staff, 1954); quotations are directly from methods 2 and 3a as described by the Salinity Laboratory Staff.

Preparation of the Saturated Soil Paste. "Prepare the saturated soil paste by adding distilled water to a sample of soil while stirring with a spatula. The soil–water mixture is consolidated from time to time during the stirring process by tapping the container on the workbench. At saturation the soil paste glistens as it reflects light, flows slightly when the container is tipped, and the paste slides freely and cleanly off the spatula for all soils but those with a high clay content. After mixing, the sample should be allowed to stand for an hour or more, and then the criteria for saturation should be rechecked. Free water should not collect on the soil surface nor should the paste stiffen markedly or lose its glistening appearance on standing. If the paste does stiffen or lose its glisten, remix with more water." The saturated soil paste is allowed to stand for 4 to 16 hours prior to extraction.

Obtaining the Saturation Extract. The extraction of soil solution from the saturated paste is typically accomplished using appropriately sized Buechner filter funnels lined with ash-free filter paper and connected to a thick-walled Erlenmeyer flask with side arm (filtration flask) which is in turn connected to a vacuum source. A test tube nested within the filtration flask can be used to collect filtrate. Alternatively, specially modified Buechner funnels may be used with a mechanical extractor (Holmgren et al., 1977).

"Transfer the saturated soil paste . . . to the filter funnel with a filter paper in place and apply vacuum. Collect the extract in a bottle or test tube. . . . If the initial filtrate is turbid, it can be refiltered through the soil or discarded. Vacuum extraction should be terminated when air begins to pass through the filter."

Comments. Methods for preparing saturation pastes often recommend that dry soil be used; because of the potential confounding effects of air-drying (section 5.1.1), it is preferable that field-moist soil be used when saturation extracts are to be used as models of soil solution composition. Soils will be most readily "worked at moisture contents near field capacity, sufficient water should be added immediately to bring the sample nearly to saturation. If the paste is too wet, additional . . . soil should be added."

"The amount of soil required depends on the measurements to be made, [that is], on the volume of extract desired. A 250-g sample is convenient to handle and provides sufficient extract for most purposes. . . . If saturation pastes are to be made from a group of soils of uniform texture, considerable time can be saved by carefully determining the saturation percentage of a representative sample. . . . Subsequent samples can be brought to saturation by adding appropriate volumes of water to known weights of soil."

"Special precautions must be taken with peat and muck soils and with soils of very fine and very coarse texture." If a peat or muck soil is dry, overnight wetting is necessary prior to preparation of the saturation paste. Following this initial equilibration, a second rewetting and mixing will produce a mixture characteristic of a saturated paste. Soils having textures that are either very fine or very coarse should be prepared with a minimal amount of mixing; this can be best achieved if the moisture required to achieve saturation has been predetermined.

5.6 WATER EXTRACTS OF SOILS

Soil–water extracts may be suitable models of in situ soil solution composition in some instances. Hoagland et al. (1920) found that, for major soil solution components occurring in soils with high buffer capacities, 1:1 and 1:5 soil–water extracts with short equilibration times were suitable models of soil solution.

Recommended Procedure. The method described here is adapted from the United States Salinity Laboratory Staff (1954). An appropriate quantity of field-moist soil (typically 10 g) is placed in a bottle or capped centrifuge tube and sufficient deionized water is added to achieve the desired soil to water ratio. (Moisture content of the soil as used is predetermined from a separate subsample, and the appropriate water to be added is determined on the basis of the soil oven-dry weight.) The soil–water mixture, in a capped container, is placed on a mechanical shaker for 15 minutes, allowed to stand for 1 hour, and shaken again for 5 minutes. The soil–water slurry is filtered through ashless filter paper in a fluted funnel, and the clear filtrate is retained for analysis.

5.7 COMPLEXATION AND EXCHANGE TECHNIQUES

Complexation and exchange techniques involve reaction of soil with chelates or ion exchangers that sequester a pool of chemicals which may be associated with an available soil pool (soil solution plus the labile chemical in the soil solid phase). The characterization of the soil pool sampled depends on the competitive strength and concentration of the sequestering sink relative to the soil solid phase pool of the ions of interest, as well as on the time of contact of the sequestering agent relative to the rate of diffusion of the ions of interest in the test system. The sequestering effect of resin pools introduced to soils—in resin-impregnated membranes, papers, or bags—can serve as an index of bioavailability (the ion sink bioavailability index) indicating the combined effects of concentration of the bioavailable pool and effective rates of diffusion through soil solution (Abrams and Jarrel, 1992; Vaidyanathan and Nye, 1966). Under the proper experimental conditions, the ion sink bioavailability index may directly correlate with soil solution concentration, but the inability to experimentally determine the effective diffusion coefficient limits the direct applicability of such methods to unambiguous measurement of soil solution composition.

Complexation and exchange techniques, however, can be modified using competitive ion equilibria to estimate ion activities in soil solutions. This approach has been used with chelating agents to measure indirectly ion activities of metals in soil solution (Baker, 1973; Norvell and Lindsay, 1982). Additionally, reactions with chelating agents have been used to determine trace cation distribution in 50 mM $Ca(NO_3)_2$ extracts (Henderson and Corey, 1983). The premise underlying metal ion chelate equilibria is that "if metal ion chelates can reach equilibrium with the metal ions in soil solution and with the metal ions in the soil labile pool, the activities of metal ions in the soil solution can be calculated" (Amacher, 1984). Two general approaches to metal ion-chelate equilibria, the variable mole fraction method and the mass balance-iteration method are described here.

5.7.1 Variable Mole Fraction Method

Norvell and Lindsay (1982) used reaction of soil with dilute EDTA solutions in which the mole ratio of Ca to Fe was varied to calculate Fe^{3+} activity in soil solution. Changes in chelated Fe occurring after equilibration with soil were determined by plotting final versus initial mole fraction of chelated Fe as the initial mole fraction of chelated Fe was varied. That point along the curve obtained where initial and final mole fractions of chelated Fe were equal represented the mole fraction of Fe that was in equilibrium with the original soil. This mole fraction of chelated Fe and the corresponding mole fraction of chelated Ca can be used along with an estimate of soil solution Ca^{2+} activity and formation constants for Fe- and Ca-EDTA complexes to calculate Fe^{+3} activity in soil solution,

$$A_{Fe^{3+}} = \frac{A_{FeEDTA}}{A_{CaEDTA}} \frac{K_{CaEDTA}}{K_{FeEDTA}} A_{Ca^{2+}}^{3/2}$$

[5.1]

5.7.2 Mass Balance/Iteration Method

Baker (1973) and Baker and Amacher (1981) devised a method for comprehensive estimation of cations in soil solution using chelation with DTPA. Soil was extracted (1:10) for 24 hours with an 0.4 mM DTPA solution with fixed initial concentrations of chelated K, Mg, and Ca. Following the equilibration period, the solution phase is analyzed for total cation concentrations. The total cation concentrations along with equilibrium solution pH and appropriate formation constants for the metal-chelate species of interest allow for computation of ion distribution between free solution and chelated phase components. True equilibrium of the solution, labile and chelate phases, does not occur after an arbitrary equilibration time; therefore, the computationally determined ion activities in the solution phase have an uncertain relationship to true soil solution cation distribution (Amacher, 1984).

5.7.3 Comments

Although complexation and exchange techniques are useful estimators of available pools of ions, they are for the most part limited for routinely modeling in situ soil solution composition. There are three major limitations to these techniques.

1. The kinetics of ion partitioning between soil solution, labile, and sequestering phases are not well defined nor are they generalizable across experimental systems.

2. Use of extracts to model the soil solution phase are most likely inappropriate, especially when trace constituents are determined.

3. Adequacy of the complexation constants, activity estimates for some components, and computational approaches have been incompletely characterized.

Thus, it is not surprising that when such methods are compared they result in grossly different estimates of trace metal activity in the soil solution (Amacher, 1984).

5.8 SUMMARY

Soil solution is operationally defined by the method whereby it is displaced or otherwise modeled. No one laboratory method is consistently superior for this purpose—the soils and analytes investigated will largely determine the adequacy of a given technique to faithfully model composition of in situ soil solution. The various column displacement methods are the most widely applicable and reliable techniques for obtaining unaltered soil solution, but their widespread adaptation is hampered by relatively high degree of operator experience required for consistent success. Saturation extracts are a useful alternative technique, particularly for moderate to high ionic strength soil solutions that are strongly buffered by the soil solid phase. Centrifugation methods are the most popular soil solution displacement techniques because of their ease and the ready availability of the requisite equipment in most laboratories. In general, preferred methods maintain soils field most at ambient temperature and pressure and entail minimal sample handling, alteration, and storage prior to displacement.

CHAPTER 6

OBTAINING SOIL SOLUTION: FIELD METHODS

The term "soil solution" is broadly—and, often, inaccurately—applied to soil waters and soil percolates sampled and analyzed in field studies. Whereas the methodologies described in Chapter 5 are intended to sample the aqueous phase of soil in quasi-equilibrium with the soil solid phase, the intent of field sampling methods is frequently indistinct. Most field sampling methods have been used to interpret soil solution chemistry from both static and dynamic standpoints without sufficient consideration of the form of soil water being sampled and its appropriateness for a given consideration of chemical reactivity in soils. Ambiguities in interpreting results of field moisture sampling, therefore, frequently arise because of a failure to clearly distinguish "soil solution" from "soil water" (section 2.2.2).

Field methods for sampling soil water are generally grouped under the rubic "lysimetry." Lysimetry is a term derived from Greek that means "the measure of what is loosened or set free." Broadly considered, lysimetry refers to any method devised to sample solvents and/or solutes occurring in or moving through the soil profile. There are four general types of lysimeters:

1. Monolith—any device utilizing an undisturbed soil block or column.
2. Filled-in—devices containing soil where the natural soil structure has been disrupted.
3. Tension—also called vacuum, suction, point, or mini- lysimeters.
4. Ebermayer—any lysimeter installation where, by access from a trench, a trough, pan, funnel, plate, or wick is placed under undisturbed soil.

The history of lysimetry research in the United States has been described by Khonke et al. (1940) and Harrold and Dreibelbis (1967). Soileau and Hauck (1987) have described lysimetry in the United States with primary emphasis on its utility for measuring leaching losses from fertilization, whereas the review by Litaor (1988) emphasizes the utility of various lysimeter designs for sampling of "unaltered" soil solution. Bergstrom (1990) has reviewed the use of lysimeters emphasizing the evaluation of pesticide leaching potential.

6.1   COMMON ATTRIBUTES OF LYSIMETERIC METHODS

6.1.1   Nature and Composition of Lysimeter Solutions

Soil water collected from lysimeters is comprised of water that has moved by diffusion and/or conduction through the soil column as well as water that has moved continuously via preferential flow.[1] Soil structure, soil moisture conditions prior to and during percolate sampling, and features of the design and operation of the lysimeter will influence the relative proportions of diffuse and conductive versus preferentially transported water obtained. Interpretation of lysimetry solution composition from a static rather than a dynamic perspective thus depends on the degree to which diffuse and mobile water are reflected in the solution collected. It also depends on the proportion of preferentially transported water represented in the mobile fraction. Differing flow velocities of water moving through intra-aggregate pores—that is, water moving diffusively—and mobile water moving through inter-aggregate pores or through preferential flow channels contribute to differing chemical compositions of soil solutions obtained by lysimetric sampling of heterogeneous soil horizons. Chemical composition of solutions obtained by lysimetry is most appropriately applied to considerations of chemical transport in soils, provided the solution obtained meaningfully samples water in a soil environment where texture and homogeneity sufficiently approximate a continuous solid matrix (section 2.2.3).

Solution Chemical Composition.   Differences among lysimeters in terms of solution composition, rate of sampling, and zone of influence are anticipated as a result of differing designs, construction materials, and methods of installation and operation. Table 6.1 presents lysimeter solution composition and recovery volumes for various types of field-installed lysimeters (Silkworth and Grigal, 1981); the significant differences in solution composition observed may indicate that different portions of in situ soil solution were sampled by the various lysimeters. Similarly, comparisons of soil solution composition for lysimeter solutions versus solutions obtained by laboratory displacement indicate the degree to which these sampling techniques sample diffuse as contrasted to mobile water.

Hantschel et al. (1988) compared soil solution chemical composition for several acid upland soils from Germany for solutions obtained as 1:0.8 soil–water saturation extracts and as percolates from 6-cm diameter undisturbed soil cores. The saturation extracts contained lower concentrations of protons and acidic cations and higher concentrations of base cations than did comparable percolate solutions. They

TABLE 6.1   Soil Solution Concentrations Extracted by Four Types of Sampler during Snow-Free Season in Northern Minnesota[a]

Element	Large Ceramic	Small Ceramic	Fritted Glass	Cellulose Fiber
		$\mu g \; mL^{-1}$		
P	0.22	0.15	0.13	0.11
K	0.91	1.38[b]	0.81	0.78
Ca	8.95	18.14[c]	7.87	19.80[c]
Mg	3.19	18.01[c]	2.86	5.44[c]
Na	4.21	5.18[b]	10.80[b]	6.50[b]
		mL		
Volume	149	54[b]	88	55
n	14	14	7	5

[a]All pairs included the large ceramic samplers. The set of tested pairs for the fritted glass and cellulose fiber samplers were primarily a subset of the large ceramic set (all samplers at 130-cm depth).
[b,c]Significant difference at 0.05 and 0.01 levels, respectively, paired t-test.
D. R. Silkworth and D. F. Grigal. *Soil Sci. Soc. Am. J.,* 1981, 45:440–442.

concluded that percolate solutions reflected mobile water flowing through inter-aggregate pores, and that the composition of exchange sites interacting with the percolating water was different than the population of exchange sites reflected in saturation extracts that modeled diffuse water in intra-aggregate pores.

Upchurch et al. (1973) and Marshall et al. (1973) conducted extensive investigations of cation exchange and pedogenesis as interpretable from soil solution composition of an Alfisol. Soil solution was modeled either as the percolate collected from filled-in lysimeters or as a 1:1 soil–water extract. Comparison of ion ratios calculated from the compositional analysis of the solutions obtained showed dramatic differences between soil–water extracts from the bottom depth of the control lysimeter and lysimeter percolates (Table 6.1). The authors attributed differences to a contaminant artifact (drainage gravel) from lysimeter installation that altered percolate composition. Preferential flow arising from the soil-filling operation may have also influenced results.

Reeve and Doering (1965) reported good correspondence in total salt content of tension lysimeter solutions and saturation extracts obtained from several depths in a sodic soil. Data were not presented to indicate whether chemical integrity of the solutions was otherwise similar between the two methods for modeling soil solution.

Zabowski and Ugolini (1990) compared composition of soil solutions obtained from monthly sampling of a Spodosol pedon. Field samples were obtained from 14 μm pore-size ceramic plates in contact with the soil and under 10 kPa tension; in the laboratory, soil solutions were displaced by centrifugation. Displaced soil solutions were characterized by distinct seasonal patterns of solution composition. In contrast, the lysimeter solutions tended to have lower concentrations of ions and exhibited less seasonal variation. Differences were attributed to seasonal changes in soil moisture and biotic activity that were reflected in displaced soil solutions that sampled water in diffuse contact with the soil. Lysimeter solutions, however, were thought to represent mobile water rapidly percolating through the soil and, thus, were less sensitive to changes in moisture and biotic activity.

O'Dell et al. (1992) noted rapid movement of Br^- and the pesticide imazethapyr when applied to an undisturbed Hapludult in column lysimeters under tensions of 10 and 33 kPa. Corresponding measurement of imazethapyr sorptivity using vacuum displacement of soil solutions suggested that inter-aggregate contact was limited in the soil columns; thus, lysimeter solutions represented significant bypass flow in the columns.

These limited comparisons indicate distinct differences in solution composition as ascertained from lysimetry, as compared to soil solution modeled by displacement or water extraction. They are not particularly informative, however, as to whether differences are attributable to sampling of different portions of soil water or to artifacts arising from the design and operation of the lysimeters.

6.1.2 Lysimeter Construction Materials

A considerable body of data documents the potential errors introduced from the materials used in construction and installation of porous cup (or porous plate) lysimeters. Porous cups or plates, and the silica grouts that are used in their installation, may release contaminants to lysimeter solutions, may exclude certain solution components by filtering, and may act as surfaces for adsorption, exchange, and precipitation of chemicals. Interpretation of the influence of lysimeter construction materials on lysimeter solution composition is complicated by the wide variation in material types, preparation methods, and evaluation methods employed by researchers (Beier and Hansen, 1992).

Common materials for the construction of porous samplers are ceramics, alundum, and Teflon. Proximate composition of these materials (expressed as percent elemental oxides) is as follows (Creasey and Dreiss, 1988): Ceramic is 55% Al_2O_3 and 35% SiO_2, with minor amounts of Fe_2O_3, TiO_2, CaO, MgO, Na_2O, K_2O and SO_3. Alundum is 90% Al_2O_3 with SiO_2, Fe_2O_3, and TiO_2. Teflon is tetrafluoroethylene containing minor organic contaminants introduced in the process of making it porous. Additionally, stainless steel, fritted glass, cellulose acetate fibers, and nylon have been used less frequently in lysimeter construction.

Chemical weathering and leaching, filtering and retention, chemical equilibration, sampling rate, and sampling variability are important factors influencing how construction materials affect changes in lysimeter solution composition (Beier and Hansen, 1992). Neary and Tomassini (1985) investigated contaminating effects from new alundum plate lysimeters (15 cm×15 cm by 13 mm thick). Washing with up to 3 L distilled water drawn through the plate released Ca, Mg, Na, K, SO_4, Cl, and Zn at levels sufficient to interfere with analysis of low-concentration soil solutions. Acidified distilled water (containing H_2SO_4) released progressively greater amounts of Fe, Mn, Al, and Zn as pH decreased below 4.5. They concluded distilled water rinsing would be sufficient to decontaminate lysimeter plates for use with soil solutions of >pH 4.5; at lower pH, special precautions for decontamination would be needed.

Reactions with Porous Surfaces. Adsorption to sampler surfaces can decrease concentrations of trace constituents as soil water is drawn into lysimeters; these trace constituents may subsequently desorb into solution at later times. Thus, variation in lysimeter solution composition with time can reflect differential reactivity of solution constituents with lysimeter materials; this may obscure changes in soil solution composition occurring as a consequence of dynamic processes in the soil environment.

Adsorption of trace metals by various porous materials was demonstrated when precleaned tension lysimeters were equilibrated with simulated soil solutions containing $\mu g \ L^{-1}$ concentrations of the radioisotopes ^{65}Zn, ^{60}Co, ^{51}Cr, and ^{109}Cd (McGuire et al., 1992). Lysimeter materials adsorbed these metals in this general order: ceramic > stainless steel $\gg$ fritted glass = Teflon. But the relationship was by no means consistent—fritted glass, for example, adsorbed substantially more Cr than other materials. The method of precleaning samplers, the number of pore volumes of simulated soil solution drawn through the samplers, and the time of equilibration with the simulated soil solution were confounding factors that influenced the magnitude of adsorption observed. Fine silica, used as packing material for seating tension lysimeters in field installations, was also a significant contributor to trace metal adsorption, removing upwards to 60% of trace metals from solution. The adsorptivity of metals to silica increased with increased pH of the equilibrating solution (Fig. 6.1).

Guggenberger and Zech (1992) investigated dissolved organic carbon reaction with both old (field-equilibrated for three years) and new (prepared by flushing with 0.1 M HCl solution) porous ceramic cups and the resulting effects on chemicals reactive with organic carbon. Solutions from zero-tension Ebermayer lysimeters draining a forest floor organic horizon were drawn through the ceramic cups; old ceramic cups reached DOC concentrations equal to the test solutions following flushing with approximately 20 pore volumes, while new cups sorbed from 20 to 30% of DOC when flushed with up to 60 pore volumes. New ceramic cups increased hydrophobic forms of DOC in lysimeter solutions, whereas old ceramic cups had no effect on qualitative distribution of DOC. Solution sulfate and pH were unaffected by the porous cups, Al was unaffected by ceramic cups once 20 pore volumes of solution were passed through either new or aged cups, Cu was variably affected, and Pb was severely reduced through flushing with 60 pore volumes. Ceramic cup reactivity with DOC precludes their use for sampling high organic matter solutions where significant temporal changes in solution DOC are anticipated; field-equilibration of ceramic cups should allow their use for sampling mineral horizons where DOC should be less variable.

Perrin–Gauler et al. (1993) found ceramic cups to reduce lysimeter solution concentrations of atrazine, isoproturon, 2,4-D, and carbofuran by up to 80% of that in test solutions for the first pore volume passed through the lysimeter. This effect

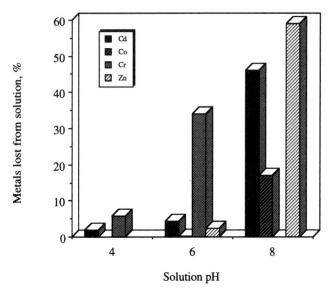

FIGURE 6.1 Trace metal adsorption from simulated soil solution by silica packing material (99.8% silica dioxide, 200 mesh) used to seat tension lysimeters. Initial trace metal concentrations: pCo = 4.0; pCd = 7.0; pCr = 5.9 (not included in pH 8 treatment); and pZn = 4.1. [After McGuire et al., 1992.]

was rapidly overcome, however, and artifacts of sampling were minimized when the initial pore volume was discarded.

Creasey and Dreiss (1988) reviewed a number of porous cup cleaning procedures. They recommended preparing ceramic cups for field installations by the method of Grover and Lamborn (1970)—rinsing with 50 to 60 pore volumes 1 M HCl followed by rinsing with 10 pore volumes distilled water. This general procedure appears suitable for preparation of ceramic, alundum, or Teflon cups or plates for sampling of many soil solution components (Ca, Mg, Na, K, Al, Zn, Mn, Fe, Cr, Cu, Cd, Cl, SO_4) provided analytes of interest do not occur at < 1 mg L^{-1}.

6.2 MONOLITH LYSIMETERS

Monolith lysimeters may be grouped as block or column lysimeters. Both types of lysimeters utilize undisturbed units (monoliths) of soil; they differ principally in size and transportability. Block lysimeters are the traditional form of monolith lysimeter that, because of their size, are typically built in place. Undisturbed soil column lysimeters are of more recent advent and are characterized by their ease of construction and ability to be transported from their point of construction to a remote location.

6.2.1 Block Lysimeters

The concept underlying the design of block lysimeters is to isolate a unique pedon of soil (a monolith) physically and equip it so soil water percolating vertically through the monolith may be collected. Additionally, block lysimeters may be constructed so as to rest on a scale allowing for determination of moisture flux through the monolith. Weighing-pain block lysimeters are particularly useful for experimental determinations of evapotranspiration (Kohnke et al., 1940).

Construction of a block lysimeter basically involves excavation of a trench surrounding the pedon to be isolated. A casing is constructed around the exposed soil block. A perforated pan is fit to the bottom of the soil block either by driving it in place from one side of the trench or lifting the block and inserting.

The first monolith lysimeters in the United States were constructed in New England in the late 1870s (Soileau and Hauck, 1987). These units were used to measure the quantity and composition of soil water percolates. Recently, modified design block lysimeter designs involving instrumentation to isolate and characterize both lateral and vertical conduits to preferential flow of solutes have been developed (Jardine et al., 1989).

6.2.2 Undisturbed Soil Column Lysimeters

Column lysimeters are constructed by driving a cylindrical casing vertically through the soil profile. Depending on the technique employed the soil column is recovered either through excavation or extraction. A variety of different methodologies have been employed to drive casings into the soil: hydraulic pressure (Walker et al., 1990; Swallow et al., 1987), progressive addition of weight (Tackett et al., 1965), driving (Bitton and Boylan, 1985), and augering (Bergstrom, 1990). The principal considerations for selection of a technique for sampling of an undisturbed soil column are availability of equipment and capability of driving the cylinder through the soil of interest without disrupting soil structure (either through fracturing or compressing the soil).

6.2.3 Operational Needs and Practical Constraints

Sidewall Effects. Monolith lysimeters differ in the forms of soil water sampled depending on design and conditions of operation. The boundaries imposed by the monolith lysimeters disrupt lateral flow and runoff of solutes.

Sidewalls of monolith lysimeters artificially truncate a portion of the vertical and lateral conduits to preferential flow of solutes that occur in natural soil bodies. These conduits may be worm or root channels; or they may be features imparted by the physicochemical properties of the soil itself. Alternatively, conduits for sidewall flow can be established as an artifact of lysimeter construction. The degree to which sidewall effects influence the yield and composition of lysimeter water depends largely on the surface area of the monolith (increased monolith surface area reduces the relative influence of sidewalls in affecting the volume and composition of percolate). Sidewall effects are influenced by design of the column sampling device, soil texture, soil moisture at sampling, and use of grouting to seal the interface between the cylinder and the contacting soil. Figure 6.2 illustrates the effect of sidewall flow on breakthrough curves, in 80-cm diameter soil columns, for a conservative tracer.

Till and McCabe (1976) applied radiotracer as an annular ring near the sidewall or as a disk avoiding the sidewalls in studies with 25-cm diameter columns; the results indicated a readily measurable sidewall influence (Fig. 6.3), but the results are somewhat ambiguous due to a fourfold greater water flow for the annular ring versus disk applications. Saffigna et al. (1977), also using disturbed soil columns, found sidewall effects insignificant for a column diameter of 87-cm diameter. Cameron et al. (1990; 1992) observed an artifact effect from their sampling method leading to significant sidewall flow in 80-cm diameter undisturbed soil columns; this effect was mitigated by injection of liquified petrolatum in the annular gap between the lysimeter sidewall and the soil column. Their approach would be useful only for studies where the grouting material did not react with analytes of interest. Bergstrom (1990) concludes that for pesticide leaching evaluations, the limiting diameter for undisturbed soil columns should be 25 cm, provided of course that the sampling and preparation method sufficiently limits the extent of the sidewall gap formed.

Column Holdup of Water. As typically used, monolith lysimeters yield percolate without application of tension—that is, the boundary of the monolith lysimeter truncates to the ambient atmosphere, a design feature referred to as "zero-tension"

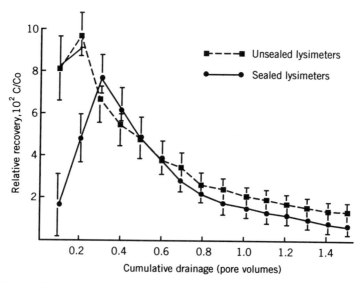

FIGURE 6.2 Solute breakthrough curves for sealed and unsealed column lysimeters. Each point represents the mean of 10 replicate lysimeters. Bars represent standard errors. [Cameron et al., 1990. *Soil Sci. Soc. Am. J.,* 1992, 56:1625–1628.]

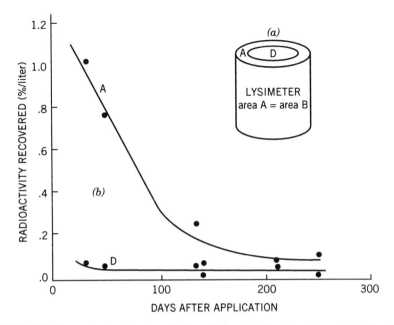

FIGURE 6.3   Example of estimation of sidewall flow. (*a*) Position of annular [A] and disc [D] zones. (*b*) Recovery in leachate of radioactivity applied in the two zones. [A.R. Till and T.P. McCabe, 1976, "Sulfur leaching and lysimeter characterization," *Soil Science* 121: 44–47.]

or "free-drainage" lysimetry. "In a free drainage system, a [near] water-saturated zone must first form at the bottom of the soil profile before the water can drain out, owing to the resistance formed by the surface tension between the soil-air boundary at the lysimeter bottom.[2] This zero-potential zone can modify the soil–water conditions throughout the profile, especially for shallow profiles" (Bergstrom, 1990). This poses problems for modeling solute transport for soil columns of finite length (Barry and Sposito, 1988). Practical interpretation of the composition of monolith percolates with time from the standpoint of the dynamics of solute flow is therefore frequently suspect for zero-tension conditions.

6.3 FILLED-IN LYSIMETERS

The earliest lysimeters were of the filled-in type—disturbed soil was introduced into a device where the movement of water and solutes could be investigated. Filled-in lysimeters—particularly columns of sieved, hand-packed soil—continue to have some degree of research utility, but in laboratory rather than field situations. These types of lysimeters are typically used in saturated flow studies designed to develop base parameters for describing transport—retardation factors, dispersion coefficients, and hydraulic conductivities. They are also used in saturated flow studies designed to give qualitative assessment of relative solute mobility, so-called "column leaching studies."

The relative uniformity of pore size and the lack of natural soil structure exhibited by disturbed soil columns precludes their use for describing field transport of solutes per se. Differences in solute transport between disturbed and undisturbed soil columns are well documented (Elrick and French, 1966; Cassel et al., 1974; McMahon and Thomas, 1974). Column breakthrough curves for undisturbed soil columns when compared to disturbed soil columns generally exhibit more rapid occurrence of solute in leachate, as the structure of undisturbed soils allows for continuous transport of solutes by way of preferential flow channels. Disturbed soil columns produce breakthrough curves more representative of the diffusive transport of solutes.

Filled-in lysimeters such as used in saturated flow studies in disturbed soil columns have little relevance to the measurement of soil solution composition. Disturbed soil columns have been used in laboratory studies where solutions modeled after rainfall, throughfall, or soil solution are leached and the leachate is analyzed and expressed as soil solution composition. Because these studies typically exhibit dynamic properties uncharacteristic of most soil environments—for example, flow velocities and water contents that are higher and less variable than in nature—extrapolation of the lysimeter solution chemistry to in situ conditions is problematic. Although not intended to obtain information on soil solution composition per se, solute breakthrough studies require consideration of the ionic strength and composition of the leaching solution relative to that of the soil solution in the soil of interest for their proper conduct and interpretation. A base leaching solution composition of 0.005 M $CaSO_4$ is frequently employed to mimic gross composition of temperate region soil solutions.

Filled-in lysimeters are not commonly employed for modern-day field lysimetry with the specialized exception of rhizotrons—root observation chambers—adapted to additionally act as lysimeters (Soileau and Hauck, 1987). These types of installations suffer from the disturbed nature of the soils studied, but they offer the benefit of specialized sampling of microsites within the rhizosphere.

6.4 TENSION LYSIMETERS

Tension, vacuum, or suction lysimeters are point or mini-lysimeters where an applied tension is used to facilitate collection of lysimeter water by way of a porous cup contacting the soil. Zone of influence, tension, time, and lysimeter construction materials are aspects of tension lysimeter installation and operation that commonly influence the solute composition of the solution obtained and the adequacy of the result obtained as a model of soil solution.

6.4.1 Zone of Influence

Zone of influence refers to the impact of lysimeter installation and operation on pathways of solute flow, the region of the soil from which lysimeter water is drawn, and the portion of soil water that is represented in lysimeter solutions. A limitation to the use of tension lysimeters is the inability to determine a priori the sampling rate and zone of influence (Morrison and Lowery, 1990). This can be addressed in part from knowledge of the water characteristic function of the soil involved and through measurement of in situ water potential. If volumes of lysimeter solution collected with time do not correlate well with rates of water percolation into the zone of influence, differences in solute composition between in situ soil solution and lysimeter solution are expected (Van der Ploeg and Beese, 1977). Rates of solution accumulation by lysimeters are related to soil heterogeneity reflected in nonuniform moisture and irregular solution flow. Applied tension increases non-uniformity of sampling, especially when lysimeter tension exceeds tension of the soil water being sampled (Cochran et al., 1970). Additional heterogeneity in moisture distribution arises as an artifact associated with lysimeter installation. When high rates of water flow are present in soil in the vicinity of tension lysimeters, the zone of influence collapses and little disturbance in the pattern of solute transport occurs. Thus, tension lysimeters are expected to realistically represent solute composition of mobile water when soil conditions favor water flow.

Morrison and Lowery (1990) presented data describing sampling rate of porous cup lysimeters receiving either constant or transient vacuum. Vacuum (for constant vacuum setups), cup radius, and water conductance of cups were important discriptors of sampling rate. Similar empirical data describing zone of influence are unavailable, but Warrick and Amoozegar-Fard (1977) have theoretically described radius of influence (r_m) as

$$r_m = (q/\pi K_o)\exp(-\alpha h_1) \tag{6-1}$$

where q is sampling rate, K_o and α are constants related to soil properties, and h_1 is the pressure head across the zone of contact when expressed at great distances. When the parameters q, K_o, and α are constant, $r_m \propto \exp -h_1$; as lysimeter tension increases relative to soil solution moisture tension the radius of influence increases.

6.4.2 Tension

Tension lysimeters are operated either as transient or constant tension devices. These two approaches to tension lysimetry result in different rates and zones of sampling and thus are expected to provide lysimeter solutions of differing composition. These differences are most clearly manifested when the lysimeters are operated at soil moisture contents near the high tension limits of the porous cups employed.

Transient tension lysimetry involves the periodic establishment of tension using pumping to evacuate the lysimeters. In the intervals between evacuation, progressive decay of the applied tension occurs; consequently, both the rate of sampling and the zone of influence vary with time. Both the volume of lysimeter solution obtained and its composition reflect in part the variation in lysimeter tension with time; this complicates the comparison of results obtained for the same lysimeter at different times as well as comparisons among different lysimeters even when they are prepared and operated in a similar manner. Careful matching of sampler cups for geometry and conductance, frequent evacuation intervals, and evacuation to uniform tensions may lessen the artifact influences imposed on transient tension lysimeter solutions.

Constant tension lysimetry utilizes a constant vacuum source to maintain the tension applied to tension lysimeters (Fig. 6.4). Sampling rate and zone of influence for lysimeters operated in this manner are less variable in time and among similarly designed and installed lysimeters; thus, the relative comparability among lysimeter solutions is improved over that achieved with transient tension. Sampling rate of a given constant tension lysimeter is mostly due to the tension applied, but comparison between lysimeters at equivalent tensions shows differences in sampling rate due to cup radius and conductance and the interaction of these with applied tension (Morrison and Lowery, 1990).

Table 6.2 compares cumulative sample volume collected from tension lysimeters operated at transient and constant tension. Cumulative volumes collected are significantly higher for constant as compared to transient tension lysimeters, irrespective of porous cup size or applied tension; however, there is a trend toward comparable cumulative volumes as cup radius and applied tension increase. This differential sampling volume will be reflected in variation in lysimeter solution composition for many situations where soil water is being sampled because differ-

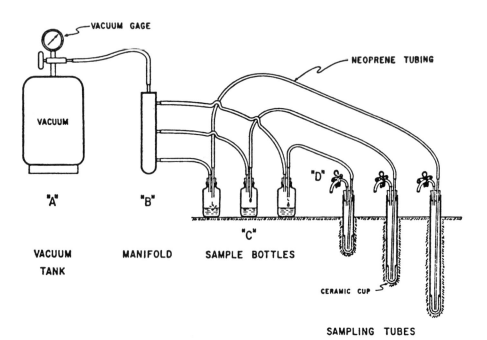

FIGURE 6.4 Apparatus for constant tension lysimetry. [R.C. Reeve and E.J. Doering, 1965, "Sampling the soil solution for salinity appraisal," *Soil Science* 99:339–344.]

TABLE 6.2 Ratios of Total Cumulative Sample for Constant (*Sc*) Versus Transient (*St*) Tension Lysimetry

Sampler and Vacuum	Cumulative Sample Collected, cm^3		St/Sc
	Constant Vacuum	Transient Vacuum	
	Sampler 1 (r_0 = 1.15 cm)[a]		
10 cbar	348 ± 15.9[b]	150 ± 29.2	0.43
30 cbar	504 ± 13.0	280 ± 15.43	0.56
50 cbar	791 ± 12.8	509 ± 3.2	0.64
70 cbar	1114 ± 12.4	639 ± 18.5	0.57
	Sampler 2 (r_0 = 2.4 cm)		
10 cbar	173 ± 2.5	93 ± 3.7	0.54
30 cbar	294 ± 2.9	192 ± 3.4	0.66
50 cbar	700 ± 10.3	389 ± 11.9	0.55
70 cbar	878 ± 16.5	513 ± 7.5	0.58
	Sampler 3 (r_0 = 3 cm)		
10 cbar	902 ± 9.2	405 ± 9.5	0.45
30 cbar	1713 ± 19.0	1432 ± 8.7	0.84
50 cbar	1828 ± 8.5	1592 ± 26.7	0.87
70 cbar	2470 ± 39.1	2076 ± 31.4	0.84

[a]r_0 = radius of porous cup.
[b]Standard deviation.

R. D. Morrison and B. Lowery, 1990, "Effect of cup properties, sampler geometry and vacuum on the sampling rate of porous cup samplers," *Soil Science,* 149:308–316.

ent proportions of soil water as well as different regions of the soil are being sampled.

6.4.3 Time

Lysimeter solution composition represents a time-averaged result for the region of soil sampled. Conditions of quasi-equilibrium between the soil solution being sampled and contacting soil compartments negate any effect of time. The norm for tension lysimeters, however, is to sample water representing transient events within the zone of influence. These dynamic events may reflect natural water flow within the soil or artifacts imposed by the sampler itself. The time-averaged lysimeter solution composition is influenced by the constancy of sampling rate. Figure 6-5 shows an example of the previously mentioned interaction of porous cup radius (r_0) and tension on cumulative volume of lysimeter solution sampled with time from constant tension lysimeters. Many combinations of r_0 result in constant rates of sampling; other combinations demonstrate reduced rates of sampling with time.

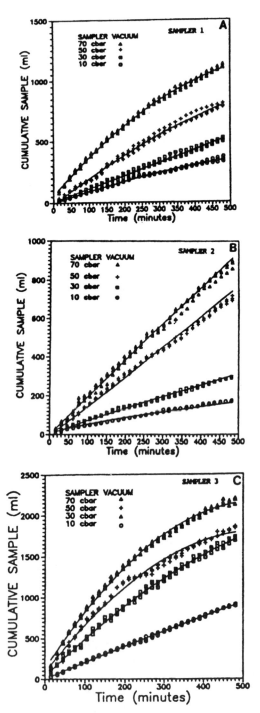

FIGURE 6.5 Interaction of porous cup radius and applied tension on cumulative volume of solution recovered for constant tension lysimetry. Porous cup radius (r_o): (*a*) 3.05 cm, (b) 2.4 cm, (*c*) 1.15 cm. [R.D. Morrison and B. Lowery, 1990, "Effect of cup properties, sampler geometry and vacuum on the sampling rate of porous cup samplers," *Soil Science,* 149: 309–316.]

Interpretation of dynamic processes reflected in lysimeter solution composition—for example, solute flow past a depth of sampling in the soil profile—is dependent on the sampling strategy employed. If sampling is continuous, resolution of the data obtained is restricted to the interval over which lysimeter solution is composited. Discontinuous sampling is even more restrictive in its interpretability because each time point sampled is only reflective of the conditions existing in the zone of influence during the time tension is applied to the lysimeter for the purpose of withdrawing solution for analysis.

6.4.4 Tension Lysimeter Construction Materials, Conditioning, and Installation

The effects of porous cup composition and silica grout on composition of lysimeter solutions are described in section 6.1.2. The potential contaminating effects of lysimeter construction materials are especially important to the interpretation of tension lysimeter data. The ratio of soil volume sampled to the surface area of the sampler is smaller for point lysimeters than for other types of lysimeters; therefore, the potential effects of construction materials are likely to be more pronounced for tension lysimeters. Tension lysimeters sample relatively small volumes of soil, so discrepancies between adjacent samplers may be due to contamination effects, installation effects, or natural soil heterogeneity.

Composition, geometry, preconditioning of porous cups, and reaction of installed lysimeters with the surrounding environment interact to influence lysimeter solution composition. Hansen and Harris (1975) noted that these factors all needed to be considered in order to obtain consistent measurement of solution NO_3^-. When tension lysimeters are first installed, soil perturbation, the establishment of a localized physicochemical environment in the region of the sampling cup compared to the region of soil sampled, and the reactivity of the porous cups may interact to contaminate lysimeter solutions obtained. For example, Rasmussen (1989) observed Al contamination of lysimeter solutions leading in shifts of predicted equilibria for nine months following porous cup lysimeter installation.

6.5 EBERMAYER LYSIMETERS

Installation of Ebermayer lysimeters involves the excavation of a trench adjacent to the intended area of sampling; lateral excavation extends from the trench below the desired depth of installation, and the lysimeter is installed in contact with the exposed base of the overburden soil. Backfilling of the excavation allows for unimpeded movement of water through the soil profile under conditions of minimal soil disturbance. Ebermayer lysimeters may be troughs, pans, funnels, plates, or capillary wicks.

Ebermayer lysimeters potentially offer the most reliable measurements of solute transport because they involve no enclosing walls and, therefore, allow unrestricted vertical and lateral flow of soil water and surface runoff in the undisturbed soil overburden that is sampled by the lysimeter. In practice, Ebermayer lysimeters may alter patterns of solute flow through soil profiles because the soil–air interface where the lysimeter and soil make contact acts as an impeding layer to hold up vertically flowing water that is then redirected as lateral flow away from the lysimeter. Ebermayer lysimeter designs have sought to overcome the effect of the soil–air interface either by facilitating the rapid gravitational flow of water away from the interface (zero-tension lysimeters) or by placing lysimeters under tension.

6.5.1   Trough, Pan, and Funnel Lysimeters

A wide variety of Ebermayer lysimeter designs have used troughs, pans, and funnels. Simple installations, such as sand-filled funnels placed beneath soil horizons and sealed in place by backfilling (Joffe, 1929) are useful for sampling water under conditions of saturated flow; but for conditions of unsaturated flow these types of collectors impede and redirect solute flow leading to biased sampling of soil water.

Zero-tension lysimeters are designed to rapidly collect soil water by gravitational flow as the water begins to accumulate at the soil–air interface; this is accomplished by eliminating as much as possible the surface tension at the interface. In soils with high water conductivities, bypass flow occurs as water accumulates at the interface leading to uneven sampling of soil water, zero-tension lysimeters selectively sample water conducted through large-size pores. A typical zero-tension lysimeter design uses a stainless steel trough or shallow pan, filled with glass wool and an array of drain rods, and fitted with fiberglass screen; the device is tightly wedged against the soil overburden and backfilled (Jordan, 1968). Figure 6.6 illustrates a pan lysimeter installation.

6.5.2   Plate Lysimeters

Plate lysimeters are constructed from porous ceramic, alundum, or occasionally sintered glass. The plate materials are typically fixed to a fiberglass resin backing connected to a drain tube; this device is connected to a source of tension such as a hanging water column or a vacuum pumping system (Cole, 1958; Cole et al., 1961). The tension achieved with such an installation is intended to assure that a uniform fraction of soil water is sampled at all times.

Operational constraints to the use of plate lysimeters are similar to those of point lysimeters: potential contamination from construction materials and inconsistent tension or loss of tension leading to nonuniform sampling in time. The comprehensive study of Neary and Tomassini (1985) describes contaminating effects of alundum plate lysimeters and methods for decontaminating and preparing such plates for field installation (see section 6.1.2).

In addition, plate lysimeters may exhibit spatial nonuniformity in sampling solutes across the plate surface. A typical lysimeter plate is drained by a single tube at its center. Such a design leads to nonuniform sampling of water at the extremities

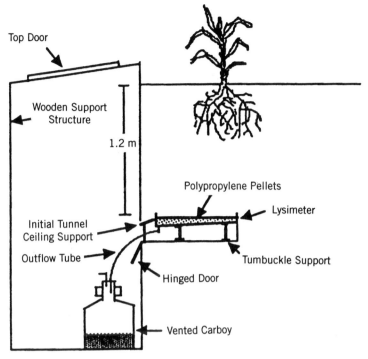

FIGURE 6.6 Example of an Ebermayer lysimeter installation. [J.M. Jemison, Jr. and R.H. Fox, 1992, "Estimation of zero-tension pan lysimeter collection efficiency," *Soil Science* 154:85–94.]

versus the center of the plate because of differing travel distances of the water sampled (Boll et al., 1992).

Haines et al. (1982) compared zero-tension trough lysimeters and ceramic plate lysimeters under tension in terms of the volume and composition of lysimeter solution recovered when the lysimeters were installed in 16 locations within a hardwood forest watershed. Zero-tension lysimeters recovered 7 times more solution than plate lysimeters when installed below the surface leaf litter, but recovered half as much solution as plate lysimeters when installed at a 30-cm depth in soil. More problematic than volumes of solution recovered were the differences in lysimeter solution composition observed between zero-tension and plate lysimeters (Table 6.3). These differences are perhaps best ascribed to the different forms of soil water reflected in the lysimeter solutions obtained; zero-tension lysimeters more effectively sampled mobile water while plate lysimeters under tension more effectively sampled diffuse soil water.

An innovative application of plate lysimetry is that of Bourgeois and Lavkulich (1972a, 1972b) who used plates attached at different positions in the floor and walls of trenches to isolate lateral versus vertical flow of solutes in forest soils on sloping topography.

TABLE 6.3 Monthly Mean Composition of Lysimeter Solutions Obtained from a Forested Watershed Using Tension-Plate and Zero-Tension Lysimeters

| | Litter | | | | Soil | | | |
| | Tension | | Zero Tension | | Tension | | Zero Tension | |
Property	$\bar{X}$	SD*	$\bar{X}$	SD*	$\bar{X}$	SD*	$\bar{X}$	SD*
Mean flow, cm	3.56	5.91 ac	26.56	25.72 ad	6.68	11.85 bc	3.19	6.07 bd
Sample numbers								
Flow	208		207		208		207	
Chemistry	57		170		86		52	
Mean µeq/liter								
H^+	5.59	10.15 ac	13.86	23.66 ad	1.17	2.24 c	1.04	1.41 d
NH_4^+	3.35	9.47	5.14	16.11	1.71	4.52 b	8.72	24.29 b
K^+	75.04	32.88 ac	63.09	40.16 ad	28.28	23.54 bc	39.03	33.09 bd
Na^+	13.71	7.98	11.16	9.09 d	21.12	31.94	19.10	13.42 d
Ca^{2+}	71.11	35.95 c	99.43	111.72 d	34.42	29.88 c	46.37	45.63 d
Mg^{2+}	56.75	35.04	73.14	61.63	57.23	24.79	65.52	30.26
NO_3^-	0.87	1.89	1.33	2.65	0.61	1.84 b	2.11	3.65 b
Cl^-	45.38	48.12 c	35.96	27.35	24.57	25.49 bc	39.84	28.50 b
$H_2PO_4^-$	0.95	1.70	0.96	4.05	0.96	1.63	1.19	1.18
SO_4^{2-}	102.51	53.25 ac	124.23	72.94 ad	67.49	23.66 bc	100.58	39.34 bd
Cation sum	225.57	93.44 c	265.83	181.61 d	143.94	64.72 bc	179.79	80.92 bd
Anion sum	149.71	85.31 c	162.47	83.63	93.64	33.26 bc	143.73	59.01 b
Cation sum − anion sum	75.86	75.48 c	103.35	172.17 d	50.30	64.76 c	36.05	77.97 d
Dissolved silica, µg/liter	2024.00	1822.60 a	618.72	393.87 ad	2519.00	1455.00 bc	1472.00 b	1020.00 bd

*Difference between (a) tension and zero-tension lysimeters in litter significant at 0.05 level; (b) tension and zero-tension lysimeters in soil significant at 0.05 level; (c) tension lysimeters in litter and soil significant at 0.05 level; and (d) zero-tension lysimeters in litter and soil significant at 0.05.

B. L. Haines, J. B. Waide, and R. L. Todd. *Soil Sci. Soc. Am. J.*, 1982, 46:658–661.

6.5.3 Capillary Wick Lysimeters

Capillary wick lysimetry is a relatively recent approach to lysimetry where wicks are used to apply tension and withdraw water from the contacting soil. Self-priming by the wick causes it to act as a hanging water column placing the unsaturated soil under tension with external application of vacuum. Figure 6.7 illustrates one type of capillary wick lysimeter comprised of a conventional plate lysimeter modified with the addition of a fiberglass wick that acts both as a source of tension and as a conduit for the removal and sampling of the lysimeter solution (Brown et al., 1989; Holder et al., 1991).

Wick-activated plate lysimeters are capable of continuous soil water sampling when soil water potentials range from 1 to 6 kPa (Holder et al., 1991). The solutions obtained are representative of mobile water flux through soil profiles for a rather narrow region of soil water potential in comparison to the tension of the hanging water column. Otherwise, lysimeter solutions represent convergent sampling of soil water at higher water potentials and divergent sampling at lower water potentials. The selective sampling of the capillary wick lysimeters is most likely related to the water tension and conductive properties of the particular lysimeter design.

Measurements of moisture-bearing properties (capillary rise and saturated hydraulic conductivity) and chemical reactivity of various wick materials—nylon and glass rope, woven fiberglass, and fiberglass wicking—were evaluated by Brown et al. (1989). Fiberglass wicks exhibit favorable properties for capillary wick lysimeters: a 54-cm capillary rise, high saturated hydraulic conductivity of 10^{-2} cm sec^{-1}, and inertness to inorganic (Br, Cd, NO_3) and organic (toluene, trichloroethylene, ethyl benzene, and naphthalene) chemicals. Further evaluations of capillary wick lysimeter design have been conducted by Boll et al. (1992) with the intent of describing travel time—that is, rate of solute movement away from the plate lysimeter surface—and matric potential developed when the cross sectional diameter of the wick is varied (Figure 6.8). Poletika et al. (1992) in laboratory evaluations of capillary wick lysimeters concluded transport properties of porous wicks would not be a confounding factor for dynamic sampling of soil water moving at flow rates from 11.1 to 745.5 mL $hour^{-1}$. Information of this nature can be used to design capillary wick lysimeters to match the anticipated water potential and flow rate to be sampled in a particular field installation, thus reducing artifact effects of wick properties—that is, induced dispersion and retardation—on the nature and composition of the lysimeter solution obtained.

6.6 WATERSHED AND FIELD-SCALE SAMPLING OF SOIL SOLUTION

Lysimetric sampling of soil solution is restricted by the limited spatial extent of sampling achieved. For example, Holder et al. (1991), determined that representative ($p > 95\%$) soil solution composition required from 2 to 31 lysimeters for 8.88 m^2 plots on soils with varied physical properties. Considering that these au-

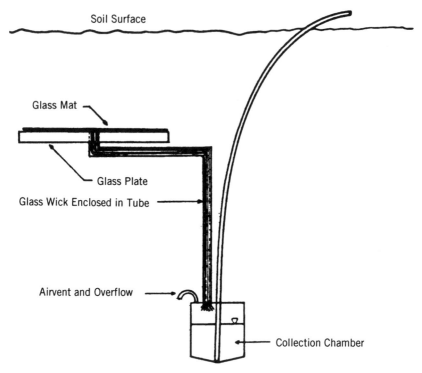

FIGURE 6.7 Schematic diagram of an unsaturated zone capillary wick sampler. [Brown et al., 1989. Reprinted with permission of author.]

thors utilized wick-activated plate lysimeters with a 0.9 m^2 sampling area, this translates to from 2 to 30% of the plot area being instrumented in order to achieve highly reliable results. This extent of sampling coverage is not remotely achievable for watershed or field-scale applications.

Two alternatives for wide area coverage investigations of soil solution chemistry are to employ sampling designs amenable to statistical evaluation of the spatial heterogeneities extant in the system being studied, or to obtain a sample integrated over this same system. Experimental design and statistics have been used to treat spatial variability in agronomic-scale studies for many decades, but the extension to watershed and regional-scale evaluations is more recent and requires design and analysis approaches markedly different from traditional field plot technique. Geostatistics (Nielson and Bouma, 1985) and probabilistic modeling (Laskowski et al., 1990) are tools that may prove useful to generalization of soil chemical processes on a broad spatial scale. A more traditional approach to the same end is to sample natural or anthropogenic sources of drainage to obtain a result integrated over the spatial area of interest. Examples of this approach are the sampling and analysis of natural seeps to elucidate mineral control of soil solution composition (Kittrick, 1969), sampling of lateral flow from shallow forest soils or streams draining high-

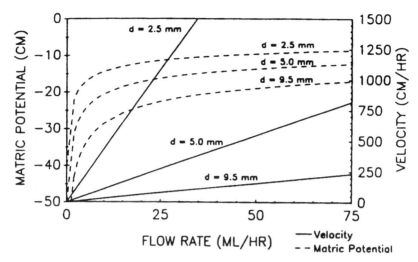

FIGURE 6.8 Velocity and matric potential (cm water tension) as related to flow rate for capillary wicks with length L=47.5 cm and diameters d=2.5, 5.0, and 9.5 mm. [Boll et al., 1992. *Soil Sci. Soc. Am. J.,* 1992, 56:701–707.]

order watersheds to evaluate transient effects on solution Al chemistry (Hendershot et al., 1992), and utilizing agricultural tile drains to assess leaching of agrochemicals (Richard and Steenhuis, 1988). The interpretation of wide-area sampling data must be carefully considered since such approaches are poorly resolved in both space and time.

6.7 SUMMARY

Field-scale sampling of soil water is much more prevalent than is laboratory-based soil solution displacement. Representation of the analytes of lysimeter solutions, tile drain effluents, and various drainage waters as "soil solution" is a common facet of research literature in many disciplines. The scientist pursuing deeper understanding of the chemistry of soil solutions must be cognizant of the limitations imposed when interpreting field-sampled waters as soil solution.

The inability to consistently define the nature of solutions sampled in space or time is a difficulty inherent with field-based sampling techniques. Field-sampling of soil solution considers spatial scales extending from points in the landscape (tension lysimeters) to fields (tile drains) to entire watersheds (drainage seeps). The results obtained may reflect the integrated effect of processes extending over time scales of hours to years. Increased utility of field-scale sampling can be achieved through experimental design (for example, consideration of the nature, number. preparation, and placement of lysimeters) and interpretation (through recognition of the spatial and temporal resolution achievable with the sampling strategy employed). More problematic is the inability to consistently link the composition of field-sampled solutions to a particular fraction of soil water because this limits the ability to distinguish clearly between transient and steady state events. Modeled approaches are an important means to overcoming this particular difficulty but at present they offer no panacea because of the inability to describe the nature and extent of bypass flow.

Porous cup tension lysimetry is clearly the most common of the vast array of field-based techniques for sampling soil solution. Despite its popularity, the method is replete with inadequacies that need be considered in the acquisition and interpretation of porous cup tension lysimeter data. Principle inadequacies are non-representative sampling of soil water occurring above the capillary fringe and potential artifact effects arising from the reaction of lysimeter materials with the surrounding soil environment.

Field-based sampling of soil solution remains an area of great opportunity for the development of innovative research approaches as typified by the development of capillary wick lysimetry and in situ monitoring of the rhizoshere.

NOTES

1. Mechanistic descriptions of water flux in soils assume a continuous solid matrix holding water in pores and films. The water content form of the Richards equation exemplifies mechanistic descriptions of vertical water transport (transport in the z-direction) for a homogeneous soil column:

$$\frac{\delta\theta}{\delta t} = \frac{\delta}{\delta z}\left(D_w(\theta)\frac{\delta\theta}{\delta z}\right) + \frac{\delta K(\theta)}{\delta z}$$

where water flow, $\delta\theta/\delta t$, is related to a diffusion term, $\delta K(\theta)/\delta z$, and a conductivity term, $(D_w(\theta)\ \delta\theta\delta z)$. Diffusion is defined by the diffusivity function, $D_w(\theta)$, which describes the increase in matric water potential per unit increase in water content. Conductivity is described by the unsaturated hydraulic conductivity as a function of water content, $K(\theta)$ (Jury et al., 1991).

Physical properties of field soils differ substantially from the homogeneous condition described from the Richards equation. Features of natural soils such as root and worm channels or cracks act as conduits to vertical water flow in a manner distinct from the surrounding soil matrix. Incomplete description of water flow from a mechanistic perspective is due to such heterogeneities in the soil column.

The matric potential-water content function (water characteristic function, section 2.2.2) provides a unifying approach for the description of soil water and soil solution. Terminology such as mobile, bypass, or diffuse water is used in this presentation as an aid in visualizing soil water relations in relation to sampling of soil solution. It must be recognized, however, that such terminology is being applied to indistinct pools of soil water.

2. Water potential in the near-saturated zone must reach *air-entry potential* before soil water will drain from the column. At the point of air-entry potential, the largest water-filled pores will drain first. The air-entry potential is a function of pore-size distribution and may range from tensions of 0.6 to 9 kPa.

CHAPTER 7

SOIL SOLUTION COMPOSITION

In Chapter 1, soil solution composition was described as a window to chemically reacting soil systems where the intensity and distribution of chemicals in the soil aqueous phase represents the integration of multiple physical, chemical, and biological processes occurring concurrently within the soil environment. Composition of displaced soil solution represents this integrated intensity and distribution of ions as modified by soil moisture content, sample handling, and displacement technique. In this chapter the ion composition of soil solution will be described in further detail with respect to the total ion composition of displaced soil solutions, the speciation of ions in soil solution, and computational approaches for expressing ion intensities and distribution in soil solution.

7.1 TOTAL ION COMPOSITION OF SOIL SOLUTIONS

Table 7.1 summarizes the major ions of interest in soil physical chemistry, along with common states of occurrence in soil solution. Comprehensive soil solution analysis typically requires measurement of soil solution pH and the total concentrations of major exchangeable cations and the anions that balance charge in the aqueous phase. A comprehensive analysis of soil solution composition is not needed to apply soil solution analysis successfully to all types of problems encountered in soil and environmental chemistry. However, comprehensive analysis of soil solution is a powerful diagnostic tool for the interpretation of many soil chemical phenomena relating to soil fertility, mineralogy, and environmental fate. Comprehensive soil solution analysis is necessary for applications where computational approaches

TABLE 7.1 Ions of Major Interest in Soil Chemistry, Grouped According to Major Behavioral Modes and Shown as Their Most Common States in Soil Solutions

Ion	Comments
Major Exchangeable Cations	
Ca^{2+} Mg^{2+} Na^+ K^+ NH_4^+ $Al^{3+}(H^+)$	Occur predominantly as exchangeable cations in soils; these ions are relatively easily manipulated by liming, irrigation, or acidification; exchangeable Al^{3+} is characteristic of, though rarely the predominant exchangeable cation in, acid soils; productive agricultural soils are rich in exchangeable Ca^{2+}.
Major Anions	
NO_3^- SO_4^{2-} Cl^- HCO_3^-, CO_3^{2-}	Present in considerably lower concentrations than the major cations in all but the most coarse-textured and strongly saline soils, where they are essentially equal; sulfate and NO_3^- are important nutrient sources for plants; sulfate, Cl^-, and HCO_3^- salts accumulate in saline soils; carbonate ions are present in appreciable amounts only in soils of pH > 9.
Weakly Soluble Anions	
$H_2PO_4^-$, HPO_4^{2-} $H_2AsO_4^-$, AsO_2 H_3BO_3, $H_2BO_3^-$ $Si(OH)_4$ MoO_4^{2-}	Strongly retained by soils; borates are the most soluble of the group; retention or fixation by soils is pH-dependent; molybdate and silica are more soluble at high pH; phosphate is more soluble at neutral or slightly acid pH.
Transition Metals and Aluminum	
Al^{3+}, $AlOH^{2+}$, $Al(OH)_2^+$ $TiOOH^+$ (?) $Fe(OH)_2^+$, Fe^{2+} Mn^{2+}	Insoluble hydroxides tending to accumulate in soils as silica and other ions weather; iron and manganese are more soluble in waterlogged or reduced soils.
Cu^{2+} Zn^{2+}	More soluble than the above cations in all but very acidic soils; availability increases with increasing soil acidity; complexed strongly by soluble organic matter.
Toxic Ions	
Cd^{2+}, Al^{3+} Pb^{2+} Hg^{2+}, Hg Be^{2+}, AsO_4^{3-}, CrO_4^{2-}	Soil behavior similar to transition metals; Al^{3+} is a hazard to plants; the others are of more concern to animals; Cd^{2+} is relatively soluble, available to plants, and its retention is relatively independent of pH; remaining ions are less available to plants with increasing pH, except perhaps for As; the last three ions have received relatively little study in soils.

TABLE 7.1 *Continued*

Ion	Comments
Active in Oxidation-Reduction Reactions	
C (organic to HCO_3^-)	Soil biochemistry revolves around the oxidation state
O (O^{2-} to O_2)	changes of soil carbon, nitrogen, and sulfur compounds;
N (—NH_2 to NO_3^-)	molecular oxygen is the main electron acceptor; Fe(III),
S (—SH to SO_4^{2-})	Mn(III-IV), nitrate, and sulfate are electron acceptors
Fe (Fe^{2+} to FeOOH)	when the oxygen supply is low.
Mn (Mn^{2+} to $MnO_{1.7}$)	
Se (organic to SeO_4^{2-})	
Hg (organic to Hg^0 or Hg^{2+})	

Bohn et al., 1985.

are used to model ion speciation and free ion concentration or activity in soil solution.

7.1.1 Major Ion Components of Soil Solution

In the majority of temperate region agricultural soils with pH near neutrality, analysis for total concentrations of Ca, Mg, K, NO_3, Cl, and SO_4, along with pH and specific additional analytes of interest, is sufficient to describe soil solution composition. Other analytes must necessarily be determined for soils that are acidic (Al becomes a consequential solution phase component) or alkaline (Na and carbonates become consequential solution phase components). Total dissolved Si determination is important when mineral stability relationships are considered (Chapter 9). Dissolved organic carbon (DOC) is an important soil solution analyte in many surface soils, especially in low management ecosystems.

Table 7.2 summarizes total ion compositions for displaced soil solution from a worldwide cross-section of soils. For most soils, Ca^{2+} is the dominant exchangeable cation; therefore, it occurs as the dominant soil solution cation as well. Nitrate is typically the dominant anion balancing ion charge in soil solution, but Cl^-, SO_4^{2-}, and carbonate can frequently be of equivalent importance. Deviations from the norm of Ca^{2+} as the dominant soil solution cation are indicative of soils with chemically restrictive limitations to plant growth. The Traver soil in Table 7.2, for example, is adversely impacted by Na^+—as indicated by the dominance of Na^+ over Ca^{2+} in soil solution and the high ionic strength of the soil solution. For other soils, the free ion concentrations or activities, or the ratios of these activities for specific ion combinations (Al^{3+} and Ca^{2+}, for example), are critical indicators of plant growth-limiting nutrient or toxic ion status of soils (Chapter 10).

The uncorrected ion strengths reported in Table 7.2 are computed as $1/2\Sigma C_i Z_i^2$, assuming the most common charge state for the ions present. Thus, uncorrected ionic strength assumes comprehensive analysis of the soil solution and ignores speciation reactions such as hydrolysis that may alter the charge distribution of soil

TABLE 7.2 Total Ion Composition of Displaced Soil Solution for Selected Soils

Soil	Taxonomic Descriptor	pH	EC dS m^{-1}	mmol L^{-1} Ca	Mg	K	Na	NH$_4$	Al	Si	HCO$_3$	SO$_4$	Cl	NO$_3$	DOC	Uncorrected Ionic Strength	Ion Difference %
Pressure Plate Extraction at 10 kPa Moisture, California USA (Eaton et al., 1960)																
Hanford	Typic Xerothents	7.52	4.96	21.1	1.80	0.80	3.00	0.40			3.5	5.9	4.6	32.0		80	−2
Escondido	Typic Xerochrepts	7.41	0.70	1.00	0.50	0.20	2.80	0.10			1.0	0.7	2.9	1.2		9	−3
Tujunga	Typic Xeropsamments	6.99	8.90	41.6	8.30	2.60	7.40	4.00			4.2	18.9	5.6	65.0		182	0
Greenfield	Typic Haploxeralfs	7.66	1.96	6.40	1.89	1.10	1.20	0.40			1.9	1.4	1.0	13.2		29	1
Yolo	Typic Xerorthents	6.89	2.74	8.70	1.95	3.70	1.00	0.50			2.3	3.0	1.9	16.9		41	−1
Corona	Pachic Agrixerolls	8.12	3.36	5.85	1.85	7.30	9.20	0.40			6.1	3.8	3.2	12.4		43	5
Chino	Aquic Haploxerolls	7.52	1.68	5.55	1.10	7.30	2.30	0.20			3.8	1.4	2.3	8.8		25	−5
Traver	Natric Haploxeralfs	7.35	143.3	118.8	105.0	5.60	1777	0.30			32.0	39.4	2205	25.0		2549	−2
Vacuum Displacement at Field Moisture Content, Tennessee USA (Qian and Wolt, 1990)																
Etowah Ap	Typic Paleudults	5.97	0.41^a	1.55	0.38	0.21	0.28		0.140	0.136		0.13	0.13	3.9	1.00	7	6
Etowah B	Typic Paleudults	5.12	0.22	0.73	0.12	0.02	0.09		0.012	0.079		0.06	0.42	1.4	1.22	3	−2
Statler Ap	Humic Hapludults	6.04	3.34	12.85	3.39	1.25	0.52		0.126	0.169		0.84	14.98	22.2	2.09	54	−6
Statler B	Humic Hapludults	6.66	0.25	0.71	0.22	0.02	0.09		0.068	0.054		0.21	0.39	1.1	0.62	3	8
Vacuum Displacement at Field Moisture Content, Tennessee USA (Wolt and Graveel, 1986)																
Memphis Ap	Typic Hapludalfs	5.91	1.19	5.10	0.88	0.62	0.15	0.035				0.27	3.34	9.48		19	−2
Memphis Bt2	Typic Hapludalfs	4.95	0.22	0.55	0.47	0.03	0.45	0.042				0.05	0.95	1.38		4	2
Lily Ap	Typic Hapludults	4.35	0.079	0.21	0.05	0.08	0.24	0.034				0.07	0.12	0.51		1	7
Lily Bt2	Typic Hapludults	4.86	0.050	0.05	0.05	0.05	0.22	0.037				0.05	0.06	0.36		1	−1
Centrifugal Displacement of Surface Soils at Field Moisture Content, United Kingdom (Kinniburgh and Miles, 1983)																
Harwell	brown earth	5.3		1.15	0.14	0.62	0.38		0.013	0.844		0.47	0.56	1.78	4.33	6	5
Ickneid	rendzina	7.8		1.23	0.07	0.04	0.03		0.004	0.075		0.37	0.54	0.19	3.83	4	29

147

TABLE 7.2 *Continued*

				Ca	Mg	K	Na	NH$_4$	Al	Si	HCO$_3$	SO$_4$	Cl	NO$_3$	DOC	Ionic Strength	Ion Difference %
																	Uncorrected
Soil	Taxonomic Descriptor	pH	EC dS m⁻¹	mmol L⁻¹													
Grove	gleyed calcareous	8.1		1.73	0.06	0.49	0.31		0.0004	0.473		0.28	0.73	0.29	2.33	5	47
Rowsham	surface water gley	8.0		1.48	0.07	0.37	0.23		0.0004	0.541		0.19	0.51	1.07	3.17	5	31
Fyfield	brown earth	7.9		1.35	0.02	0.41	0.26		0.003	0.121		0.25	0.51	0.47	3.00	4	40
Thames	ground water gley			3.13	0.25	0.37	0.23		0.001	0.285		0.34	1.41	0.44	3.33	9	53
Marcham	brown calcareous	8.3		1.73	0.07	0.33	0.20		0.006	0.128		0.09	0.24	0.66	2.17	5	59
Denchworth	surface water gley	7.0		1.25	0.26	0.53	0.33		0.021	0.413		0.59	1.66	0.12	4.67	6	14
Sothhampton	podzol	4.3		0.35	0.08	0.21	0.13		0.025	0.142		0.31	0.82	0.02	7.17	2	−7
Berkhamsted	gleyed brown earth	7.1		1.15	0.21	1.52	0.95		0.011	0.384		0.31	0.51	1.86	4.17	6	27

Centrifugal Displacement of Surface Soils at Field Moisture Content, Ireland (Curtin and Smilie, 1983)

Soil	Taxonomic Descriptor	pH	EC dS m⁻¹	Ca	Mg	K	Na	NH$_4$	Al	Si	HCO$_3$	SO$_4$	Cl	NO$_3$	DOC	Ionic Strength	Ion Difference %
Clonroche	Dystric Eutrochrepts	4.7	1.18^a	3.69	1.06	0.55	1.08		0.006	0.206			0.94	10.56		16	−2
Mortarstown	Typic Hapludalfs	5.8	0.42	1.28	0.21	0.03	0.63		0.002	0.153	0.04		0.85	2.54		5	3
Castlecomer	Aeric Haplaquepts	5.6	0.59	1.87	0.37	0.09	0.68		0.002	0.139	0.04		0.99	4.39		8	−1

Centrifugal Displacement of Surface Soils at 33 kPa, Virginia, USA (Elkhatib et al., 1987)

Soil	Taxonomic Descriptor	pH	EC dS m⁻¹	Ca	Mg	K	Na	NH$_4$	Al	Si	HCO$_3$	SO$_4$	Cl	NO$_3$	F	Ionic Strength	Ion Difference %
Gilpin	Ultic Hapludalfs	5.15	0.53^a	1.61	0.75	1.00			0.13			1.40	1.04	0.16	1.16	9	9
Hayesville	Typic Hapludults	4.95	2.84	8.02	3.97	3.45			0.07			0.10	0.84	30.75	0.17	42	−7
Porters	Umbric Dystrocrepts	3.96	1.31	2.49	1.12	0.83			1.32			0.19	0.00	11.90	0.27	15	−3

Centrifugal Displacement at 7.5 kPa Moisture, Georgia USA (Gillman and Sumner, 1987)

Soil	Taxonomic Descriptor	pH	EC dS m⁻¹	Ca	Mg	K	Na	NH$_4$	Al	Si	HCO$_3$	SO$_4$	Cl	NO$_3$	DOC	Ionic Strength	Ion Difference %
Cecil Ap	Typic Hapludults	4.7		0.21	0.14	0.12	0.17		0.0004	0.14		0.22	0.28	0.31		2	−2
Cecil B	Typic Hapludults	4.6		0.21	0.14	0.64	0.21		0.0107	0.10		0.17	0.54	0.30		2	15
Davidson Ap	Rhodic Hapludults	6.3		0.06	0.09	0.05	0.21		0.0004	0.06		0.16	0.08	0.05		1	11
Davidson B	Rhodic Hapludults	6.8		0.02	0.02	0.03	0.70		0.0074	0.08		0.31	0.17	0.05		1	0
Dyke Ap	Typic Rhodudults	7.0		2.60	0.73	0.10	0.10		0.0004	0.09		2.86	0.19	0.04		13	7
Dyke B	Typic Rhodudults	6.3		1.20	0.53	0.08	0.24		0.0004	0.11		0.26	0.93	0.92		5	23
Townley Ap	Typic Hapludults	6.6		0.70	0.21	0.02	0.14		0.0004	0.10		0.78	0.28	0.01		4	3
Townley B	Typic Hapludults	6.2		0.60	0.18	0.03	0.15		0.0004	0.15		0.10	0.53	0.46		2	19

TABLE 7.2 *Continued*

			EC	mmol L⁻¹												Uncorrected	
Soil	Taxonomic Descriptor	pH	dS m⁻¹	Ca	Mg	K	Na	NH₄	Al	Si	HCO₃	SO₄	Cl	NO₃	DOC	Ionic Strength	Ion Difference %
Centrifugal Displacement at 10 kPa Moisture, Queensland, Australia (Gillman and Bell, 1978)																	
surface soil	Kransnozem	5.9	6.00	0.06	0.08	0.27	0.25	5.50			0.76	1.28	2.12	0.63		7	2
subsoil	Kransnozem	5.5	0.64	0.01	0.01	0.04	0.16	0.22			0.06	0.02	0.04	0.05		<1	41
surface soil	Euchrozem	7.9	2.90	0.64	0.44	0.30	0.07	0.03			2.12	0.11	0.22	0.01		3	0
subsoil	Euchrozem	6.6	0.57	0.05	0.05	0.04	0.14	0.01			0.14	0.02	0.04	0.01		<1	25
surface soil	Xanthozem	4.1	8.20	0.28	0.86	0.39	0.69	4.03			0.01	1.99	3.97	0.68		11	−8
subsoil	Xanthozem	4.6	2.45	0.02	0.10	0.16	0.38	0.92			0.08	0.17	0.34	0.47		2	16
surface soil	Red Pozolic	5.0	10.00	0.40	0.73	0.79	0.86	5.22			1.62	1.53	3.06	0.53		11	5
subsoil	Red Pozolic	5.8	3.20	0.01	0.09	0.18	0.30	1.52			0.12	0.23	0.45	0.05		2	34
surface soil	Red Earth	7.4	2.22	0.22	0.21	0.45	0.29	0.17			0.46	0.23	0.45	0.04		2	12
subsoil	Red Earth	5.3	1.40	0.10	0.07	0.17	0.22	0.13			0.10	0.17	0.34	0.01		1	4
surface soil	Yellow Earth	4.2	1.90	0.01	0.06	0.10	0.27	0.67			0.10	0.10	0.20	0.05		1	36
subsoil	Yellow Earth	5.1	0.71	0.01	0.01	0.03	0.22	0.02			0.02	0.02	0.04	0.05		<1	33
Centrifugal Displacement of Surface Soils at a Field Moisture Content, New Zealand (Edmeades et al., 1985)																	
Egmont	Yellow-brown loam	6.3		0.82	0.37	1.03	1.72	0.39	0.019	0.33	0.12	0.44	0.31	2.79		7	15
Stratford	Yellow-brown loam	6.0		0.65	0.15	0.45	0.64	0.19	0.019	0.22	0.08	0.50	0.28	1.10		4	9
Tirau	Yellow-brown loam	6.1		1.12	0.29	1.02	0.53	0.24	0.022	0.38	0.16	0.56	0.20	1.64		6	20
Horotiu	Yellow-brown loam	6.2		1.18	0.29	0.85	0.52	0.30	0.021	0.39	0.12	0.46	0.22	2.29		6	14
Kaharoa	Yellow-brown pumice	6.1		0.65	0.21	0.45	0.63	0.22	0.027	0.33	0.17	0.54	0.15	0.94		4	14
Taupo	Yellow-brown pumice	6.2		0.52	0.19	0.83	0.43	0.31	0.071	0.36	0.24	0.46	0.14	0.64		4	26
Matapiro	Yellow-grey earth	6.1		0.34	0.18	0.24	0.98	0.09	0.096	0.53	0.11	0.45	0.30	0.21		3	29
Tokomaru	Yellow-grey earth	6.3		0.75	0.29	0.52	0.89	0.15	0.048	0.32	0.12	0.58	0.26	1.09		5	18
Waikare	Northern Yellow-brown earth	6.9		1.12	0.17	0.38	0.74	0.17	0.014	0.25	0.67	0.59	0.25	0.85		5	14

[a] EC is a correction of that appearing in original citation.

solution components. Adjustment of ion strength for changes in solution composition due to hydrolysis, complexation, and ion pairing will reduce the calculated ion strength of soil solutions. A correction for ion complexes will be unnecessary for dilute soil solutions; Gillman and Bell (1978) found that correcting ionic strength for ion pair formation resulted in values that were approximately 98% of the uncorrected value for soils with ionic strengths < 12 mmol L^{-1}. Correlations of ionic strength on electrical conductivity (Table 4.8), however, suggest that corrected ion strengths will average $\approx 75\%$ of uncorrected ionic strengths. Therefore, ion pair correction may be important to ionic strength computation for higher ionic strength soil solutions. An example of this effect can be shown for the Statler Ap soil solution from Table 7.2; the uncorrected ionic strength of 54 mmol L^{-1} reduces to 50 mmol L^{-1} when corrections for ion pair formation are made. When the same comparison is made for a higher ionic strength soil solution, such as the Tujunga soil in Table 7.2, the uncorrected and corrected ion strengths differ to a greater degree (182 and 145 mmol L^{-1}, respectively).

The uncorrected ion differences [$= (\Sigma C_{i+} - \Sigma C_{i-}) \times 100/(\Sigma C_{i+} + \Sigma C_{i-})$] in Table 7.2 are similarly based on the most common charge state of the total ions and therefore do not reflect the effects of ion pairing or complexation that may alter distributions of charged species in solution. Despite the inexactness of the uncorrected ion difference, it is a good indicator of the degree that a particular soil solution analysis is comprehensive. Large charge balance deviations indicate failure to account for an important charge contributing component of soil solution. For example, the data of Kinniburgh and Miles (1983) occurring in Table 7.2 indicate negative change deficits of up to 60%. This is because the negative charge assignable to DOC is excluded from the ion balance, as is HCO_3^- which is likely substantial in those soils with pH > 7.

It is commonplace in many soil and environmental studies to utilize 5 mmol L^{-1} (0.01 N)Ca^{+2} solutions—where the counter ion is NO_3^-, Cl^-, or occasionally, SO_4^{2-}—for batch equilibration with soils to mimic soil solution ionic strength and composition. This is mainly an historical artifact arising from earlier studies of soil exchangeable cations and the dissolved salt content of soils supporting calcicole plants. The uncorrected ionic strength of these equilibration solutions (15 mmol L^{-1} when the counter ion is monovalent and 20 mmol L^{-1} when the counter ion is divalent) indeed appears representative of the average case for surface soils listed in Tables 7.2 and 7.3 ($= 16$ mmol L^{-1} when the Na-impacted Traver soil is excluded). The average case Ca^{2+} concentration for this same set of soils is 3.5 mmol L^{-1} and is also reasonably approximated by the equilibration solutions. Many of the soils summarized in Table 7.2 are acid or represent low-intensity fertility management, thus the median ionic strength and Ca^{2+} concentration (6 and 1.27 mmol L^{-1}, respectively) are significantly lower than average values. This points to a problem arising from the uncritical application of 5 mmol L^{-1} Ca^{2+} solutions for batch equilibration studies on weakly buffered soils exhibiting low ionic strengths in soil solution. Depending on the soil-to-solution ratio, the equilibration solution may reflect the ionic strength and ion composition of the added solution more than it does that of the in situ soil solution.

TABLE 7.3 Total Ion Composition of Displaced Soil Solution as Affected by Fertility Management

Soil	Treatment	pH	EC $dS\ m^{-1}$	Ca	Mg	K	Na	NH₄	Al	Fe	Mn	Zn	HCO₃	SO₄	Cl	NO₃	Si	Ionic Strength	Ion Difference %
				\multicolumn{5}{c}{$mmol\ L^{-1}$}					\multicolumn{5}{c}{$\mu mol\ L^{-1}$}					\multicolumn{4}{c}{$mmol\ L^{-1}$}		\multicolumn{2}{c}{Uncorrected}
\multicolumn{20}{l}{*Column Displacement at Field Moisture ontent, California USA (Vlamis, 1953)*}																			
Siskiyou Co. loam	Unlimed	4.2		0.50	0.70	0.40	0.40	0.40	237.2	1.8	291.2			0.40	0.20	3.80		7	1
	Limed	5.8		3.00	0.35	0.10	2.80	0.10	7.4	0.9	142.0			0.45	0.20	5.30		12	22
\multicolumn{20}{l}{*Centrifugal Displacement of Surface Soils at Field Moisture Content, Ireland (Curtin and Smillie, 1983)*}																			
Clonroche	Unlimed	4.7		3.69	1.06	0.55	1.08		5.6	0.4	1.3	0.8			0.94	10.56	0.21	16	−2
	Limed	7.7		6.33	0.72	0.26	0.68		0.4	0.2	0.0	0.2	1.65		1.08	12.20	0.02	22	0
Mortarstown	Unlimed	5.8		1.28	0.21	0.03	0.63		2.2	0.5	1.6	1.4	0.04		0.85	2.54	0.15	5	3
	Limed	8.0		2.92	0.27	0.03	0.63		2.2	1.3	0.2	0.2	2.55		1.23	3.21	0.02	10	0
Castlecomer	Unlimed	5.6		1.87	0.37	0.09	0.68		2.2	0.7	1.5	1.2	0.04		0.99	4.39	0.14	8	−1
	Limed	7.8		3.80	0.46	0.06	0.49		1.5	0.9	0.4	0.3	2.14		1.28	4.87	0.03	13	4
\multicolumn{20}{l}{*Centrifugal Displacement at 7.5 kPa Moisture, Georgia USA (Gillman and Sumner, 1987)*}																			
Cecil	Unamended	4.7		0.21	0.14	0.12	0.17		0.4					0.22	0.28	0.31	0.14	2	−2
	Amended	4.7		3.60	0.09	0.33	0.23		4.4					2.14	1.35	1.17	0.21	13	8
Davidson	Unamended	6.3		0.06	0.09	0.05	0.21		0.4					0.16	0.08	0.05	0.06	1	11
	Amended	6.4		3.80	1.99	0.15	0.18		1.1					5.42	0.33	0.11	0.08	23	3
Dyke	Unamended	7.0		2.60	0.73	0.10	0.10		0.4					2.86	0.19	0.04	0.09	13	7
	Amended	7.1		8.60	1.46	0.19	0.23		1.5					9.58	0.28	0.02	0.07	40	3
Townley	Unamended	6.6		0.70	0.21	0.02	0.14		0.4					0.78	0.28	0.01	0.10	4	3
	Amended	6.6		5.35	1.35	0.03	0.15		4.4					6.96	0.32	0.02	0.09	28	−2

TABLE 7.3 Continued

			EC	mmol L⁻¹					μmol L⁻¹				mmol L⁻¹					Uncorrected	
Soil	Treatment	pH	dS m⁻¹	Ca	Mg	K	Na	NH₄	Al	Fe	Mn	Zn	HCO₃	SO₄	Cl	NO₃	Si	Ionic Strength	Ion Difference %

Residue Management

Vacuum Displacement at Field Moisture Content, Tennessee USA (Qian et al., 1994)

			EC	Ca	Mg	K	Na	NH₄	Al	Fe	Mn	Zn	HCO₃	SO₄	Cl	NO₃	Si (DOC)	Ionic Strength	Ion Diff %
Loring silt loam	Disk, chisel plow, and harrow	5.8	0.34	1.13	0.38	0.05	0.14		59.0					0.05	0.13	3.03	0.38	5	2
	No tillage	6.0	0.16	0.58	0.30	0.02	0.11		60.2					0.07	0.17	1.44	0.09	3	8

Nitrogen Fertilization

Vacuum Displacement at Field Moisture Content, Tennessee USA (Qian et al., 1994)

| Loring silt loam | None | 6.0 | 0.34 | 2.56 | 0.85 | 0.12 | 0.21 | | | | | | 0.08 | 6.43 | 0.49 | 0.63 | | 11 | 0 |
| | 168 kg N ha⁻¹ | 4.9 | 0.75 | 7.37 | 1.88 | 0.26 | 0.12 | | | | | | 0.03 | 28.24 | 0.21 | 0.52 | | 33 | −20 |

Phosphorus Fertilization

Vacuum Displacement at Field Moisture Content, Tennessee USA (Harden, 1988)

			EC	Ca	Mg	K	Na	NH₄	Al	Fe	Mn	Zn	HCO₃	SO₄	Cl	NO₃	P, μmol L⁻¹	Ionic Strength	Ion Diff %
Memphis silt loam	None	5.3	2.41	6.96	2.69	0.35	0.62		411					0.09	0.18	27.36	0.9	36	−12
	90 kg Pha⁻¹	5.5	1.77	7.71	2.07	0.21	0.58		971					0.36	1.59	21.80	1.5	37	−2
	180 kg Pha⁻¹	5.5	1.54	6.57	1.75	0.49	0.67		534					0.53	1.73	19.20	8.1	31	−6

7.1.2 Soil Fertility Management Affects on Soil Solution Composition

Table 7.3 highlights examples of the effects of fertility management on soil solution composition. Amendment of soil that is acid and Ca-deficient with either limestone or gypsum has comprehensive influence on the nutrient intensities in soil solution. Fertilization with a specific nutrient source can beneficially influence the availability of that nutrient in soil solution but may have less desirable effects on other soil solution components. An example is the fertilization with an acid-forming N source as illustrated for the Loring silt loam in Table 7.3. The beneficial effect of increased soil solution NO_3^- is accompanied with a decline in soil solution pH.

The example of P fertilization of an initially P-deficient soil (Memphis silt loam, Table 7.3) shows an effect of biotic response on soil solution composition. Phosphorous fertilization overcomes a growth-limiting nutrient deficiency. The resulting enhanced plant growth and nutrient uptake depletes soil solution NO_3^- and accompanying cations.

Some impacts of soil fertility management may have counter-intuitive effects on soil solution composition. The effect of long-term no-tillage management on the Loring silt loam in Table 7.3 is to reduce nutrient intensities in soil solution over that of a conventionally tilled soil. In this case, buildup of the organic carbon pool of the surface soil may act as a sink for nutrients (the labile or potentially available pool is increased and the readily available pool decreased).

These examples serve to indicate how soil solution compositional analysis provides an immediate insight into soil nutrient status. The diagnostic application of such information is discussed in greater detail from the standpoint of bioavailability in Chapter 8.

7.1.3 Minor Ion Components of Soil Solution

The occurrence and distribution of trace metals in soil solution has been usefully employed for understanding the environmental soil chemistry of trace metals occurring from natural soil sources or from the introduction of metal-bearing wastes to soils. This is discussed in detail in Chapter 12.

Examples of total trace metal concentrations in soil solution displaced from agricultural soils are shown in Table 7.3. Typically, concentrations are in the micromolar to submicromolar range. Increased total trace metal concentrations in soil solution may occur as soils become more acid, as soil organic matter is mineralized, or for redox sensitive metals when rapid cycling occurs between aerobic and anaerobic conditions.

7.2 ION SPECIATION IN SOIL SOLUTION

Even though total ion composition of soil solution is useful for discerning gross effects on available pools of soil ions, the diagnostic utility of total ion concentrations in soil solution is limited, especially when comparisons are made across different soils and environments (see section 2.1.2, Sampling the "True" Soil Solution). The most powerful interpretive utility of soil solution compositional analysis for descriptions of chemical reactivity in soils is attained when free ion concentrations or ion activities of soil solution components are determined. This requires either that individual ion species be measured or computed for soil solution components. Ion speciation of total ions in soil solutions occurs through hydrolysis reactions or through the formation of ion pairs and ion complexes.

7.2.1 Hydrolysis

Association and reactions of solutes with water will significantly alter the distribution and intensity of soil solution components. Hydration describes the tendency of solutes to alter the structure of water physically through energetically favorable orientation of dipolar water about solute particles (Fig. 7–1). Hydrolysis describes the chemical reaction of water leading to the splitting of the water molecule. The attraction of water for some solutes is so strong as to favor chemical hydrolysis leading to deprotonation. That is, in response to solution alkalinity hydrated solutes repel H^+ to solution leaving hydroxyl ligands in association with the central ion of the hydrated complex (Fig. 7–1).

Thermodynamic Descriptions of Hydration and Hydrolysis. The net effect of hydration, hydrolysis, and deprotonation on a solute can be expressed as an enthalpy of hydration. Highly negative enthalpies of hydration indicate an energetically favorable formation of hydrated species in solution (Table 7.4).

Entropies of hydration yield further thermodynamic information regarding ion-solvent interactions. Large positive entropies of hydration are associated with large

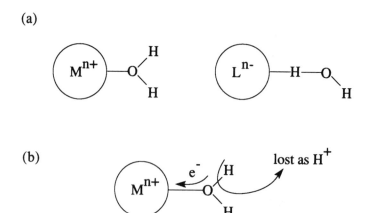

FIGURE 7.1 (*a*) Geometric relations between a solvating water molecule and a cation and an anion. (*b*) Acidity of a hydrated cation. [Reproduced with permission from *Ions in Solution* by J. Burgess published in 1988 by Ellis Horwood limited, Chichester.]

ions of low charge (compare $Cr_2O_7^{2-}$ with CrO_4^{2-} in Table 7.4). The introduction of these ions to aqueous solution acts to disrupt the structure of water with a net increase in the entropy of the system. In contrast, small ions with charges of $2\pm$ or larger have negative entropies of hydration. They act to order water through transfers from bulk solution into their solvent shell, resulting in a net decrease in entropy of the system. Due to the effect of the ion electric field, partial molar volumes correlate well with partial molar entropies of hydration (Table 7.4). Thus, reduced solvent volumes are a result of ion size and charge relationships, just as are entropies of hydration (Burgess, 1988).

Reactions of Aquo-Metal Ions. Proton loss from hydrated ions in solution to form a variety of hydroxyaquo species is a common hydrolysis reaction of importance to soil solution composition. The general reaction is described for a metal, M, as

$$[M(H_2O)_x]^{n+} \rightleftharpoons [M(OH)_y(H_2O)_{x-y}]^{n-y+} + H^{y+}$$

Trivalent aluminum is the most hydrolytically reactive ion component of typical

TABLE 7.4 Enthalpies, Partial Molar Entropies, and Partial Molar Volumes Relative to H$^+$ for Ion Hydration at 298.15 °K and 101.3 kPa

Ion	$\Delta H_{hydration}$, kJ mol^{-1}	Si, J °K^{-1} mol^{-1}	$\partial V/\partial n_i$, L mol^{-1}
H$^+$	-1091	0	0
Na$^+$	-405	59	-1.2
K$^+$	-321	101	9.0
Ca^{2+}	-1592	-53	-17.9
Mg^{2+}	-1922	-183	-21.2
Mn^{2+}	-1845		
Fe^{2+}	-1920		
Cd^{2+}	-1806	-76	-20
Hg^{2+}	-1823	-36	
Cu^{2+}	-2100		
Zn^{2+}	-2044	-110	-21.6
Fe^{3+}		-300	-44
Al^{3+}	-4660	-322	-42.2
F$^-$	-503	-10	-1.1
Cl$^-$	-369	55	17.8
OH$^-$		-11	-4.0
NO$_{3-}$	-328	125	29.0
SO$_4^{2-}$	-1145	17	14.0
CrO$_4^{2-}$		50	
Cr$_2$O$_7^{2-}$		262	

Reproduced with permission from *Ions in Solution* by J. Burgess published in 1988 by Ellis Horwood limited, Chichester.

soil solutions, exhibiting a $\Delta H_{hydration} = -4640$ kJ mol^{-1}. The size of Al^{3+} relative to water leads to the association of six water ligands with a central Al^{3+} ion to form Al(H$_2$O)$_6^{3+}$. The hexaquoaluminum ion can undergo a series of deprotonation reactions leading to a suite of hydrolysis species in soil solution:

$$Al(H_2O)_6^{3+} + H_2O \rightleftharpoons AlOH(H_2O)_5^{2+} + H^+$$

$$AlOH(H_2O)_5^{2+} + H_2O \rightleftharpoons Al(OH)_2(H_2O)_4^+ + H^+$$

$$Al(OH)_2(H_2O)_4^+ + H_2O \rightleftharpoons Al(OH)_3(H_2O)_3^\circ + H^+$$

$$Al(OH)_3(H_2O)_3^\circ + H_2O \rightleftharpoons Al(OH)_4(H_2O)_2^- + H^+$$

This scheme of hydrolysis is further complicated by the possible occurrence in some soil solutions and natural waters of dimers and polymers of aquoaluminum in addition to the monomers illustrated here (Chapter 11). For example, the bridging of hydroxyaquoaluminum monomers can lead to the following hydroxyaquoaluminum dimer.

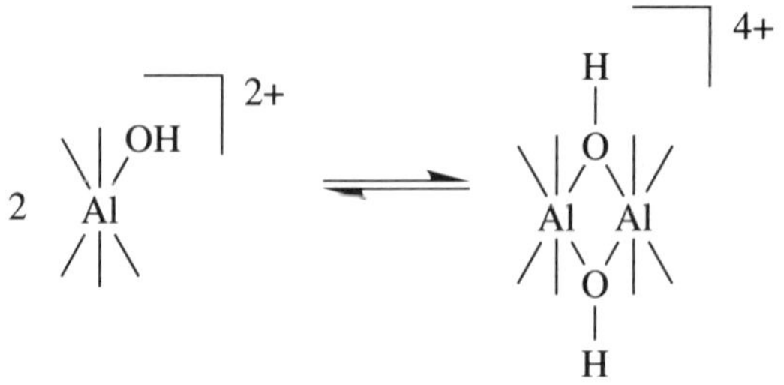

The interrelationships amongst monomeric and polymeric hydrolysis species of aquoaluminum can lead to quite complex speciation relationships in solution as illustrated in Figure 7-2.

   Prevalent hydrolysis and deprotonation reactions important to ion distribution in soil solution are summarized in Table 7.5 for common soil solution components. Following conventional chemical notation, the reactions described in Table 7.5 do not include the associated water in their formulae (for example, Al(H$_2$O)$_6^{3+}$ is expressed as Al^{3+}). These reactions are described by a hydrolysis or acid dissociation constant (expressed as log K$^\circ$ in Table 7.5).[1] The smaller the value of K$^\circ$ for a given hydrolysis/deprotonation reaction, the weaker acid is the reactant.

   Certain hydrated ions exhibit *amphoteric behavior* in soil solution. Hexaquoaluminum, for example, may react hydrolytically to form mixtures of positively charged species under acidic or weakly alkaline pH, but at strongly alkaline pH the negative ion [Al(OH)$_4$(H$_2$O)$_2^-$] may be increasingly consequential (Fig. 7–2).

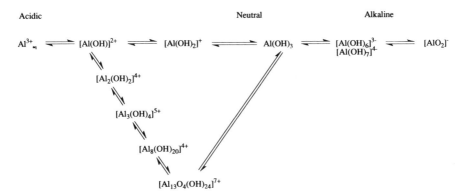

FIGURE 7.2 Interrelationships among monomeric and polymeric hydroxaquoaluminum ions as influenced by solution acidity. [After Burgess, 1988. Reproduced with permission from *Ions in Solution* by J. Burgess published in 1988 by Ellis Horwood limited, Chichester.]

7.2.2 Ion Complexation and Pairing

Tertiary species of aquo-metal ions, $[ML_y(H_2O)_{x-y}]^{n-y}$, represent the more generalized case of hydroxyaquo metal speciation described in section 7.2.1. Here, L may represent an oxyanion, such as OH^-, COO^-, CO_3^{2-}, or PO_4^{3-}, or may represent organic N, S, and P electron donors. The replacement of OH^- with other ligands in *ion complexes* results in sometimes significant effects on the acidity of water ligands within the complex. Consider, for example, deprotonation from the following complexes of Cr^{3+} (Burgess, 1988).

	$-\log K°$
$Cr(H_2O)_6^{3+} \rightleftharpoons CrOH(H_2O)_6^{2+} + H^+$	3.9
$CrCl(H_2O)_5^{2+} \rightleftharpoons CrClOH(H_2O)_4^{+} + H^+$	5.2
$cis{-}Cr(O_2CCO_2)_2(H_2O)_2^{-} \rightleftharpoons cis{-}Cr(O_2CCO_2)_2OH(H_2O)^{2-}$	5.6
$+ H^+$	
$Cr(CN)_5(H_2O)^{2-} \rightleftharpoons Cr(CN)_5OH^{3-} + H^+$	≈ 9.0

Complexation reactions refer to specific types of substitution reactions described for the general reaction of a metal (M^{n+}) and monodentate ligands (L and L′) as,

$$ML'_y(H_2O)_{x-y}^{n-y} + L \rightleftharpoons ML_y(H_2O)_{x-y}^{n-y} + L'$$

These reactions can be described by equilibrium association constants as

$$(ML_y(H_2O)_{x-y}^{n-y})(L')/(ML'_y(H_2O)_{x-y}^{n-y})(L) = K°$$

TABLE 7.5 Hydrolysis and Deprotonation Equilibria for Soil Solution Components

Equilibrium Reaction	log $K°$
$Ca^{2+} + H_2O \rightleftharpoons CaOH^+ + H^+$	-12.70
$Ca^{2+} + 2H_2O \rightleftharpoons Ca(OH)_2^0 + 2H^+$	-27.99
$Mg^{2+} + H_2O \rightleftharpoons MgOH^+ + H^+$	-11.45
$K^+ + H_2O \rightleftharpoons KOH^0 + H^+$	-14.50
$Na^+ + H_2O \rightleftharpoons NaOH^0 + H^+$	-14.20
$Al^{3+} + H_2O \rightleftharpoons AlOH^{2+} + H^+$	-5.02
$Al^{3+} + 2H_2O \rightleftharpoons Al(OH)_2^+ + 2H^+$	-9.30
$Al^{3+} + 3H_2O \rightleftharpoons Al(OH)_3^0 + 3H^+$	-14.99
$Al^{3+} + 4H_2O \rightleftharpoons Al(OH)_4^- + 4H^+$	-23.33
$Al^{3+} + 5H_2O \rightleftharpoons Al(OH)_5^{2-} + 5H^+$	-34.24
$2Al^{3+} + 2H_2O \rightleftharpoons Al_2(OH)_2^{4+} + 2H^+$	-7.69
$Fe^{3+} + H_2O \rightleftharpoons FeOH^{2+} + H^+$	-2.19
$Fe^{3+} + 2H_2O \rightleftharpoons Fe(OH)_2^+ + 2H^+$	-5.69
$Fe^{3+} + 3H_2O \rightleftharpoons Fe(OH)_3^0 + 3H^+$	-13.09
$Fe^{3+} + 4H_2O \rightleftharpoons Fe(OH)_4^- + 4H^+$	-21.59
$2Fe^{3+} + 2H_2O \rightleftharpoons Fe_2(OH)_2^{4+} + 2H^+$	-2.90
$Fe^{2+} + H_2O \rightleftharpoons FeOH^+ + H^+$	-6.74
$Fe^{2+} + 2H_2O \rightleftharpoons Fe(OH)_2^0 + 2H^+$	-16.04
$Fe^{2+} + 3H_2O \rightleftharpoons Fe(OH)_3^- + 3H^+$	31.99
$Fe^{2+} + 4H_2O \rightleftharpoons Fe(OH)_4^{2-} + 4H^+$	-46.38
$3Fe^{2+} + 4H_2O \rightleftharpoons Fe_3(OH)_4^{2+} + 4H^+$	-45.39
$Mn^{2+} + H_2O \rightleftharpoons MnOH^+ + H^+$	-10.95
$Mn^{2+} + 2H_2O \rightleftharpoons Mn(OH)_2^0 + 2H^+$	
$Mn^{2+} + 3H_2O \rightleftharpoons Mn(OH)_3^- + 3H^+$	-34.00
$Mn^{2+} + 4H_2O \rightleftharpoons Mn(OH)_4^{2-} + 4H^+$	-48.29
$2Mn^{2+} + H_2O \rightleftharpoons Mn_2OH^{3+} + H^+$	-10.60
$2Mn^{2+} + 3H_2O \rightleftharpoons Mn_2(OH)_3^+ + 3H^+$	-23.89
$Mn^{3+} + 3H_2O \rightleftharpoons MnOH^{2+} + 3H^+$	0.40
$Zn^{2+} + H_2O \rightleftharpoons ZnOH^+ + H^+$	-7.69
$Zn^{2+} + 2H_2O \rightleftharpoons Zn(OH)_2^0 + 2H^+$	-16.80
$Zn^{2+} + 3H_2O \rightleftharpoons Zn(OH)_3^- + 3H^+$	-27.68
$Zn^{2+} + 4H_2O \rightleftharpoons Zn(OH)_4^{2-} + 4H^+$	-38.29
$Cu^{2+} + H_2O \rightleftharpoons CuOH^+ + H^+$	-7.70
$Cu^{2+} + 2H_2O \rightleftharpoons Cu(OH)_2^0 + 2H^+$	-13.78
$Cu^{2+} + 3H_2O \rightleftharpoons Cu(OH)_3^- + 3H^+$	-26.75

TABLE 7.5 *Continued*

Equilibrium Reaction	log $K°$
$Cu^{2+} + 4H_2O \rightleftharpoons Cu(OH)_4^{2-} + 4H^+$	-39.59
$2Cu^{2+} + 2H_2O \rightleftharpoons Cu_2(OH)_2^{2+} + 2H^+$	-10.68
$H_4SiO_4^\circ \rightleftharpoons H_3SiO_4^- + H^+$	-9.71
$H_4SiO_4^\circ \rightleftharpoons H_2SiO_4^{2-} + 2H^+$	-22.98
$H_4SiO_4^\circ \rightleftharpoons HSiO_4^{3-} + 3H^+$	-32.85
$H_4SiO_4^\circ \rightleftharpoons SiO_4^{4-} + 4H^+$	-45.95
$4H_4SiO_4^\circ \rightleftharpoons H_6Si_4O_{12}^{2-} + 2H^+ + 4H_2O$	-13.32
$CO_2(g) + H_2O \rightleftharpoons H_2CO_3^\circ$	-1.46
$H_2CO_3^\circ \rightleftharpoons H^+ + HCO_3^-$	-6.36
$HCO_3^- \rightleftharpoons H^+ + CO_3^{2-}$	-10.33
$CO_2(g) + H_2O \rightleftharpoons H^+ + HCO_3^-$	-7.82
$CO_2(g) + H_2O \rightleftharpoons 2H^+ + CO_3^{2-}$	-18.15
$NH_4^+ \rightleftharpoons NH_3^\circ + H^+$	-9.28
$NH_4^+ + H_2O \rightleftharpoons NH_4OH^\circ + H^+$	-9.25
$H_2SO_4^\circ \rightleftharpoons H^+ + HSO_4^-$	1.98
$HSO_4^- \rightleftharpoons H^+ + SO_4^{2-}$	-1.98
$H_3PO_4^\circ \rightleftharpoons H^+ + H_2PO_4^-$	-2.15
$H_2PO_4^- \rightleftharpoons H^+ + HPO_4^{2-}$	-7.20
$HPO_4^{2-} \rightleftharpoons H^+ + PO_4^{3-}$	-12.35
$H_2PO_4^- \rightleftharpoons 2H^+ + PO_4^{3-}$	-19.55
$2H_2PO_4^- \rightleftharpoons (H_2PO_4)_2^{2-}$	-0.35

where the parentheses denote activities. For the most prevalent complexation reactions in soil solution, where L' is H_2O, water of hydration is excluded from the formulae and the activity of water equals 1, this expression becomes

$$\frac{A_{ML}}{A_M A_L} = K°$$

An example of this type of substitution reaction is the complexation of Fe^{3+} with glycinate.

$$Fe(H_2O)_6^{3+} + H_2NCH_2CO_2^- \rightleftharpoons Fe(H_2NCH_2CO_2)(H_2O)_5^{2+} + H_2O$$

Common complexation reactions involving metals and oxyanions in soil solution are summarized in Table 7.6.

TABLE 7.6 Complexation Equilibria for Soil Solution Components

Equilibrium Reaction	$\log K^\circ$
$Ca^{2+} + H_2PO_4^- \rightleftharpoons CaH_2PO_4^+$	1.40
$Ca^{2+} + H_2PO_4^- \rightleftharpoons CaHPO_4^\circ + H^+$	-4.46
$Ca^{2+} + H_2PO_4^- \rightleftharpoons CaPO_4^- + 2H^+$	-13.09
$Ca^{2+} + SO_4^2 \rightleftharpoons CaSO_4^\circ$	2.31
$Ca^{2+} + Cl^- \rightleftharpoons CaCl^+$	-1.00
$Ca^{2+} + 2Cl^- \rightleftharpoons CaCl_2^\circ$	0.00
$Ca^{2+} + CO_2(g) + H_2O \rightleftharpoons CaHCO_3^+ + H^+$	-6.70
$Ca^{2+} + CO_2(g) + H_2O \rightleftharpoons CaCO_3^\circ + 2H^+$	-15.01
$Ca^{2+} + NO_3^- \rightleftharpoons CaNO_3^+$	-4.80
$Ca^{2+} + 2NO_3^- \rightleftharpoons Ca(NO_3)_2^\circ$	-4.50
$Mg^{2+} + 2H_2O \rightleftharpoons Mg(OH)_2^\circ + 2H^+$	-27.99
$Mg^{2+} + H_2PO_4^- \rightleftharpoons MgHPO_4^\circ + H^+$	-4.29
$Mg^{2+} + SO_4^{2-} \rightleftharpoons MgSO_4^\circ$	2.23
$Mg^{2+} + 2Cl^- \rightleftharpoons MgCl_2^\circ$	-0.03
$Mg^{2+} + CO_2(g) + H_2O \rightleftharpoons MgHCO_3^+ + H^+$	-6.76
$Mg^{2+} + CO_2(g) + H_2O \rightleftharpoons MgCO_3^\circ + 2H^+$	-14.92
$Mg^{2+} + 2NO_3^- \rightleftharpoons Mg(NO_3)_2^\circ$	-0.01
$K^+ + SO_4^{2-} \rightleftharpoons KSO_4^-$	0.85
$K^+ + Cl^- \rightleftharpoons KCl^\circ$	-0.70
$2K^+ + CO_2(g) + H_2O \rightleftharpoons K_2CO_3^\circ + 2H^+$	-18.17
$Na^+ + Cl^- \rightleftharpoons NaCl^\circ$	0.00
$Na^+ + CO_2(g) + H_2O \rightleftharpoons NaCO_3^- + 2H^+$	-16.89
$2Na^+ + CO_2(g) + H_2O \rightleftharpoons Na_2CO_3^\circ + 2H^+$	-18.14
$Na^+ + CO_2(g) + H_2O \rightleftharpoons NaHCO_3^\circ + H^+$	-7.58
$Al^{3+} + F^- \rightleftharpoons AlF^{2+}$	6.98
$Al^{3+} + 2F^- \rightleftharpoons AlF_2^+$	12.60
$Al^{3+} + 3F^- \rightleftharpoons AlF_3^\circ$	16.65
$Al^{3+} + 4F^- \rightleftharpoons AlF_4^-$	19.03
$Al^{3+} + 3NO_3^- \rightleftharpoons Al(NO_3)_3^\circ$	0.12
$Al^{3+} + SO_4^{2-} \rightleftharpoons AlSO_4^+$	3.20
$Al^{3+} + 2SO_4^{2-} \rightleftharpoons Al(SO_4)_2^-$	1.90
$2Al^{3+} + 3SO_4^{2-} \rightleftharpoons Al_2(SO_4)_3^\circ$	-1.88
$Fe^{3+} + Cl^- \rightleftharpoons FeCl^{2+}$	1.48
$Fe^{3+} + 2Cl^- \rightleftharpoons FeCl_2^+$	2.13
$Fe^{3+} + 3Cl^- \rightleftharpoons FeCl_3^\circ$	0.77
$Fe^{3+} + Br^- \rightleftharpoons FeBr^{2+}$	-0.60

TABLE 7.6 *Continued*

Equilibrium Reaction	log K°
$Fe^{3+} + 3Br^- \rightleftharpoons FeBr_3^\circ$	0.04
$Fe^{3+} + F^- \rightleftharpoons FeF^{2+}$	6.00
$Fe^{3+} + 2F^- \rightleftharpoons FeF_2^+$	9.20
$Fe^{3+} + 3F^- \rightleftharpoons FeF_3^\circ$	11.70
$Fe^{3+} + NO_3^- \rightleftharpoons FeNO_3^{2+}$	1.00
$Fe^{3+} + SO_4^{2-} \rightleftharpoons FeSO_4^+$	4.15
$Fe^{3+} + 2SO_4^{2-} \rightleftharpoons Fe(SO_4)_2^-$	5.38
$Fe^{3+} + H_2PO_4^- \rightleftharpoons FeH_2PO_4^{2+}$	5.43
$Fe^{3+} + H_2PO_4^- \rightleftharpoons FeHPO_4^+ + H^+$	3.71
$Fe^{2+} + 2Br^- \rightleftharpoons FeBr_2^\circ$	0.00
$Fe^{2+} + 2Cl^- \rightleftharpoons FeCl_2^\circ$	−0.07
$Fe^{2+} + 2F^- \rightleftharpoons FeF_2^\circ$	0.03
$Fe^{2+} + H_2PO_4^- \rightleftharpoons FeH_2PO_4^+$	2.70
$Fe^{2+} + H_2PO_4^- \rightleftharpoons FeHPO_4^\circ + H^+$	−3.60
$Fe^{2+} + SO_4^{2-} \rightleftharpoons FeSO_4^\circ$	2.20
$Mn^{2+} + Cl^- \rightleftharpoons MnCl^+$	0.61
$Mn^{2+} + 2Cl^- \rightleftharpoons MnCl_2^\circ$	0.04
$Mn^{2+} + CO_2(g) + H_2O \rightleftharpoons MnHCO_3^+ + H^+$	−6.02
$Mn^{2+} + CO_2(g) + H_2O \rightleftharpoons MnCO_3^\circ + 2H^+$	−18.87
$Mn^{2+} + SO_4^{2-} \rightleftharpoons MnSO_4^\circ$	2.26
$Zn^{2+} + H_2PO_4^- \rightleftharpoons ZnH_2PO_4^+$	1.60
$Zn^{2+} + H_2PO_4^- \rightleftharpoons ZnHPO_4^\circ + H^+$	−3.90
$Zn^{2+} + NO_3^- \rightleftharpoons ZnNO_3^+$	0.40
$Zn^{2+} + 2NO_3^- \rightleftharpoons Zn(NO_3)_2^\circ$	−0.30
$Zn^{2+} + SO_4^{2-} \rightleftharpoons ZnSO_4^\circ$	2.33
$Zn^{2+} + Cl^- \rightleftharpoons ZnCl^+$	0.43
$Zn^{2+} + 2Cl^- \rightleftharpoons ZnCl_2^\circ$	0.00
$Zn^{2+} + 3Cl^- \rightleftharpoons ZnCl_3^-$	0.50
$Zn^{2+} + 4Cl^- \rightleftharpoons ZnCl_4^{2-}$	0.20

The products of complexation reactions (Table 7.6) are frequently, and somewhat vaguely, described as ion complexes or ion pairs. The distinction lies in the degree that ion hydration spheres are affected as metals and ligands electrostatically interact. In the example of Fe^{3+} reacting with glycinate, a stable complex ion is formed as glycinate ion replaces water from the solvation shell of Fe^{3+}. *Ion complexes*—stable entities found in solution that are formed largely via covalent metal-ligand bonds—are also referred to as *inner sphere complexes*. Hydroxyaquo-metal ions are a specific type of ion complex where OH^- is the complexing ligand. *Ion pair* or *outer sphere complexes* involve electrostatically interacting metals and ligands where the interaction occurs over a measurable time interval but where spheres of hydration are not altered.

The original concept of ion pairing advanced by Bjerrum (1926) considered columbic interactions and used a probability density function to calculate the distribution of bulk solution anions with distance from a cation. Ion pairs were defined as the population of pairs of cations and anions whose calculated interaction energy was > 2 times the thermal energy. Despite concerns over some aspects of the theoretical development this approach, calculated ion association constants arising from Bjerrum's theory of ion association give fair agreement with measured values (Johnson and Pytkowicz, 1979). The use of a probability density function to describe ion pairs indicates that for any metal–ligand complexation reaction, the distinction of inner versus outer sphere complexation is one of the relative proportion of pairs of cations and anions that are interacting electrostatically in their outer versus inner solvation shell.

The Eigen–Wilkins mechanism affords a dynamic theory for complexation whereby outer versus inner solvation shell interactions can be kinetically probed. Schematically, for a mondentate ligand, complexation can be viewed as a two-step process.

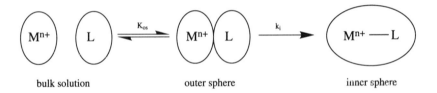

bulk solution   outer sphere   inner sphere

Mechanistically, metals and ligands first distribute between distinct ions in bulk solution and outer sphere complexes described by an equilibrium *outer sphere association constant* (K_{OS}) developed on the basis of electrostatic interactions. The second step involves displacement of a coordinated water molecule to form an inner sphere complex described by the *interchange rate constant*, k_i.

$$M(H_2O)_x^{n+} + L^{y-} \xrightleftharpoons{K_{OS}} M(H_2O)_x, L^{n-y} \xrightarrow{k_i} ML(H_2O)_{x-1}^{n-y} + H_2O$$

The general rate law for this reaction describes the formation of the metal–ligand inner sphere complex (ML).

$$\frac{d[ML^{n-y}]}{dt} = \frac{K_{OS}k_i[M^{n+}][L^{y-}]}{1 + K_{OS}[L^{y-}]} \qquad [7\text{-}1a]$$

Experimental determinations of complexation rates are conducted under conditions where $[M^{n+}] \gg [L^{y-}]$ so that equation 7-2b simplifies to

$$\frac{d[ML^{n-y}]}{dt} = K_{OS}k_i[M^{n+}][L^{y-}] \qquad [7\text{-}2b]$$

Experimentally, the apparent rate constant $k_f = K_{OS}k_i$ is measured using a technique such as the absorbance trace developed with stopped-flow kinetics. Rates of outer sphere complex formation are too rapid to measure; therefore, K_{OS} is calculated from electrostatics. Values of k_i thus equal k_f/K_{OS}.

Table 7.7 summarizes measured k_f and derived k_i for the formation of Mg^{2+}_{aq} inner sphere complexes with ligands of biological consequence. Differences in k_f for Mg^{2+}_{aq} complexation with various ligands are due mostly to K_{OS} that are sensitive mainly to ligand charge. The values of k_i are of the same relative order of mag-

TABLE 7.7 **Measured Apparent Rate Constants (k_f), Outer Sphere Association Constants (K_{OS}), and Derived Interchange Rate Constants (k_i) for Reactions of Biologically Important Ligands with Mg^{2+}_{aq} at 298.2 K**

Ligand	Measured $10^{-5}k_f(M^{-1}\,s^{-1})$	Estimated K_{OS}	Derived $10^{-5}k_i\,(s^{-1})$
Oxine[a]	0.13	0.2	0.7
Oxinate[-b]	6.0	2.1	2.9
Fluoride[-]	0.55	1.6	0.4
5-Nitrosalicylate[-]	7.1	2	3.6
Bicarbonate[-]	5.0	0.9	5.6
Carbonate[2-]	0.15	3.5	0.4
PyrophosphateH$_2^{2-}$	5.4	13	0.4
ADPH[2-c]	10	9	1.1
ATPH[3-d]	30	30	1.0
ATP[4-d]	130	120	1.1
Cf. water exchange			1.0

[a]Oxine =   [b]Oxinate =

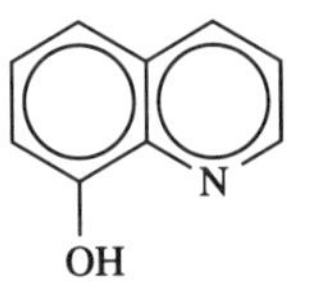

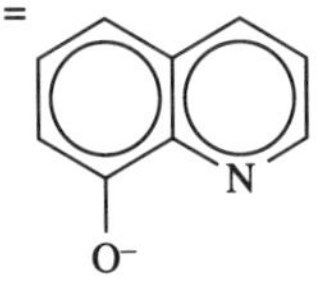

OH   O[-]

[c]ADP = adenosine diphosphate.
[d]ATP = adenosine triphosphate.

[c]ADP = adenosine diphosphate.
[d]ATP = adenosine triphosphate.

nitude as the exchange rate of H_2O (k_{ex}) from the Mg^{2+} solvation shell. This indicates that dissociation reactions—the expelling of water from the inner solvation shell—control the inner sphere complexation reaction. This is in contrast to the associative nature of the initial step where outer sphere complexes form.

The nature of the solvent exchange mechanism varies dependent on metal ion charge, ion radius, and degree of acid "hardness" (Burgess, 1988). Thus, the inner sphere complexation reactions of ligands with Fe^{3+}_{aq} and $Fe(OH)^{2+}_{aq}$ occur by differing mechanisms—associative and dissociative, respectively.

Measurement of Association Constants. Measurements of ion association in solution are limited mostly to synthetic solutions of relatively simple composition and high ion concentration relative to natural waters. Differing methods for measurement of ion association constants frequently do not yield equivalent results when applied to the same system. Reliable measurement of ion association constants require validation by more than one analytical technique and/or agreement with dynamic or static predictions.

A variety of methods have been employed for the measurement of ion association. Conductivity, solubility, and potentiometric methods have broad application, whereas numerous other methods such as physicochemical separation, competitive equilibria, or electrode kinetics have useful specific applications. Very seldom do these methods afford direct measurements of the ion complex of interest; rather, the nature of these postulated complexes is inferred from failure of synthetic solutions to behave as predicted by simple solution theory. Most complexes do not have distinct electromagnetic spectra; therefore, direct spectroscopic evidence for their existence is lacking. Further discussion of limitations to measurements of complexes are presented in section 13.2.1 for the specific case of metal complexation with organic ligands.

Most methods for measurement of ion association are not well suited to multiple complexes occurring in multi-ion mixtures and at micromolar concentrations—thus, directly applying such approaches to analyze soil solutions is not possible. Even when it is possible to measure the concentrations of associated ions in solution and when these measurements are integral to studies designed to elucidate the environmental significance of particular ion complexes, the problems associated with such measurements—particularly shifts in ion species distributions when soil solutions or natural waters are sampled, displaced, and analyzed—preclude their widespread use. This is especially true when a comprehensive analysis of soil solution composition is desired. Measurements of the various complexation products (hydrolytic species, ion complexes, and ion pairs) relevant to soil solution composition are not possible. One way to circumvent this problem is to use analytical methods that do not distinguish individual ion species but group ion species on the basis of their reactivities with ligands. Examples of this approach are described for Al in section 11.1

7.3 COMPUTATIONAL MODELING OF SOIL SOLUTION COMPOSITION

Because it is largely not possible to analyze comprehensively and unambiguously for the full complement of ion species that may be important to applied soil solution chemistry, the prevalent approach utilizes computational modeling of ion distribution in soil solution. This approach requires comprehensive characterization of total ion composition of soil solution (section 7.1) and an appropriate ion speciation model. Ion speciation models basically consist of schemes for dealing with materials balance, computational algorithms for solving the multiple simultaneous equations that describe equilibrium distribution of ions in aqueous solution, and thermodynamic data files describing equilibrium partitioning of ions.

7.3.1 Model Comparisons and Choice of a Model

The kinds of ion speciation models available vary significantly in their design and intended use. The models in most widespread current use in soil solution chemistry—models such as GEOCHEM (Mattigod and Sposito, 1980) and MINEQL+ (Schecher and McAvoy, 1992)—represent second- and third-generation versions built upon the numerical approaches outlined by Morel and Morgan (1972) for the model REDEQL (Nordstrom et al., 1979). The principal distinction of these various models is the thermodynamic data files that reflect the type of aqueous systems the model is most intended to simulate (for instance, sea water, mine water, soil water). These models offer a considerable degree of flexibility in terms of the ability to consider subsets of chemicals occurring in aqueous environments (in contrast to comprehensive ion analyses), computation of pH, precipitation of minerals for which the system is oversaturated, and consideration of complex redox equilibria.

In contrast to these complex models are the less computationally intensive models that assume comprehensive total ion analysis for soil solutions under typical field conditions. The approach used in the design of this nature of model was described by Adams (1971) for CALPHOS and has been used in latter generations of this model, such as SOILSOLN (Wolt, 1989). These types of models give results equivalent to those of more comprehensive models, but are limited with regard to the number of complexes and species considered and the variety of auxiliary operations that may be performed on the speciation results. They are a useful tool, however, for many problems typical to soil fertility assessment, mineral stability, and environmental soil chemistry. Table 7.8 compares the free ion concentrations computed by two ion speciation models differing in their computational intensity, when applied to natural water composition. The results differ principally in the numbers of metals and ligands treated, not in the free ion concentrations computed for ions shared in common.

A common feature of these various models is the use of an equilibrium-constant approach for description of ion distribution and computation of ionic activities. A contrasting approach is the use of Gibbs free energy minimization, such as in the various versions of WATEQ (Ball et al., 1979). The equilibrium-constant approach

TABLE 7.8 Comparison of Output for Two Ion Speciation Models

	Redding Saturation Extract			Cumberland Plateau Stream Water		
	Total Ion Concentration	Free Ion Concentration		Total Ion Concentration	Free Ion Concentration	
		GEOCHEM	SOILSOLN		GEOCHEM	SOILSOLN
			$-$log Concentration, mol L^{-1}			
Metals						
H	5.60	5.60	5.60	6.80	6.80	6.80
Ca	3.15	3.22	3.22	3.08	3.15	3.12
Mg	2.82	2.89	2.87	3.12	3.18	3.16
K	3.40	3.40	3.40	4.20	4.20	4.20
Na	2.55	2.55	2.56	3.84	3.85	3.84
Fe(III)				5.51	13.89	
Mn				3.62	3.69	3.65
Cu(II)				6.80	7.12	
Zn				7.12	7.22	7.22
Ni				7.07	7.16	
Hg				7.00	15.34	
Pb				6.71	6.96	
Cd(II)				5.99	6.09	
Cr				7.02	16.37	
Al				5.35	10.72	10.45
TiO				6.22	6.22	
Ligands						
CO$_3$	3.00	8.29	8.33	4.09	7.53	7.16
SO$_4$	2.65	2.66	2.72	2.65	2.70	3.05
Cl	2.68	2.63	2.68	4.07	4.07	4.41
P				5.03	5.36[a]	5.25[a]
Si				4.14	4.14[b]	4.14[b]
NO$_3$				6.09	6.09	6.42
Ionic strength, mmol L^{-1}	10.44		10.40		7.32	6.25

[a]H_2PO_4
[b]H_4SiO_4

is more flexible in terms of the thermodynamic data than is Gibbs free energy minimization as the appropriate thermodynamic data for Gibbs free energy minimization of geochemical problems is frequently lacking.

Developments in chemical modeling of aqueous systems are well summarized in two symposia series (Jenne, 1979; Melchior and Bassett, 1990). Nordstrom et al. (1979) present an especially useful comparison of the structure and outputs of common ion speciation models. In recent years, emphasis has shifted from the development of new ion speciation models to the refinement of existing models.

7.3.2 Materials Balance

Materials balance is addressed at two levels in ion speciation models. Charge balance is used to gauge the adequacy of the input data set, while mole balance equations comprise the core computational algorithms (section 7.3.3). Few ion speciation models require charge balance (Baham, 1984). This allows for applications of the model to problems that may comprise incomplete characterization of a system and may therefore lead to erroneous results. A more rigorous approach requires charge balance for input data of soil solution composition. When charge is not balanced, a charge deficit can be overcome either through normalization of the input data or through assignment of the deficit charge against an ion species that does not markedly interact to speciate in the system considered (frequently this assignment is made against NO_3^-) or by adjusting the proton balance for the system. Initial charge balance can be computed based on assignment of the free ion charge to corresponding total ion concentration in the model input data, as net charge balance will be conserved when ions interact to speciate. Ion difference (section 7.1.1) is useful in evaluation of input data sets for adequately representing electroneutrality for treatment of materials balance by this approach.

7.3.3 Computational Algorithms

Materials balance is additionally important to the computational algorithms describing speciation of total ion concentrations among the free ions and various hydrolytic species, ion complexes, and ion pairs. The computational algorithms are nonlinear expressions for each metal and ligand occurring in soil solution,

$$C_{M_{total}} = \frac{A_{M^{n+}}}{f_{M^{n+}}} + \sum_i \alpha K_i^0 \frac{A_{M^{n+}}^\alpha + A_{L^{y-}}^\beta}{f_{M^{n+}}^\alpha + f_{L^{y-}}^\beta} \qquad [7\text{-}1a]$$

$$C_{L_{total}} = \frac{A_{L^{y-}}}{f_{L^{y-}}} + \sum_i \beta K_i^0 \frac{A_{M^{n+}}^\alpha + A_{L^{y-}}^\beta}{f_{M^{n+}}^\alpha + f_{L^{y-}}^\beta} \qquad [7\text{-}1b]$$

where K_i^0 describes the ion association constant for a metal–ligand complex, $M_\alpha L_\beta^{\alpha n - \beta y}$. Each expression distributes a metal or ligand among its free ionic form in solution and its various metal–ligand complexes (these may be hydrolytic species, ion complexes, or ion pairs) while maintaining the number of moles input to the problem.

The core of the ion speciation model is, therefore, a series of simultaneous equations of the sort shown by equations 7-1a and b describing the equilibrium distribution of metals and ligands in aqueous solution. These equations are solved iteratively using successive approximation. The Newton–Raphson algorithm (that is, Newton's method) or a simplification thereof is the usual iterative approach, but the secant method and simple bisection have also been successfully used. Choice of an iteration algorithm will influence the rate of convergence and the likelihood of divergence. Convergence rate and divergence are most frequently of concern when the input data incompletely describe the aqueous system of interest leading to a significant imbalance in the charge distribution that is inadequately corrected with charge balance normalization.

Ancillary calculations in the model code may allow for correction of K^0 for temperature and pressure, concentration conversions, and redox distributions. The core ion speciation model may be augmented with subroutines that allow for interactions with other soil phases (degassing, adsorption, precipitation–dissolution), computation of ion activity products and disequilibrium ratios (Chapter 9), kinetics, mixing, and reaction path simulation.

7.3.4 Thermodynamic Data Files

As mentioned in the foregoing discussion, the most important feature leading to differences in the output from ion speciation models is the thermodynamic data file (TDF) that accompanies the computational algorithms. These files are comprised of the ion size parameters necessary for activity coefficient computations, the stoichiometry of metal–ligand complexes, and appropriate thermodynamic data. For the equilibrium-constant approach, these data are ion association constants; for the Gibbs free energy minimization approach, they are free energies, enthalpies, and entropies. Selection of appropriate data for thermodynamic data files is very much related to the judgments of individual investigators who should best know which data are suited for a given application. Because of this, those models with user-definable TDF tend to be the most useful.

Thermodynamic data are derived from a multitude of sources and selection of the most appropriate data for inclusion in TDF requires that data sources are critically reviewed. Review of thermodynamic data is a formidable task, and the soil chemist is well advised to utilize critically reviewed compilations of thermodynamic data as an aid to data selection. Primary data compilations such as IUPAC and NBS tabulations (Högfeldt, 1982; Perrin, 1979; Wagman et al., 1982) are the most reliable source of peer-reviewed thermodynamic data. Secondary compilations contain selected thermodynamic data of use for specific applications. An example of secondary compilations specific to soil chemistry is that of Sadiq and Lindsay (1979) used in the equilibrium computations of Lindsay (1979). Additional secondary compilations are the TDF accompanying most ion speciation models. Secondary compilations are the obvious starting point for the soil chemist undertaking ion speciation modeling, but a more in-depth examination of primary compilations or of original data sources may be required for those thermodynamic data of greatest importance to a given application. This is because even comparisons among well-researched secondary thermodynamic data compilations will show orders of magnitude variation in certain of the thermodynamic data selected.

NOTE

1. Unless otherwise indicated, the thermodynamic data occurring throughout this text are from the thermodynamic data file compilation of Sadiq and Lindsay (1979).

CHAPTER 8

QUANTITY-INTENSITY RELATIONSHIPS

The intimate interation of chemicals in soil solution with chemicals associated with the soil solid phase is a fundamental aspect of all soil and environmental chemistry research dealing with availability, mobility, and distribution of chemicals in soils. The multiple types of chemical reactions governing soil solid-solution phase distribution are generically described as *quantity-intensity* (Q/I) relationships, where quantity refers to a solid phase pool of a chemical and intensity refers to the actual concentration (or activity) of the chemical in solution phase at any time. Much of soil chemistry involves the attempt to understand and express these Q/I relationships from a variety of perspectives.

The types of quantity-intensity relationships important to a given problem of soil chemical reactivity can vary broadly. They may involve cation and anion exchange between the quantity of a cation or anion residing on the soil exchange complex and a charged species in the soil solution; precipitation–dissolution of solid phases, as governed by the solubility product principle; or chemisorptive process involving both adsorption and partition. This chapter deals principally with the generalized Q/I relationship described by ratio law and common models for cation exchange and chemisorption. Precipitation–dissolution as a Q/I relationship is discussed in detail in Chapter 9 as it relates to mineral stability relationships. Collectively, the interest of the soil solution chemist in these Q/I relationships is the ability to predict the intensity of chemicals in soil solution and, thus, to better evaluate biogeochemical availability (Chapter 10).

8.1 RATIO LAW

The equilibrium distribution of two cations (M^{m+} and N^{n+}) between soil solution and an exchangeable phase (—X) is described

$$nM\text{—}X + mN^{n+} \rightleftharpoons mN\text{—}X + nM^{m+}$$

The equilibrium condition for this reaction is expressed

$$K_{ex} = \frac{A_{M^{m+}}^n}{A_{N^{n+}}^m} \frac{AX_{N^{n+}}^m}{AX_{M^{m+}}^n} \qquad [8\text{-}1]$$

where K_{ex} describes an exchange coefficient. The determination of K_{ex} is complicated by the inability to measure or appropriately express activities of cations in the exchange phase, AX_i. Thus, a variety of empirical expressions have been developed for the description of ion exchange on the basis of *selectivity coefficients*.

The concept of Q/I relationships derives from the work of Schofield (1947) who observed that nutrient status was better described by the equilibrium distribution of ions in solution (intensity) than it was by the solid phase pool (quantity) of these ions. Schofield's *ratio law* states, with respect to exchangeable cations, that a negatively charged surface (cation exchange complex) containing a mixture of cations will support in solution in intensity (activity) of these cations in proportion to the ratio of Z_i-root of their concentrations on the exchange complex, where Z_i is the valence of a cation i. For the case of binary exchange of cations M^{m+} and N^{n+}, where CX_i is the concentration (mol kg^{-1}) of i in the exchangeable phase and K is a selectivity coefficient,

$$K = \frac{A_{M^{m+}}^{1/m} \gamma_{N^{n+}}^{1/n}}{A_{N^{n+}}^{1/n} \gamma_{M^{m+}}^{1/m}} \frac{CX_{N^{n+}}^{1/n}}{CX_{M^{m+}}^{1/m}} \qquad [8\text{-}2]$$

$$\frac{A_{M^{m+}}^{1/m}}{A_{N^{n+}}^{1/n}} = \frac{\gamma_{M^{m+}}^{1/m}}{\gamma_{N^{n+}}^{1/n}} K \frac{CX_{M^{m+}}^{1/m}}{CX_{N^{n+}}^{1/n}} \qquad [8\text{-}2]$$

The relationship described by equation 8-2a can be developed from either the Kerr or Gapon equations. The Kerr equation as derived from the mass action expression for the binary exchange of M^{m+} and N^{n+} is

$$K_K = \frac{C_{M^{m+}}^n}{C_{N^{n+}}^m} \frac{CX_{N^{n+}}^m}{CX_{M^{m+}}^n} \qquad [8\text{-}3]$$

When equation 8-3 is raised to the n/m power (when the m/n root is taken) and when solution phase concentrations are corrected to activities, equation 8-2a is obtained ($K = K_{K^{n/m}}$). The Gapon equation expresses exchange phase concentration in terms of equivalents kg^{-1} ($= CX_{i^{z+}}/Z_i$), thus, the equilibrium distribution described is

$$K_G = \frac{C_{N^{n+}}^n}{C_{M^{m+}}^m} \frac{\left(\dfrac{1}{n} CX_{N^{n+}}\right)^m}{\left(\dfrac{1}{m} CX_{M^{m+}}\right)^n} \tag{8-4}$$

Taking the m/n root of equation 8-4 and rearranging results in equation 8-2a ($K = K_{G^{n/m}}(n^n/m^m)$).

Equation 8-2a is most readily applicable to monovalent–divalent exchange, such as between K^+—X and Ca^{2+} and Mg^{2+} where the resulting ratio law is

$$CX_{K^+} = \frac{\gamma_{(Ca^{2+}+Mg^{2+})}^{1/2}}{\gamma_{K^+}} K \frac{A_{K^+}}{A_{(Ca^{2+}+Mg^{2+})}^{1/2}} CX_{(Ca^{2+}+Mg^{2+})}^{1/2} \tag{8-5}$$

Defining the activity ratio, $AR \equiv A_{K^+}/(A_{Ca^{2+}} + A_{Mg^{2+}})^{1/2}$, equation 8-5 can be rewritten as

$$CX_{K^+} = \frac{\gamma_{(Ca^{2+}+Mg^{2+})}^{1/2}}{\gamma_{K^+}} K \ AR \ CX_{(Ca^{2+}+Mg^{2+})}^{1/2} \tag{8-6}$$

The utility of this relationship arises from the observation that K^+ exchange with Ca^{2+} and Mg^{2+} (expressed as ΔK_{exch}^+ or change in CX_{K^+}) is nonlinear as AR varies in the range of low to moderate values characteristic of normal soil solutions. Since Ca^{2+} and Mg^{2+} will dominate the typical soil solution at high concentrations relative to K^+, change in AR is mostly a function of change in soil solution K^+ activity.

A Q/I plot (Fig. 8-1) of ΔK_{exch}^+ on AR provides useful information as to K^+ availability in soils. The slope of the linear proportion of the Q/I curve is the potential K^+ buffering capacity of the soil. Potassium release for the soils described in Figure 8-1 varies significantly. The linear portion of the Q/I curve for the Montell soil is steep and indicates that while the initial pool of exchangeable K^+ is large, this pool is largely nonlabile to exchange. The Grenada soil has a relatively flat linear portion, and even though the exchangeable K^+ content is low, the majority of this pool is labile to exchange.

8.1.1 Dilution Effects on Soil Solution Composition

The Schofield ratio law is a generalized form of equation 8-2b,

$$\frac{C_{M^{m+}}^{1/m}}{C_{N^{n+}}^{1/n}} = K' \tag{8-7a}$$

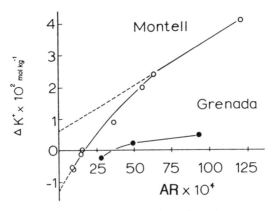

FIGURE 8.1 Quantity-intensity relationships for K^+ exchange on two soils. [After Thomas, 1974. Permission granted by the University Press of Virginia].

Equation 8-7a will hold true if fluxes of cations i through the system considered are small, since $Q_i \gg I_i$ and $CX_{N^{n+}}^{1/n}/CX_{M^{m+}}^{1/m} \approx$ constant, and where $\gamma_{N^{n+}}^{1/n}/\gamma_{M^{m+}}^{1/m} \approx$ constant as well. Equation 8-7a can be recast in a form useful to considerations of dilution effects on soil solution composition as

$$\frac{N_{M^{m+}}^{1/m}}{N_{N^{n+}}^{1/n}} = V^{m/n}K' \qquad [8\text{-}7b]$$

where N_i is the number of moles of i.

When equation 8-7b holds, the cation concentration measured in water extracts of soil can be corrected to the cation concentration extant in soil solution at field moisture content.

The relationship described by equation 8-7b does not hold widely for prediction of mixed valence cation concentrations in soil solution. This may be due to differential effects of dilution on the ratio of activity coefficients; when soil solutions are diluted with addition of water, the ionic strength decreases ($I \rightarrow 0$) and $\gamma_i \rightarrow 1$. The ratio of activity coefficients, $\gamma_{N^{n+}}^{1/n}/\gamma_{M^{m+}}^{1/m}$ is not constant for large changes in V in mixed valence systems, because the change in γ_i with a change in I will be greater as ion valence increases.

8.2 ION EXCHANGE

The experimental observation of ion exchange first described by Way in 1850 has been developed empirically by ratio law as well as through simple exchange relationships such as the Gapon and Kerr equations (section 8.1). Theoretical description of ion exchange has been approached from both static and dynamic perspectives. While "more precise models of . . . [ion] exchange are academically interesting . . . it must be remembered that, in practice, the situation is complicated by competing effects of microbial and plant uptake and inputs from geochemical weathering and other sources" (Cresser et al., 1993). For completeness, some of the more widely used exchange relationships are summarized here, but their comprehensive development lies outside the intent of this chapter. An expanded discussion of ion exchange and selectivity can be found in White and Zelazny (1986) and an interesting contemporary treatment of ion exchange from the standpoint of statics and statistical mechanics can be found in Sposito (1981).

Again, consider the case of binary exchange of cations M^{m+} and N^{n+},

$$nM\text{---}X + mN^{n+} \rightleftharpoons mN\text{---}X + nM^{m+}$$

The Vanselow Equation. The Vanselow expression assumes that the mole fraction of cations on the exchange complex is a truer estimate of exchange phase cation activity than is the assumption of unit activity implicit in the Kerr equation.

$$K_V = \frac{C_{N^{n+}}^n}{C_{M^{m+}}^m} \frac{CX_{N^{n+}}^m}{CX_{M^{m+}}^n} (CX_{N^{n+}}^m + CX_{M^{m+}}^n)^{n/m} \qquad [8\text{-}8]$$

The Vanselow equation reduces to the Kerr equation for considerations of homovalent cation exchange.

The Expression of Krishnamoorthy and Overstreet. This expression uses correction factors for valence developed from a molecular approach that accounts for the lattice properties of the exchange phase.

$$K_{KO} = \frac{C_{N^{n+}}^n}{C_{M^{m+}}^m} \frac{CX_{N^{n+}}^m}{CX_{M^{m+}}^n} (p_{n+}CX_{N^{n+}}^m + p_{m+}CX_{M^{m+}}^n)^{n/m} \qquad [8\text{-}9]$$

The values of p_{Z_i} are 1, 1.5, and 2 for $Z_i = 1, 2,$ and 3, respectively.

8.2.1 Dynamic Expression of Ion Exchange

The forgoing expressions for ion exchange have been described from a static perspective. The conditional distribution coefficients (selectivity coefficients) describing ion exchange equilibria differ only in the way that exchange phase ion concentrations are expressed and the assumptions underlying correction of exchange phase concentrations to activities.

These various equations may be developed from a dynamic perspective as well. For the binary exchange relationship previously described, the rate of exchange of cation N^{n+} is expressed

$$\frac{dC_{N^{n+}}^m}{dt} = -k_f C_{N^{n+}}^m CX_{M^{m+}}^n + k_b C_{M^{m+}}^n CX_{N^{n+}}^m \qquad [8\text{-}10]$$

Note that the rate of exchange is first order with respect to a monovalent ion and fractional order for multivalent ions; the rate of exchange is fractional order overall. At equilibrium,

$$\frac{dC_{N^{n+}}^{m}}{dt} = 0$$

Thus,

$$k_f C_{N^{n+}}^{m} C X_{M^{n+}}^{n} = k_b C_{M^{m+}}^{n} C X_{N^{n+}}^{m} \qquad [8\text{-}11a]$$

$$\frac{k_f}{k_b} = \frac{C_{M^{m+}}^{n} C X_{N^{n+}}^{m}}{C_{N^{n+}}^{m} C X_{M^{m+}}^{n}} \qquad [8\text{-}11b]$$

where $k_f/k_b = K$, the selectivity coefficient, and equation 8-11b reduces to equation 8-2b.

Cation selectivity can be measured kinetically using miscible displacement (Sparks and Jardine, 1981). The rate of adsorption is determined by reacting an exchange complex with a solution of constant concentration N^{n+} and where $C_{M^{m+}} = 0$. The efflux concentration of N^{n+} is monitored with time, and k_f is developed. Subsequently, N^{n+} is desorbed from the exchange complex by miscible displacement with a solution where $C_{M^{m+}}$ is constant and $C_{N^{n+}} = 0$. Again the efflux concentration of N^{n+} is measured with time, and this information is used to develop k_b. When activities are determined for both solution and exchange phases, a true equilibrium constant, K_{ex}, is obtained (equation 8-1).

8.3 CHEMISORPTION

Chemisorption describes indeterminate processes leading to the distribution of chemicals between solution and solid phases. Equilibrium distribution of a chemical, Y, between soil solution and a sorbing phase is described by a distribution coefficient, K_d.

$$Y_{solution} \rightleftharpoons Y_{sorbed}$$

$$K_d = \frac{C_{Y_{sorbed}}}{C_{Y_{solution}}} \qquad [8\text{-}12]$$

The K_d is a ratio of disparate concentrations and is therefore a conditional equilibrium constant. Failure of K_d to describe partitioning has lead to a variety of empirical sorption relationships. Empirical relationships describing chemisorption may be applied to well-defined processes—such as exchange or adsorption to a surface, or precipitation–dissolution. They are most usefully employed, however, for describing the net distribution observed for sorption occurring along multiple paths or along pathways that are indistinct.

8.3.1 The Freundlich Equation

The Freundlich equation is a specific application of the power function, $f(x) = ax^k$, to the distribution of chemicals between solution and solid phases. The equation takes the form

$$K_f = \frac{C_{Y\,sorbed}}{C_{Y\,solution}^{1/n}}$$ [8-13]

where K_f and n are empirical fitting constants. The Freundlich equation is of greatest utility for linearization of sorption data that shows concentration-dependence when concentrations are varied over orders of magnitude.

Quasi-theoretical significance is imparted to the interpretation of Freundlich isotherms in soil systems, where the exponential coefficient $1/n$ is thought to describe the fraction of sorption sites that are available. When this interpretation is used, sorptive sites are viewed as being heterogeneous, such that sorption energy varies with the degree of site coverage. In actuality, the Freundlich equation fits most sorption measurements well, regardless of the mechanism involved. Its utility, therefore, arises from ability to generalize data rather than from any ability to correctly infer mechanism. As $n \rightarrow 1$, equation 8-13 reduces to equation 8-12 and $K_f \rightarrow K_d$.

8.3.2 The Langmuir and Multisite Langmuir Equations

The Langmuir equation describes sorption as a site specific phenomena involving monolayer coverage of homogeneous sites where sorbing molecules do not interact. The distribution of a sorbate molecule, Y, between the sorbing and solution phases is described,

$$K_1 = \frac{C_{Y\,sorbed}/(C_{Y\,max\ sorbed} - C_{Y\,sorbed})}{C_{Y\,solution}}$$ [8-14a]

The quantity of Y sorbed is expressed in equation 8-14a as the fraction of sorptive sites covered, $C_{Y\,sorbed}/(C_{Y\,max\ sorbed} - C_{Y\,sorbed})$, where $C_{Y\,max\ sorbed}$ is the concentration of Y sorbed at monolayer coverage. The coefficient K_1 is an empirical fitting parameter with units of L mol^{-1}. At any point along the isotherm, the distribution coefficient $K_d = K_1(C_{Y\,max\ sorbed} - C_{Y\,sorbed})$. Equation 8-14a can be rearranged to yield the more conventional form

$$C_{Y\,sorbed} = \frac{K_1 C_{Y\,solution} C_{Y\,max\ sorbed}}{1 + K_1 C_{Y\,solution}}$$ [8-14b]

The multisite Langmuir equation applies to the situation where two or more populations of noninteracting homogeneous sites participate in the observed sorption process. For the case of two populations of sorption sites, from equation

8-14b,

$$C_{Y_{sorbed}} = \frac{K_{li}C_{Y_{solution}}C_{Y_{i\ max\ sorbed}}}{1 + K_{li}C_{Y_{solution}}} + \frac{K_{lii}C_{Y_{solution}}C_{Y_{ii\ max\ sorbed}}}{1 + K_{lii}C_{Y_{solution}}} \qquad [8\text{-}15a]$$

For sites i and ii, respectively, $C_{Y_{i\ max\ sorbed}}$ and $C_{Y_{ii\ max\ sorbed}}$ represent the sorption maxima and K_{li} and K_{lii} are empirical fitting parameters having units of L mol^{-1}.

The various adjustable parameters of the multisite Langmuir equation describe the overall distribution coefficient, K_d, for the system. This can be shown from equation 8-15a by substituting $C_{Y_{sorbed}}$ K_d for $C_{Y_{solution}}$ and rearranging. This yields the following expression for sorptive distribution of Y in terms of K_d,

$$K_d^2 + (K_{li} + K_{lii})K_dC_{Y_{sorbed}} + K_{li}\ K_{lii}C_{Y_{sorbed}^2}$$

$$- (K_{li}\ C_{Y_{li\ max\ sorbed}} + K_{lii}\ C_{Y_{lii\ max\ sorbed}})K_d$$

$$- K_{li}\ K_{lii}C_{Y_{sorbed}}(C_{Y_{li\ max\ sorbed}} + C_{Y_{lii\ max\ sorbed}}) = 0 \qquad [8\text{-}15b]$$

Consider two conditions for the sorption isotherm described by the multisite Langmuir model: First, as $C_{Y_{sorbed}} \rightarrow 0$,

$$K_d^2 - (K_{li}\ C_{Y_{li\ max\ sorbed}} + K_{lii}\ C_{Y_{lii\ max\ sorbed}})K_d = 0 \qquad [8\text{-}15c]$$

and $K_d \rightarrow (K_{li}C_{Y_{li\ max\ sorbed}} + K_{lii}C_{Y_{lii\ max\ sorbed}})$. Second, as $C_{Y_{sorbed}} \rightarrow (C_{Y_{li\ max\ sorbed}} + C_{Y_{lii\ max\ sorbed}})$, $K_d \rightarrow (K_{li} - K_{lii})C_{Y_{li\ max\ sorbed}} - C_{Y_{sorbed}}$. Sorption data may be fit to the multisite Langmuir equation through a Scatchard plot of K_d on $C_{Y_{sorbed}}$ (Holford et al., 1974) or through nonlinear model fitting and optimization (Wolt et al., 1992).

The linear ($C_{Y_{sorbed}} = K_d\ C_{Y_{solution}}$, from equation 8-12), Langmuir (equation 8-14b), and multisite Langmuir (equation 8-15a) isotherms comprise a series of nested models describing sorptive partition. The models differ only in the number of parameters invoked to describe the distribution of data. L-curve isotherms that deviate significantly from linearity will be better described by the Langmuir and multisite Langmuir models than by the simple linear model. Selection of the appropriate nonlinear isotherm to describe L-curve isotherms is complicated, however, by the improved fit to data that occurs with increasing degree of model parameterization. Thus, simple goodness of fit of sorption data to these models cannot be used to infer mechanism. This problem can be approached in part through use of statistical measurements of improvement in model fit with increased model parameterization to select the most appropriate isotherm description (Wolt et al., 1992). Figure 8-2 illustrates the difficulty in using model fitting as a tool for interpretation of sorption mechanism when the sorption processes involved are not clearly understood. In cases such as this, model selection should be based largely on the ability to adequately describe the trend in data with the minimum numbers of parameters.

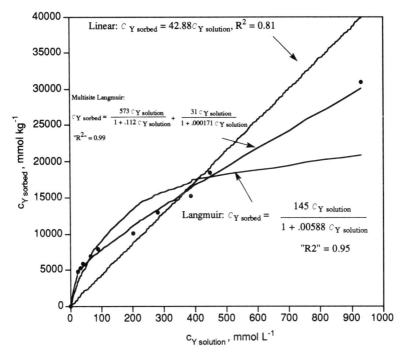

FIGURE 8.2 Chemisorption of SO_4^{2-} (●) by a thixotropic Typic Hydrandept subsoil as described by three different sorption isotherms.

8.3.3 Dynamic Description of Chemisorption

The use of equilibrium partitioning coefficients to describe chemical distribution in field environments is often inadequate for complex organic molecules. This is largely due to the failure for natural systems to approach equilibrium with respect to desorption from the soil solid phase. Examples of kinetically restricted sorption are discussed for xenobiotics in Chapters 3 and 14. Figure 14.2 presents an example of how K_d derived from conventional batch equilibrium measurements fails to adequately describe field observations of partition.

Kinetically restricted desorption from a slowly desorbed to a rapidly desorbed pool can be described in dynamic terms using a *bicontinuum approach*—this entails a two-compartment model describing the differentially reacting pools of a sorbed chemical. The bicontinuum approach hypothesizes a rapidly reacting compartment consisting of a fraction of the total sorbed phase pool ($fC_{Ysorbed}$) that is in instantaneous equilibrium with the soil solution. A second, slowly reacting compartment undergoes mass transfer with the rapidly reacting pool; this transfer is described by simple first-order kinetics. The overall reaction is

$$Y_{solution} \underset{k_1}{\overset{k_2}{\rightleftharpoons}} Y_{rapid} \underset{k_4}{\overset{k_3}{\rightleftharpoons}} Y_{slow}$$

The distribution between the rapidly reacting compartment of the sorbed Y and Y in soil solution can be described by an apparent distribution coefficient, K_d',

$$K_d' = \frac{C_{Y_{rapid}}}{C_{Y_{solution}}} \qquad [8\text{-}20]$$

Since k_1, $k_2 >> k_3$, k_4, the rate law for the rapidly reacting compartment is

$$\frac{\partial C_{Y_{rapid}}}{\partial t} = K_d' \frac{\partial C_{Y_{solution}}}{\partial t} \qquad [8\text{-}21]$$

The rate law for the reaction with respect to Y_{slow} is

$$\frac{\partial C_{Y_{slow}}}{\partial t} = k_3 C_{Y_{rapid}} - k_4 C_{Y_{slow}} \qquad [8\text{-}22]$$

The net partition of S between sorbed and solution phases is described by a partition coefficient, K_p, as

$$K_p = \frac{(C_{Y_{rapid}} + C_{Y_{slow}})}{C_{Y_{solution}}} = \frac{C_{Y_{sorbed}}}{C_{Y_{solution}}} \qquad [8\text{-}23]$$

Substituting equation 8-23 into equation 8-22, recalling f is defined as the fraction of sorbed Y in the rapidly reacting compartment, and rearranging results in an expression for the rate of change in the slowly reacting compartment,

$$\frac{\partial C_{Y_{slow}}}{\partial t} = k_3 K_p f C_{Y_{solution}} - k_4 C_{Y_{slow}} \qquad [8\text{-}24]$$

When the system is at equilibrium with respect to $C_{Y_{slow}}$,

$$\frac{k_3}{k_4} = \frac{C_{Y_{slow}}}{C_{Y_{rapid}}} = \frac{(1 - f)}{f} \qquad [8\text{-}25]$$

Substitution into equation 8-24 yields

$$\frac{\partial C_{Y_{sorbed}}}{\partial t} = k_4 [K_p (1 - f) C_{Y_{solution}} - C_{Y_{slow}}] \qquad [8\text{-}26]$$

Combining equations 8-22 and 8-26 provides an expression for the rate of change in total sorbed S,

$$\frac{\partial C_{Y_{sorbed}}}{\partial t} = K_d' \frac{\partial C_{Y_{solution}}}{\partial t} + k_4[K_p(1 - f)C_{Y_{solution}} - C_{Y_{slow}}] \qquad [8\text{-}27]$$

Two cases are instructive of the sorptive behavior described by the biocontinuum model. The first describes the pseudo-equilibrium observed when freshly added Y distributes between soil solution and the labile phase (that is, the rapidly reacting compartment). In this instance, $f \to 1$, $C_{Y_{slow}} \to 0$, and equation 8-27 reduces to equation 8-21. This is the condition when batch equilibration is performed for a chemical freshly added to soil.

The second case describes the approach to true sorptive equilibrium within the system. As $f \to 0$, $C_{Y_{rapid}} \to 0$, and, thus, $K_d' \to 0$. Equation 8-27 reduces to

$$\frac{\partial C_{Y_{sorbed}}}{\partial t} = k_4[K_p C_{Y_{solution}} - C_{Y_{slow}}] \qquad [8\text{-}28]$$

When the system is at equilibrium,

$$\frac{\partial C_{Y_{sorbed}}}{\partial t} = 0$$

and equation 8-28 yields equation 8-23. The partition coefficient K_p expresses the hypothetical desorptive K achieved as a chemical exhibiting significant nonequilibrium sorptivity approaches stasis in a soil. As Y_{sorbed} distributes from the rapid to slow reacting sorptive compartment with time of chemical aging in soil, f decreases and sorptive control shifts from the labile pool (described by K_d') to the slowly reacting pool (described by K_p) giving rise to the so-called "aged soil effect" on chemical sorptive partition.

Linear Free Energy Relationship (LFER). Equation 8-20 and 8-24 can be combined and rearranged to produce the log-transformed expression.

$$\log K_p = \log\left[\frac{C_{Y_{sorbed}}}{C_{Y_{solution}}}(1 - f)\right] - \log k_4 \qquad [8\text{-}29]$$

Equation 8-29 is a linear free energy relationship. A LFER is any relationship between logarithms of rate and equilibrium constants. More broadly defined a LFER is any extrathermodynamic relationship between thermodynamic quantities that does not follow from the fundamental relationships of thermodynamics (Exner, 1972).

The relationship between soil–water partitioning coefficient (K_p) and rate of desorption has been expressed as a LFER to describe nonequilibrium sorptive behavior of xenobiotics in soil as related to families of chemistry (Brusseau and Rao,

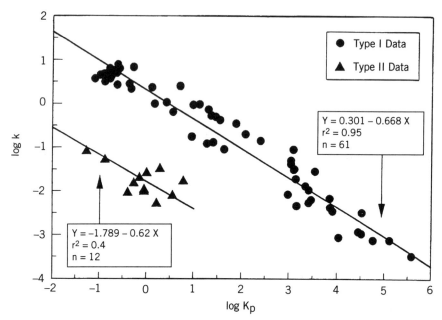

FIGURE 8.3 Relationship between sorption rate constant (k) and equilibrium sorption constant (K_p). [Reprinted from *Chemosphere*, 18, Brusseau and Rao, "The Influence of sorbate-organic matter interactions on sorption nonequilibrium," pp. 1691–1706, 1989, with kind permission from Pergamon Press Ltd, Headington Hill Hall, Oxford OX3 0BW, UK.]

1989, 1991). The LFER presented in Figure 8-3 was constructed from K_p and desorption rate constants (k) for a diverse range of xenobiotics. The data fall within two classes: Type I represents nonpolar, hydrophobic chemicals and Type II represents polar/ionizable, hydrophilic chemicals. The differing nature of LFER obtained for these two classes may be indicative of differing rate-controlling factors to the sorption/desorption process. Although such approaches offer some insight into processes governing chemical reactivity, it is important to recognize that the LFER represents linearization of the nonlinear expression developed by the biocontinuum approach. Thus, the insights offered by the LFER are only resolvable to within an order of magnitude.

CHAPTER 9

MINERAL STABILITY AND PEDOGENESIS

Soil mineral stability relationships and perturbations of these relationships through chemical weathering reactions are critical facets of pedogenesis (the formation of soils). Static approaches to the interpretation of mineral stability, and consequently pedogenesis, entail applications of idealized stationary state thermodynamic equilibrium models to predict solid phase minerals controlling composition of soil solution and the direction of weathering processes. If indeed soil systems were in a stationary state, composition of the soil solution would be invariant and mineral weathering would not occur. Over appropriate time scales, however, soil solution transfer rates are sufficiently small that a steady state approximation is applicable for thermodynamic equilibrium modeling of mineral stability relationships.

The usefulness of idealized stationary state thermodynamic equilibrium models to depict real soil systems "should be weighed by the extent to which the open system has been ... characterized ... in terms of solid and solution phase composition and the availability of reliable thermodynamic data." Discrepancies between observations for real soil systems and predicted relationships describing these systems "can be useful for evaluating the prevalence and direction of specific weathering reactions ... and mineral stability relationships" or identifying limitations of analytical data or thermodynamic predictors (Karanthanasis, 1989). The principal limitations to unambiguous application of equilibrium models to considerations of mineral stability and pedogenesis are improper or incomplete definition of consequential species in the solution and soild phases, failure to achieve steady state, and failure to account for effects of temperature and pressure.

Dynamic approaches to pedogenesis consider the rates of mineral weathering and attempt to determine if rates of mineral weathering are physically or chemically controlled. Physical weathering is a mechanical process where minerals are reduced to smaller sized particles without alteration in their chemical composition. Chemical weathering entails reactions (oxidation–reduction, hydrolysis, chelation) of minerals with air and water that result in chemical alteration of the mineral phase. Thus, chemical weathering reactions influence the mineral assemblages dominating soil bodies within landscapes. Chemical weathering occurs because equilibrium is seldom achieved between minerals and contacting aqueous phases in near-surface environments. The effects of chemical weathering on pedogenesis are manifested over time scales of months to years. At the level of clay mineral suites and the contacting soil solution within a soil body, the rates of chemical reactions are quite rapid and perturbations in soil solution composition arising from water flux are often quite minor. This allows for the assumption of localized equilibrium. Additionally, these conditions are achievable in laboratory settings, allowing for development of experimental models for pedogenic processes. For these instances, soil solution composition is a sensitive dynamic indicator of deviations in the extent and frequency of chemical reactions important to pedogenesis.

9.1 ION ACTIVITY PRODUCTS

The free energy relationship describing the dissolution of a solid phase mineral and composition of the contacting soil solution is described from Chapter 3 equations 3-31 and 3-47 at constant T and P as

$$\Delta F_r = \Delta F_r^\circ + \Sigma(RTln\ A_i)_{products} - \Sigma(RTln\ A_i)_{reactants} \qquad [9\text{-}1]$$

For example, considering the dissolution of gypsum

$$CaSO_4 \cdot 2H_2O_s \rightleftharpoons Ca^{2+}_{aq} + SO^{2-}_{4aq} + 2H_2O_l$$

equation 9-1 becomes

$$\Delta F_r = \Delta F_r^\circ + RT\ ln\ \frac{A_{Ca^{2+}_{aq}}A_{SO^{2-}_{4aq}}}{A_{CaSO_4 \cdot 2H_2O_s}} \qquad [9\text{-}2]$$

where

$$A_{Ca^{2+}_{aq}}A_{SO^{2-}_{4aq}}/A_{CaSO_4 \cdot 2H_2O_s}$$

is the reaction quotient, Q. The reaction quotient reduces to the ion activity product (IAP) of soil solution components [$= A_{Ca^{2+}_{aq}}A_{SO^{2-}_{4aq}}$] when the aqueous phase is infinitely dilute and the activities of the pure solid phase component and water equal 1. The IAP is calculated solely from ion speciation in the aqueous phase and can be expressed regardless of whether or not the solution components are in equilibrium with the solid phase. At equilibrium, $\Delta F_r = 0$ and equation 9-2 becomes

$$\Delta F_r^\circ = -RT \ln \frac{A_{Ca_{aq}^{2+}} A_{SO_{4aq}^{2-}}}{A_{CaSO_4 \cdot 2H_2O_s}} \qquad [9\text{-}3]$$

The right side of equation 9-3 can be expressed in terms of an equilibrium constant K° which is described as a solubility product (K_{sp}) for the case of a dissolution reaction. Equation 9-2 can therefore be expressed as

$$\Delta F_r = RT\ln(Q/K^\circ) = RT\ln(IAP/K_{sp}) \qquad [9\text{-}4]$$

Equation 9-4 assumes homogeneity of mineral species (ideal behavior in the standard state of the pure solid phase component). Therefore, activities of solid phase components $= 1$. For an ideal system approximating this assumption and achieving steady state equilibrium, $\Delta F_r = 0$ and $K_{sp} = IAP$ (the ratio $IAP/K_{sp} = 1$). If, for a real system, $IAP/K_{sp} > 1$, the solution is oversaturated relative to the mineral of interest and if $IAP/K_{sp} < 1$, the solution is undersaturated relative to the mineral of interest.

In a real system of soil and its contacting solution phase, deviation of IAP/K_{sp} (the *disequilibrium ratio*)[1] from 1 may indicate any of the following.

1. Failure of the system to achieve equilibrium
2. Inadequate expression of ion activities of soil solution components contributing to IAP
3. Improper selection of a reference K_{sp}
4. Failure of the postulated solid phase mineral to control ion composition of soil solution
5. Nonhomogeneity of the solid phase mineral such that the activity of solid phase components $\neq 1$

Generally, for considerations of soil mineral stability, disequilibrium ratios ranging from 0.1 to 1 are considered to confirm control of the postulated mineral over soil solution composition. This consideration is appropriate because of uncertainty in measurements of IAP and the appropriateness of applying K_{sp} obtained from chemical and geological literature to soils (Bohn and Bohn, 1987). Impurities and crystal defects are seldom of consequence for interpretation of mineral stability relationships for the dominant mineral components of soils because transition to new composition controlling minerals will occur before ion substitution can have a measurable effect on IAP.

9.1.1   Solid Phase Activities

Consideration of the nonideality (heterogeneity) of soil mineral phases is an approach to mineral stability evaluation where deviations from thermodynamic predications are considered to represent incomplete description of solid solution composition. The assumptions here are that the proper stasis of the system is achieved and soil solution IAP are correctly represented; thus, deviations in the ratio IAP/K_{sp} arise not from disequilibrium but from failure to account for deviations in solid solution activities from ideality. The consideration of solid phase activities appears most appropriate for trace mineral components of soils where nonhomogeneous mixing may lead to significant deviations from ideality (Bohn and Bohn, 1987).

If experimental conditions are such that the forgoing assumptions are valid, deviations of the disequilibrium ratio from 1 are indicative of deviations from ideality of the activities of solid phase components. The reaction quotient for this case incorporates the activity of solid phase components. The example of gypsum dissolution (equation 9-2) becomes

$$\Delta F_r = \Delta F_r^{\circ} + RT \ln \frac{A_{Ca_{aq}^{2+}} A_{SO_{4aq}^{2-}}}{C_{CaSO_4 \cdot 2H_2O_s} g_{gypsum}} \qquad [9\text{-}5]$$

where g_{gypsum} is the activity coefficient for solid phase gypsum. The analogous expression for equation 9-4 is

$$\Delta F_r = RT \ln \frac{Q}{K^{\circ}} = RT \ln C_i g_i \qquad [9\text{-}6]$$

where C_i and g_i represent the concentrations and activity coefficients, respectively, of the solid phase molecule of interest. Estimation of g_i is therefore achieved for a solid phase in equilibrium with a contacting aqueous phase by the relationship $g_i = IAP/(C_i K_{sp})$.

The solid phase activity represents the change in chemical potential resulting from transfer of a solute from an ideal to a real solid solution of equimolar composition. Factors affecting the solid phase activity of a mineral species are the lattice energy, the degree of hydration, and the homogeneity of the solid solution. Values of $g_i > 1$ for an ion species i indicate higher energy states for the solid solution than for the pure mineral, and a greater tendency for solution into the aqueous phase than for the ion in an ideal solid solution (Fu-Yong et al., 1992).

A further assumption necessary for determining solid solution activities is the consideration of what portion of the solid phase materially participates in the solution process with the contacting aqueous phase. For example, differing pools of solid phase aluminum and phosphate ions have been used for C_i in estimations of solid phase $AlPO_4$ activity. Blancher and Stearman (1984, 1985) used a measurement of labile phosphate to estimate the pool of P participating in chemisorptive processes at the soil surface. Fu-Yong et al. (1992) instead considered total solid phase aluminum and phosphate to participate materially in transfers between the solid and aqueous phases.

9.1.2 IAP and K_{sp} as Affected by Variation in Temperature and Pressure

A common simplifying assumption in thermodynamic representations of the near-surface environment—encompassing soils, the vadose zone, and surface and ground waters—is that variation in T and P from standard state conditions of 298.15 °K and 101.33 kPa is negligible. This appears reasonable for first approximations of mineral stability relationships in near-surface environments where uncertainties associated with experimental conditions, analytical sensitivity, and appropriateness of thermodynamic data probably introduce errors more substantial than the error associated with slight variations in T and P.

Indeed, the effect of variation in P is negligible at the near surface, as exemplified by the discussion of P effects on pH in section 4.1.2. Temperature variations in near-surface environments, however, are more substantial and will lead to shifts in chemical potential. This effect can be described as

$$\frac{\delta\mu_i/\delta T}{\delta T} = -\frac{H_i}{T^2} \qquad [9\text{-}7]$$

Therefore, refined considerations of mineral stability relationships need to correct measured IAP and reference K_{sp} to an equivalent basis of T (generally, to standard $T = 298.15°K$).

Consider, for example, the dissolution of solid phase $Al(OH)_3$ in a surface soil at 298.15 °K and in associated shallow groundwater at 285.15 °K.

$$Al(OH)_{3s} \rightleftharpoons Al^{3+}_{aq} + 3OH^-_{aq}$$

where pressure is constant at 101.33 kPa. Selection of a reference K_{sp} representative of $Al(OH)_{3s}$ in this environment could result in values ranging from $10^{-32.34}$ to $10^{-36.30}$, depending on the assumption of crystalline gibbsite or amorphous $Al(OH)_{3s}$ as the quasi-stable solid phase. The effect of T on K_{sp} for these species is presented in Figure 9-1.[2] The effect of temperature on K_{sp}, although significant, is secondary to the selection of the reference value of K_{sp} itself.

Ion distribution in soil solution—and thus, IAP—is similarly influenced by T. The distribution of total soil solution Al (Al_T) amongst free Al^{3+} and hydrolysis species is expressed

$$Al_T = A_{Al^{3+}} \left[\frac{1}{\gamma_{Al^{3+}}} + \frac{K_{AlOH^{2+}}}{\gamma_{AlOH^{2+}} A_{H^+}} + \frac{K_{Al(OH)_2^+}}{\gamma_{Al(OH)_2^+} A_{H^+}^2} + \frac{K_{Al(OH)_3^\circ}}{\gamma_{Al(OH)_3^\circ} A_{H^+}^3} + \frac{K_{Al(OH)_4^-}}{\gamma_{Al(OH)_4^-} A_{H^+}^4} \right] \qquad [9\text{-}8]$$

If the term in brackets is defined as R, equation 9-8 can be expressed as

$$A_{Al^{3+}} = Al_T R^{-1} \qquad [9\text{-}9]$$

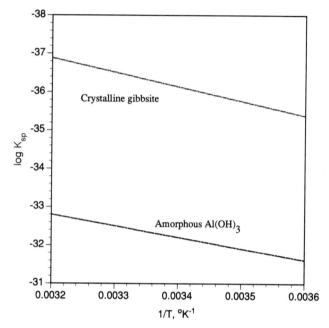

FIGURE 9.1 Effect of temperature variation (5 to 35°C) on the reference solubility product of $Al(OH)_{3s}$.

The value of R is sensitive to pH as well as T, since K_i, γ_i, and A_{H^+} are all temperature sensitive. Figure 9-2 illustrates the relative magnitude of the effect of pH and T on R when γ_i is constant and T affects only K_i. As pH increases, significant effects of T on R are observed for a given solution pH. Thus, sampling, analysis, and thermodynamic expression of $A_{Al^{3+}}$ should allow for temperature dependence on the result obtained.

9.2 STABILITY DIAGRAMS

Graphical representation of mineral stability relationships is a valuable tool for elucidating the nature and controlling influence of a soil solid phase in contact with soil solution. Additionally, mineral stability diagrams may provide insight as to mineral weathering sequences and pedogenic processes.

In order to find a means of graphical expression of mineral stability relationships, the solubility relationships among mineral phases of interest often must be determined in a manner that allows for the summarization of multiple variables using either two or three coordinate axes. Rigorous interpretation of stability diagrams requires that axes are independent. Occasionally, the construction of stability diagrams with dependent axes may offer useful interpretative insights regarding the status of a system, but caution must be used when interpreting diagrams prepared

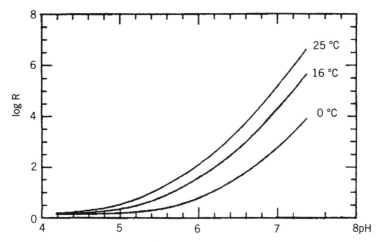

FIGURE 9.2 The effect of pH and temperature variation on log R, where R is the composite term correcting total solution Al to Al^{3+} activity. [After Neal, 1988. *J. Hydrology* 104:141–159. With permission of Elsevier Science Publishers.]

in this manner such that artifacts of the graphing procedure are avoided. The text by Garrels and Christ (1965) describes development of mineral stability diagrams in considerable depth. In this section, selected examples of stability diagram construction and interpretation specific to soil mineral relationships are presented.

The simplest type of stability diagram involves two variables and two axes. Stability relationships for $Al(OH)_{3s}$ are a good example of this type of diagram. The reaction of interest is

$$Al(OH)_{3s} \rightleftharpoons Al^{3+}_{aq} + 3OH^-_{aq}$$

which at equilibrium is described by the expression

$$K^\circ = A_{Al^{3+}} A^3_{OH^-} \qquad [9\text{-}10a]$$

This can be expressed in terms of negative logarithms ($p \equiv -\log$) as

$$pK^\circ = pAl^{3+} + 3pOH^- \qquad [9\text{-}10b]$$

If this expression is recast in terms of pH ($= pK_w - pOH^-$), the result is more readily recognizable in terms of soil pH.

$$pK^\circ = pAl^{3+} + 3(pK^\circ_w - pH) \qquad [9\text{-}10c]$$

$$pAl^{3+} = pK^\circ - 3pK^\circ_w + 3pH \qquad [9\text{-}10d]$$

A graph of pAl^{3+} on pH relative to stability lines describing the domain of stable

$Al(OH)_3$ minerals indicates whether free Al^{3+} and H^+ are in equilibria with respect to solid phase $Al(OH)_3$ having $pK_{sp} = pK°$. A graphical representation of pAl^{3+} versus pH is presented in Figure 9-3 for natural waters as compared to $Al(OH)_3$ minerals postulated as quasi-stable species in the systems sampled.

9.2.1 Layer Silicates and Crystalline Oxyhydroxides

Compositional complexity of the soil system precludes graphical summarization of mineral stability relationships for soil solids in toto; rather, selection of useful subsets of soil components is employed, allowing for a focus on specific mineral groups of interest. A focus on Al and Si, as major structural components of soil minerals, and pH, as an important influence over soil weathering reactions, allows for useful graphical representation of layer silicate and oxyhydroxide mineral species that predominate moderately to highly weathered soils.

Al_2O_3—SiO_2—H_2O System. The system comprised of the proximate components Al_2O_3—SiO_2—H_2O was originally applied by Kittrick (1969) to describe mineral relationships in a two-phase system of soil mineral and soil solution. From the phase rule, $F = C + 2 + R - P$, there is one degree of freedom—three components (C), one reaction (R, precipitation–dissolution), two phases (P), and

(b)

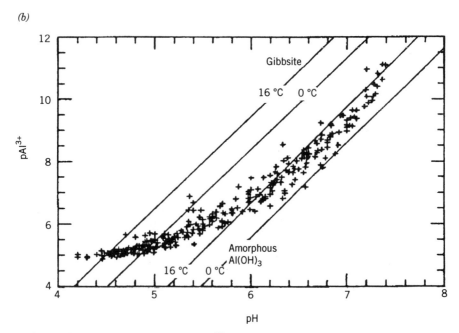

FIGURE 9.3 Stream water pH and pAl^{3+} in comparison to the stability of gibbsite and amorphous $Al(OH)_3$ at two temperatures. [Neal, 1988. *J. Hydrology* 104:14–159. With permission of Elsevier Science Publishers.]

temperature, pressure, and the chemical potential of water are fixed. Thus, the status of the system at equilibrium is fully described by one component (Al_2O_3 or SiO_2).

For graphical expression, Al_2O_3 or SiO_2 are recast in terms of $Al(OH)_3$ and H_4SiO_4. The selection of $Al(OH)_3$ and H_4SiO_4 is based on the addition of H_2O (constant chemical potential) to Al_2O_3 and SiO_2, respectively,

$$Al(OH)_3 = 1/2\ Al_2O_3 + 3/2\ H_2O$$

$$H_4SiO_4 = SiO_2 + H_2O$$

In terms of negative logarithms, $Al(OH)_3$ and H_4SiO_4 are expressed as $[pAl^{3+} + 3pOH^-]$ and pH_4SiO_4, respectively. The term $[pAl^{3+} + 3pOH^-]$ is frequently recast in the traditional notation $pH - 1/3\ pAl$ (aluminum hydroxide potential; Schofield and Taylor, 1955). From equation 9-10d,

$$pH - 1/3pAl = pK_W^\circ - 1/3pK^\circ \qquad [9\text{-}10e]$$

It is possible to express a number of aluminosilicate and aluminum oxyhydroxide minerals as a function of $[pH - 1/3pAl]$ and pH_4SiO_4. These may then be graphically expressed as in Figure 9-4.

Figure 9-4 graphically summarizes a great deal concerning soil mineral weathering. Since soil weathering results in Si mobilization and Al accumulation, pH_4SiO_4 is selected as the independent variable. The activity of H_4SiO_4 in soil solution is controlled by dissolution of unstable primary minerals, formation of stable secondary minerals such as montmorillonite and kaolinite, and rate of H_4SiO_4

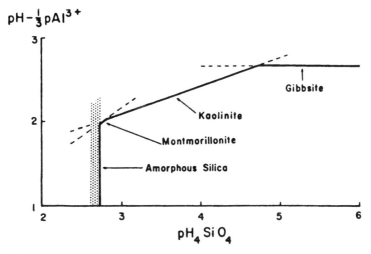

FIGURE 9.4 Composite stability diagram for some minerals in the Al_2O_3—SiO_2—H_2O system at 298.15°K and 101.3 kPa. The stability line of each mineral is solid where it is the most stable mineral of the group [Kittrick, 1969. Permission of Clay Minerals Society.]

removal from the soil environment via leaching and plant uptake. The pH-dependent accumulation of Al-enriched minerals occurs as a consequence of weathering of Si-bearing minerals and H_4SiO_4 removal from the system; thus, the pH-linked Al activity appears as the dependent variable in Figure 9-4. The stability diagram illustrates a weathering sequence typical of many soils where as pH_4SiO_4 increases (decreasing activity of H_4SiO_4 in soil solution), control of soil solution Si shifts from amorphous Si to secondary aluminosilicate minerals. Aluminum hydroxide potential in soil solution increases with decreasing H_4SiO_4 activity to the point where gibbsite occurs as a stable mineral species. Minerals adjacent to one another within the stability field can form assemblages. Points of intersection of stability lines represent an invariant condition where two associated minerals occur as a stable mixture at equilibrium. Shifts to either side of this point indicate one mineral in the association is dissolving while the other is precipitating.

Figure 9-5 is an expanded version of Figure 9-4 where the minerals represented are amorphous SiO_2, quartz, soil montmorillonite (SM), hydroxyinterlayered soil montmorillonite (HISM), kaolinite (KLN), hydroxyinterlayered vermiculite (HIV), and gibbsite (GIB). Inclusion of SM, HISM, and HIV in this graph requires the assumption of fixed soil solution composition for certain minor ion constituents of these minerals such as K^+, Ca^{2+}, Mg^{2+}, and Fe^{3+}. Overlain on the stability diagram are points representing analysis of soil solutions displaced from Ultisols. The concurrence of these data points with the stability fields of the minerals represented is instructive of the solid phase mineralogy supporting the observed soil solution composition. For example, the North Carolina Ultisols (■) cluster along the inter-

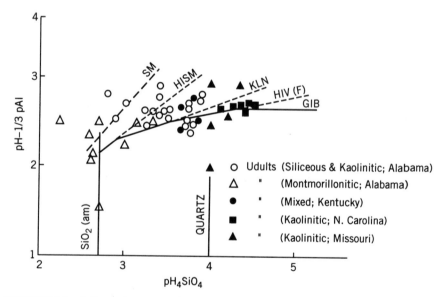

FIGURE 9.5   Soil solution compositions superimposed on a stability diagram of representative soil minerals identified in Ultisols with variable mineralogical composition. [Karanthanasis, 1989]

section of the HIV-gibbsite stability lines, which indicates these two minerals are most likely in stable association in these soils. On the other hand, Alabama Ultisols from siliceous or kaolinitic taxonomic classes ($\bigcirc$) tend to be oversaturated with respect to aluminosilicate minerals (aluminum hydroxide potentials are higher than can be supported by stable mineral phases).

K_2O—Al_2O_3—SiO_2—H_2O System. Consideration of the ratio of A_{H^+} to the activities of base cations in soil solution is instructive of the mineral status of soils, as soil weathering will result in a loss of base cations and an increase in H^+ during pedogenesis. The stability relationships among potassium-aluminum-silicates, gibbsite, and kaolinite are represented in Figure 9-6. For these minerals, a system comprised of four components (K_2O—Al_2O_3—SiO_2—H_2O) can be defined where precipitation–dissolution is the only reaction considered and where T, P, and the chemical potential of H_2O are constant. Additionally, for this system, the chemical

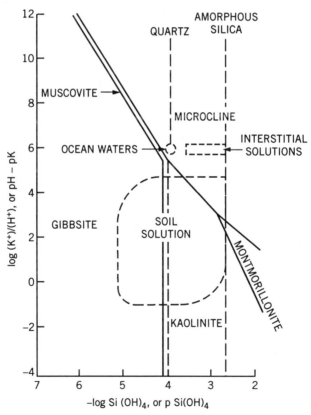

FIGURE 9.6 The K_2O—Al_2O_3—SiO_2—H_2O system at 298.15°K and 101.3 kPa, superimposed on the ranges of compositions of soil solutions, oceans, and rock interstitial solutions. [From Bohn et al., 1985.]

potential of Al_2O_3 is constant as well, as Al is conserved in the mineral phase and does not materially participate in the dissolution–precipitation reactions. The proximate components SiO_2 and K_2O react with water as follows:

$$5/2H_2O + SiO_2 + K^+ \rightleftharpoons H_4SiO_4 + H^+ + 1/2K_2O$$

$$pH_4SiO_4 = -(pH - pK^+) + \text{constant}$$

Thus, mineral relationships are described as a consequence of Si solubilization and the transfer of H^+ and K^+ between solid and solution phases as a function of pH_4SiO_4 and $[pH - pK^+]$.[4]

Superimposed on the stability relationships of Figure 9-6 are representative ranges in composition of soil solutions, ocean waters, and rock interstitial solutions. Weathering of soils results in Si mobilization from soils and favors accumulation of base-depleted secondary minerals. The divergence of soil solution composition from that of natural waters in contact with primary minerals is evidence of the unique chemical nature of soils imparted by the prevalence of secondary minerals as the products of pedogenesis.

Figure 9-6 also illustrates the ability for many soil solutions to exhibit activities of silicic acid in soil solution well below that predicted by quartz solubility. Quartz, however, dominates many soils where it is the major component of the sand-size fraction (0.05 to 2 mm diameter particles). This seeming discrepancy arises because reduced surface-to-volume ratios with increasing particle size physically restrict the rates of mineral dissolution. Thus, large portions of quartz minerals in soils are not reacting with soil solution (see section 9.3, Rates of Mineral Weathering).

9.2.2   Carbonates and Soluble Salts

Carbonates and soluble salts occur as consequential components of soils in arid regions. When evaporation exceeds rainfall, the mobilization of bases from the soil profile is restricted and carbonates and soluble salts accumulate. Soil solutions and natural waters contacting a solid phase carbonate mineral and open to the atmosphere will be buffered against shifts in pH through complex interactions among soil solution components, solid phase carbonates, and dissolved CO_{2_g}.

The complexity of carbonate buffering frequently results in use of simplified systems to describe a specific case. For example, in Chapter 4 (section 4.1.3) several cases were described relating soil solution displacement technique, dissolved CO_{2_g}, and soil solution pH. In these cases, an important simplifying assumption was that the effect of a solid-phase carbonate mineral was not considered. The approach presented here follows that of Robbins (1985) and provides a more generalizable description of soil carbonate equilibria.

CaO—CO₂—H₂O System. This system is comprised of three components in three phases and involves two reactions (precipitation–dissolution occurring between solid and liquid phases, and gas dissolution occurring between gas and liquid phases). The system has two degrees of freedom when T and P are fixed. Thus, any one component is dependent on the other two in the system for its description.

The key equations describing CO_2 partitioning between gas and liquid phases and the distribution of carbonate species in the liquid phase are summarized in Table 4.4. Assuming the system behaves ideally, the equations in Table 4.4 and equation 4-17 can be expressed in terms of activities and rearranged to yield the following expression that relates $P_{CO_{2g}}$, A_{H^+}, and $A_{CO_3^{2-}}$ for the system at equilibrium.

$$P_{CO_{2g}} = \frac{A_{H^+}^2 A_{CO_3^{2-}}}{K_H K_1 K_2}$$ [9-11]

Carbonate in solution will also be in equilibrium with solid phase $Ca(CO)_3$.

$$Ca(CO)_{3s} \rightleftharpoons Ca^{2+} + CO_3^{2-}$$

$$K_{sp} = A_{Ca^{2+}} A_{CO_3^{2-}}$$ [9-12]

Combining equations 9-11 and 9-12,

$$P_{CO_{2g}} = \frac{K_{sp}}{K_H K_1 K_2} \frac{A_{H^+}^2}{A_{Ca^{2+}}}$$ [9-13]

If the various equilibrium K are grouped as K_C,

$$P_{CO_{2g}} = K_C \frac{A_{H^+}^2}{A_{Ca^{2+}}}$$ [9-14a]

Rearranging and expressing in terms of negative logarithms,

$$pH = 1/2 pP_{CO_{2g}} + 1/2 pCa^{2+} - 1/2 pK_C$$ [9-14b]

where $pK_C = pK_{sp} - [pK_H + pK_1 + pK_2]$. Using values of pK_H, pK_1, pK_2 (Table 4.6), and pK_{sp} ($= 8.42$, Stumm and Morgan, 1981) at 298.15 °K, the value of pK_C is -14.737. Equation 9-14b becomes

$$pH = 1/2 pP_{CO_{2g}} + 1/2 pCa^{2+} + 7.3685$$ [9-15]

The stability relationship among pH, $pP_{CO_{2g}}$, and pCa^{2+} can be expressed graphically as the surface shown in Figure 9-7. A change in any one variable results in changes in the two related variables in order that equilibrium be maintained in the system. If any one variable is held constant then interrelationships between the other two become apparent. For example, if the system is held in equilibrium with atmospheric CO_{2g} (33 Pa, $pP_{CO_{2g}} = -1.52$) and $A_{Ca^{2+}}$ from dissolution of pure

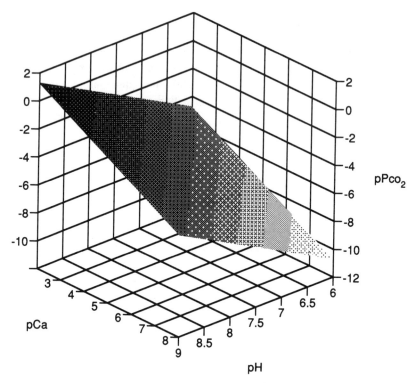

FIGURE 9.7 Stability relationships in the CaO—CO_2—H_2O system at 298.15°K and 101.3 kPa.

calcite (pCa^{2+} = 4.21), the pH 8.3 will be observed. If pCa^{2+} is instead 2.30 (reflecting typical $C_{Ca^{2+}}$ = 0.01 mol L^{-1} and $\gamma_{Ca^{2+}}$ = 0.5 L mol^{-1} for calcareous soils), soil pH will be 7.8 when in equilibrium with atmospheric CO_{2g}. Next, if respiration by soil biota raises CO_{2g} in the soil atmosphere to 300 Pa ($pP_{CO_{2g}}$ = −2.48), pH will either decrease to 7.3 when pCa^{2+} remains constant, or pCa^{2+} will increase to 5.74 ($A_{Ca^{2+}}$ decreases to 1.82×10^{-6}) when pH remains constant. As Figure 9-7 indicates, the actual response of the system may likely be a shift in both pH and pCa^{2+} somewhat intermediate between these two extreme cases.

When soil solution IAP > K_{sp} (oversaturation with respect to calcite), the plot of pH, $pP_{CO_{2g}}$, and pCa^{2+} for that solution will be a point falling below the surface described in Figure 9-7. If the system is undersaturated with respect to calcite, IAP < K_{sp}, and the plotted point will be above the surface.

Na_2O—CO_2—H_2O System. Soluble Na-salts may accumulate in arid and semi-arid soils where Na^+ exceeds the soil cation exchange capacity. Soil pH in these soils may be controlled by the hydrolysis of Na^+ released from exchange sites or by solubilization. The upper limit to soil pH that may be achieved in these saline-sodic soils is defined by Na_2CO_3 in the Na_2O—CO_2—H_2O system. Accumulation of Na_2CO_3 in saline-sodic soils is possible due to reactions of Na^+ with atmospheric CO_2. Additionally, sodium- and carbonate-rich solutions may occur from sulfate reduction in anaerobic groundwaters, and these may result in Na_2CO_3 evaporites following seepage of these solutions to surfaces. Environmentally significant amounts of $NaCO_3$ may also accrue from disposal of "red muds" that occur as by-products of bauxite ore processing.

Equilibria relationships in this system can be described in the same manner as described for the $CaO-CO_2-H_2O$ system. By analogy with equation 9-15,

$$pH = 1/2pP_{CO_2} + pNa - 1/2pK_C \qquad [9\text{-}16]$$

where $pK_C = -23.886$, reflecting a pK_{sp} for Na_2CO_{3s} of 0.729. For the system at equilibrium with atmospheric CO_2 and pure Na_2CO_{3s}, $pH = 10.6$.

In carbonate-rich environments, a number of hydrated species of Na_2CO_{3s} may form. The activity of water departs significantly from unity in strongly saline systems and defines the species of sodium carbonate present (Garrels and Christ, 1965),

$$Na_2CO_3 \cdot xH_2O_s + yH_2O \rightleftharpoons Na_2CO_3 \cdot (x + y)H_2O_s$$

$$K^\circ = 1/A_{H_2O^y} \qquad [9\text{-}17]$$

Figure 9-8 illustrates stability relationships among sodium carbonate minerals as a function of A_{H_2O}.

9.2.3 Amorphous Minerals and Sparingly Soluble Salts

The use of soil solution composition to discern mineral stability relationships of trace components of soils is more tenuous than for the forgoing cases of aluminosilicate minerals and carbonates. This is because physical evidence for the presence of amorphous minerals and sparingly soluble salts in the soil solid phase is frequently lacking. Herein lies both a strength and weakness of extending soil solution compositional analysis to trace constituents of soils. Soil solution compositional analysis can be used to make useful predictive insights concerning the composition of the soil solid phase where other analytical approaches are lacking, but these predictions frequently cannot be validated other than through the frequency with which they appear to hold true.

$Al_2O_3-H_2SO_4-H_2O$ System. The chemical control of sulfate retention-release in acid soil environments is important for understanding control of soil solution SO_4^{2-} under conditions where S may be limiting for plant growth, and the control of solution Al^{3+} and H^+ for soils impacted by anthropogenic sources of acid sulfate. Both Fe- and Al-hydroxy-sulfate minerals are important in this regard, but the occurrence of these sparingly soluble trace minerals can only be indirectly inferred through soil solution chemistry.

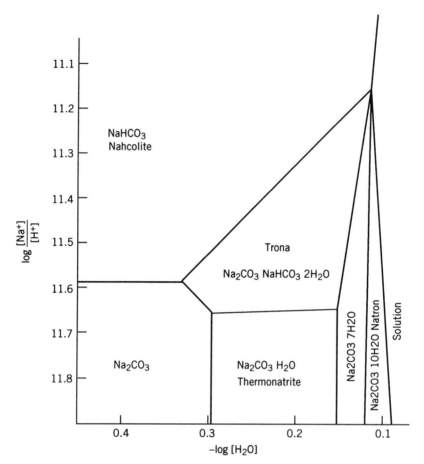

FIGURE 9.8 Stability relationships in the Na_2O—CO_2—H_2O system at 298.15°K and 101.3 kPa. [Garrels and Christ, 1965]

The mineral stability relationshps of Al-hydroxy-sulfate minerals can be developed for a three-component system (Al_2O_3—H_2SO_4—H_2O) comprising one reaction (precipitation–dissolution) and two phases (soil mineral and soil solution). When temperature, pressure, and chemical potential of water are constant, the system has one degree of freedom and can be defined from H_2SO_4 and Al_2O_3.

Choice of the independent variable is dictated by the question being addressed experimentally. In the following example, the effects of anthropogenic acid sulfate on soil Al are considered. Therefore, H_2SO_4 is selected as the dependent variable because its fluctuation as a consequence of natural or anthropogenic inputs of acid and sulfate will cause shifts in the status of solution-solid phase Al distribution. Conversely, for considerations of sulfate availability in acid soils, it may be desirable to consider pH-dependent Al instead as the independent variable controlling SO_4^{2-} availability in soil solution.

Following the convention established for description of alminosilicate stability, minerals of interest can be resolved relative to acid sulfate potential, $2pH + pSO_4$, and aluminum hydroxide potential, $pH - 1/3pAl$. Three Al-hydroxy-sulfate minerals postulated to be of consequence with respect to anthropogenic acid sulfate are alunite [$KAl_3(OH)_6(SO_4)_2$], basaluminte [$Al_4(OH)_{10}SO_4$], and jurbinite [$AlSO_4OH \cdot 5H_2O$]. Equilibrium phase distribution of these minerals as functions of acid sulfate potential and aluminum hydroxide potential are expressed as

alunite

$$KAl_3(OH)_6(SO_4)_2 \rightleftharpoons K^+ + 3Al^{3+} + 6OH^- + 2SO_4^{2-}$$

$$pH - 1/2pAl = -1/9([pK° - 5pK_w] - [pK^+ + pOH] - 2[2pH + pSO_4])$$

[9-18]

When $pK° = 85.4$, $pK_w = 14$, and $[pK^+ + pOH] = 14$,

$$pH - 1/3pAl = -0.16 + 0.22[2pH + pSO_4]$$ [9-19]

basaluminite

$$Al_4OH_{10}SO_4 \rightleftharpoons 4Al^{3+} + 10OH^- + SO_4^{2-}$$
$$pH - 1/3pAl = -1/12 ([pK° - 10pK_w] - [2pH + pSO_4])$$ [9-20]

When $pK° = 117.6$ and $pK_w = 14$,

$$pH - 1/3pAl = 1.87 + 0.08[2pH + pSO_4]$$ [9-21]

jurbanite

$$AlOHSO_4 \cdot 5H_2O \rightleftharpoons Al^{3+} + OH^- + SO_4^{2-} + 5H_2O_l$$
$$pH - 1/3pAl = -1/3 ([pK° - pK_w] - [2pH + pSO_4])$$ [9-22]

When $pK° = 3.8$ and $pK_w = 14$,

$$pH - 1/3pAl = 1.27 + 0.33[2pH + pSO_4]$$ [9-23]

Two additional minerals of interest in describing soil $Al(OH)_3$ potential are gibbsite and kaolinite.

gibbsite

$$Al(OH)_3 \rightleftharpoons Al^{3+} + 3OH^-$$

$$pH - 1/3pAl = -1/3(pK^\circ - 3pK_W) \qquad [9\text{-}10f]$$

When $pK^\circ = 34.0$ and $pK_W = 14$,

$$pH - 1/3pAl = 2.66 \qquad [9\text{-}24]$$

kaolinite

$$Al_2Si_2O_5(OH)_4 + 5H_2O \rightleftharpoons 2Al^{3+} + 2H_4SiO_4 + 6OH^-$$

$$pH - 1/3pAl = -1/6 ([pK^\circ - 6pK_W] - 2pH_4SiO_4) \qquad [9\text{-}25]$$

When $pK^\circ = 76.4$, $pK_W = 14$, and $pH_4SiO_4 = 3.5$,

$$pH - 1/3pAl = 2.421 \qquad [9\text{-}26]$$

These relationships are graphically presented as a stability diagram in Figure 9-9. Superimposed on this stability diagram are compositional data for soil solutions displaced from soils exposed to high levels of natural and/or anthropogenic sulfate.

Soils from the Belgian Ardennes represent forested sites where spruce and beech exhibit decline (van Praag and Weissen, 1984), perhaps in part due to acid sulfate deposition. Soils from Kentucky are sampled from watersheds where coal was surface mined in the 1960s and 1970s (Karanthanansis et al., 1988). The soils from the Cooper Basin region of Tennessee (Wolt, 1981) received historically high inputs of acid sulfate deposition from open air smelting of copper ore in the period 1854 through 1910. There is a striking similarity in the relationship of sulfate potential to aluminum hydroxide potential for many of the displaced soil solutions from these various sites. These data suggest that alunite may be a stable mineral phase in these soils acting to control the soil solution aluminum hydroxide potential.

Al_2O_3—CaO—P_2O_5—H_2O System. The reactions governing phosphate retention and availability in soils can be described by the solubility of sparingly soluble metal-hydroxyphosphate minerals that extend pH-dependent control over soil solution phosphate activity. In acid soils, iron and aluminum phosphates appear to control soil solution phosphate; in slightly acid to alkaline soils, control shifts to calcium phosphates. The highest availability of soil P occurs at slightly acid soil pH where control over soil solution phosphate activity shifts from iron and aluminum phosphate to calcium phosphates.

An example of pH-dependent P availability can be developed for the system Al_2O_3—CaO—P_2O_5—H_2O where varisicite [$AlH_2PO_4(OH)_2$] and hydroxyapatite [$Ca_5(PO_4)_3OH$] are postulated as the solid phase minerals controlling P solubility.

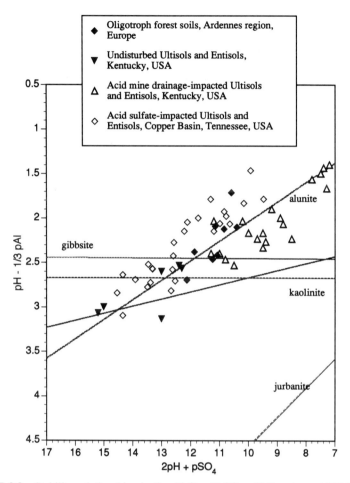

FIGURE 9.9 Stability relationships in the Al_2O_3—H_2SO_4—H_2O system at 298.15°K and 101.3 kPa, superimposed on compositions of soil solutions exposed to anthropogenic acid sulfate. Soil solution compositions are from Wolt (1981), Van Praag and Weissen (1985), and Karanthanansis et al. (1988).

For variscite,

$$AlH_2PO_4(OH)_{2s} + 2H^+ \rightleftharpoons Al^{3+} + H_2PO_4^- + 2H_2O$$

$$pK° = pAl + pH_2PO_4 - 2pH \qquad [9\text{-}27]$$

and for hydroxyapatite,

$$Ca_5(PO_4)_3OH_s + 7H^+ \rightleftharpoons 5Ca^{2+} + 3H_2PO_4^- + 2H_2O$$

$$pK° = 5pCa + 3pH_2PO_4 - 7pH \qquad [9\text{-}28]$$

In acid soils, the dominant species of soil solution phosphate is $H_2PO_4^-$; therefore, stability of variscite and hydroxyapatite in acid soils can be expressed graphically with pH as the independent variable and pH_2PO_4 as the dependent variable (Fig. 9-10). The stability relationship in Figure 9-10 uses $pK°$ of 2.5 and -14.46 for variscite and hydroxyapatite, respectively. This system can be described by two variables when Al_2O_3 and CaO are held constant ($pAl - 3pH = 9.3$ and $pCa = 2.5$, representing values characteristic of the soil being described). Thus, for variscite,

$$pH_2PO_4 = 11.8 - pH \qquad\qquad [9\text{-}29]$$

and for hydroxyapatite,

$$pH_2PO_4 = -8.99 + 2.33pH \qquad\qquad [9\text{-}30]$$

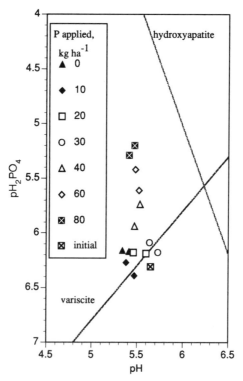

FIGURE 9.10   Soil solution P in an acid Hapludalf in relation to phosphate mineral stability and rate of applied P for samples obtained 25 days following application to cropped soils at an average temperature of 26°C. [After Hardin et al., 1989]

Solving equations [9-29] and [9-30] for pH,

$$11.8 - pH = -8.99 + 2.33pH \qquad [9\text{-}31]$$

represents the case where pH_2PO_4 is maximized in the system (pH6.2).

Overlain on the stability diagram in Figure 9-10 are measurements of soil solution pH and pH_2PO_4 for an acid Hapludalf fertilized with monocalcium phosphate, a highly soluble form of P. The pH_2PO_4 in soil solutions displaced 25 days following low rates of P fertilization show rapid equilibration with native P (variscite). In contrast, higher rates of P fertilization (> 30 mg kg^{-1}) result in progressively higher levels of soil solution pH_2PO_4 where added P has yet to achieve equilibrium with soil P.

Manganese Solubility Relationships. The final example of mineral stability relationships involves the prediction of solid phase control over soil solution Mn^{+2}. Soil Mn exhibits a number of valence states and, consequently, a variety of solution and solid phase species (Chapter 12). Thermodynamic relationships for manganese minerals potentially important for the control of soil solution Mn^{2+} are well defined and involve a variety of oxides, carbonates, silicates, and sulfates (Lindsay, 1979). It is difficult to describe adequately, or control experimentally, the many variables in the soil environment that may influence Mn solubility relationships because of the host of various Mn mineral types that must be considered. For example, Figure 9-11 graphically represents stability relationships of Mn minerals as a function of pH and pMn^{2+}. In order to represent the minerals of interest graphically, values of pe, $pP_{O_{2g}}$, $pP_{CO_{2g}}$, and pH_4SiO_4 must be held constant. The result is that the solution composition represented by Figure 9-11 is rigidly defined and therefore has restricted utility for comparison of soil solutions obtained from diverse environments.

Overlain on the mineral stability relationships of Figure 9-11 are values of soil solution pH and pMn^{2+} for limed Inceptisols and Alfisols. Soil solution pMn^{2+} is highly correlated on pH, but in a manner clearly unrelated to any of the Mn minerals postulated to be controlling solution Mn^{2+}. The system has not been adequately described to address the complexity of the solubility relationships being considered; organic carbon, the exchange complex, soil biota, solution phase complexation reactions, or an alternative Mn mineral may be responsible for the observed effect.

Figure 9-12 illustrates another format for graphical resolution of Mn solubility relationships. Here, distribution of Mn among soil solution Mn^{2+} and solid phase Mn oxide, oxyhydroxide, carbonate, and sulfide are described as a function of the master variables pH and E_H ($= 2.303$ RTpe/$\mathcal{F}$). This form of representation has the advantage of describing Mn distribution in soil as a function of redox status. Oxidized forms of Mn dominate the system at low pH or high E_H, while reduced forms of Mn solids dominate the system as pH and E_H decline.

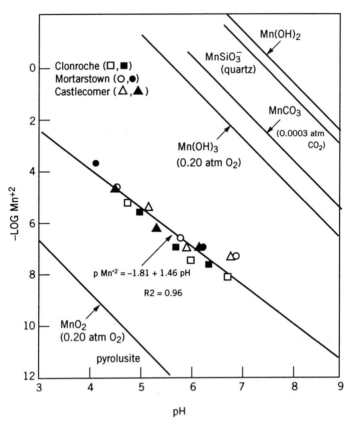

FIGURE 9.11   Stability relationships for selected Mn minerals at 298.15°K and 101.3 kPa, superimposed on compositions of acid Inceptisol and Alfisol soil solutions treated with lime. [Curtin and Smillie, 1983. Published in *Soil Sci. Soc. Am. J.* 47:701–707. 1983. Soil Sci. Soc. Am.]

9.3   RATES OF MINERAL WEATHERING

Soil mineral weathering results from release of ions to soil solution and the subsequent transport of ions away from mineral surfaces. Physicochemical processes that may control rate of mineral weathering are temperature, leaching and internal drainage, soil solution pH and redox potential, soil biota, particle size and specific surface, and intrinsic mineral weatherability (Jackson and Sherman, 1953). The physical manifestation of these physicochemical processes is the suite of minerals extant in a given soil system. Mineral stability relationships (section 9.2) describe the direction and extent of mineral weathering; *ion mobility*—the relative ability for ions to release into soil solution and subsequently leach—describes the rate of mineral transformations. The relative mobility of ions important to weathering of

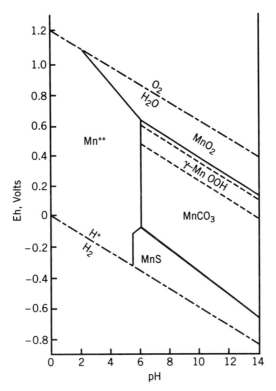

FIGURE 9.12 E_H—pH diagram for Mn minerals and Mn^{2+} at $P_{CO_2} = 33$ Pa (dashed line) and 20.26 kPa. [Bohn et al., 1985.]

dominant soil materials is

$$Na, K, Mg > Si > Fe, Al$$

with Ca being either mobile or immobile dependent on CO_{2g} as defined by calcium carbonate equilibria (see CaO—CO_2—H_2O System in section 9.2) (Chesworth, 1973; Birkeland, 1974).

Studies of mineral weathering rate, therefore, focus on methods for estimation of ion mobility in soil systems. A few examples of these methods will be presented here.

9.3.1 Approaches to Measurement of Mineral Weathering Rates

As described by White et al. (1990) absolute determinations of mineral weathering rates are not possible due mostly to the time scale involved. Various indirect methods are used instead that describe mineral weathering from the standpoint of occurrence of the products of weathering (ions) in solution over time. The methods used range from conceptual approaches—based on mineral solubility and observations of mineral suites occurring in nature—to laboratory investigations with isolated minerals, column leaching studies, and to greenhouse and field investigations. The various approaches used allow for varying degrees of experimental control. Thus, for example, laboratory investigations with pure minerals allow for determination of rates of dissolution as controlled by physicochemical variables to weathering, and greenhouse studies allow for inclusion of biotic effects in these considerations.

Because of the complexity of the processes involved, field studies are the only true measurement of natural weathering rates. However, field studies suffer from high variability, lack of control, and lack of sensitivity. Soil solution or leachate composition reflects the net weatherability of a suite of minerals. The rate of change in soil solution or leachate composition in an open system is the integrated result over the entire soil system. For natural systems, such as a watershed, the change in leachate composition with time may reflect control by any of a variety of sinks/ sources within the system. In the upper soil profile, organic matter may control mineral release and this control may be evidenced in the composition of soil water draining laterally from the watershed. Leachates percolating through the subsoil, however, may be more influenced by soil mineralogy and may thus better reflect mineral release due to weathering.

Watershed scale studies of nutrient release use mass-balance approaches describable as

$$\text{Weathering rate} = (\text{Efflux} - \text{Influx}) + \Delta \text{ Net storage in plant and soil pools}$$

Measurements of flux or storage are hampered by great uncertainty. White et al. (1990) show, for example, that an imperceptible change in one pool (a change in exchangeable Ca^{2+} of 0.02 cmol kg^{-1}) is sufficient to account for the annual Ca weathering rate (24 kg Ca ha^{-1}) for a typical watershed.

Because of the problems inherent in various experimental approaches to measuring mineral weathering rates, knowledge concerning soil mineral weathering rates in nature requires that these various approaches be combined. Thus, concepts and results of controlled laboratory and greenhouse investigations are confirmed against field observations.

9.3.2 Interpretations of Mineral Weatherability

The rate of weathering of a particular mineral under a specified set of environmental conditions depends on the nature of the mineral crystalline structure. Properties of a mineral indicative of its susceptibility to weathering are lattice structure, ion composition, bond energies, ion substitutions leading to altered lattice structure and bond energies, and the prevalence of cleavage planes and fractures allowing for access by weathering agents. In soils a mineral will be one component of a mineral assemblage, so weatherability under a specified set of environmental conditions will be complicated by the properties of the mineral relative to those of other minerals within the assemblage.

Mineral dissolution may be either congruent or incongruent. Congruent dissolution results in ion release to soil solution in the same ratios as they are represented in the mineral being dissolved. Mineral weathering proceeds toward congruent dissolution but may be incongruent over time scales of importance to soil mineral weathering.

Incongruent dissolution is postulated to occur due to differential ion mobilities of mineral components. Initially, a freshly exposed mineral surface will become depleted of the more mobile ion components, relative to the less mobile components that enrich the surface layer that is formed as a consequence of weathering. At later times, the relative rates of ion loss from the weathered surface depend on the rate of solid-state diffusion of the more mobile ions through the weathered surface and subsequent loss to soil solution in comparison to the rate of loss to soil solution of the less mobile elements enriching the weathered surface. Preparation of minerals can produce various-sized particles that have different degrees of reactivity. Surface reaction control over the dissolution process results from the differential reactivity of particles exhibiting different surface-to-volume ratios. This could be misinterpreted as solid-phase diffusion control of dissolution. Spectroscopic evidence for the presence of base-depleted surfaces on minerals leaves the possibility for diffusion-control versus surface reaction-control of incongruent dissolution unresolved.

Kinetic studies of primary mineral dissolution have shown linear (that is, apparent first-order) dissolution of all mineral components or linear dissolution of

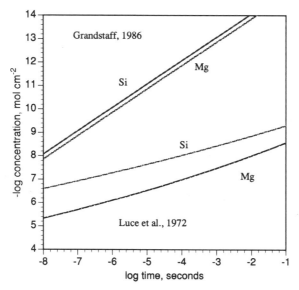

FIGURE 9.13 Fosterite dissolution kinetics measured in two experimental systems. (Data of Luce et al., 1972 and Grandstaff, 1986.]

basic cations and parabolic (that is, fractional order) dissolution of Si. The modeling of ion-release rates and, consequently, the ascribing of differing reaction mechanisms for mineral solubility on the basis of the models are largely artifacts of experiment design. Solution composition, temperature, sample preparation, sample source, sampling intervals, and the use of batch versus flow reactor designs will alter greatly the experimental outcomes. For example, batch reactors maintain solutions at elevated concentrations and allow for secondary precipitates to form. In contrast, flow reactors allow for lower solution concentrations and flow rate can be adjusted upward to restrict secondary precipitate formation. Thus, batch reactors may result in apparent first-order descriptions of a mineral dissolution whereas fractional order kinetic may result from flow reactor experiments.

Figure 9-13 illustrates the differing results obtainable in laboratory mineral dissolution studies performed under relatively similar conditions. Luce et al. (1972) described fosterite $[(Mg_{1-x}Fe(II)_x)_2SiO_4]$ dissolution in stirred batch reactors and fit data for Mg and Si loss with time to a half-order kinetic model. Approximately tenfold higher Mg than Si was released to solution at early sampling times with the difference decreasing with time. Grandstaff (1986) described fosterite dissolution kinetics for experimental batch systems maintained at constant pH and ionic strength in terms of a zero-order model. Slight differences in Mg and Si release to solution were observed under experimental conditions that were apparently less favorable to dissolution than those of Luce et al. (1972).

NOTES

1. The disequilibrium ratio may also be expressed as a *disequilibrium index,* DI $=$ pK$_{sp}$ $-$ pIAP. In this instance, a disequilibrium index < 0 indicates the solution is undersaturated with respect to the mineral of interest and a DI > 0 indicates oversaturation with respect to the mineral of interest.

2. The effect of T on K$_{eq}$ is described by the van't Hoff equation (section 3.4.6) as

$$d\ln K_{eq} = \frac{\Delta H^\circ}{RT^2} \, dT \qquad \text{[119c]}$$

Integration between limits of K$_{eq1}$, K$_{eq2}$ and T$_1$, T$_2$, with the assumption that ΔH° is constant (that is, independent of T) over the range of T considered, results in

$$\ln \frac{K_{eq2}}{K_{eq1}} = -\frac{\Delta H^\circ}{R} \left[\frac{1}{T_2} - \frac{1}{T_1} \right]$$

The general case for a dissolution reaction, r, with a solubility product, K$_{sp}$, is

$$\ln K_{sp} = -\frac{\Delta H_r^\circ}{R} \frac{1}{T} + c$$

A plot of ln K$_{sp}$ against 1/T will have a slope of $-\Delta H_r/R$.

For $Al(OH)_{3s}$,

$$Al(OH)_{3s} \rightleftharpoons Al^{3+}_{aq} + 3OH^-_{aq}$$

where $Al(OH)_{3s}$ may represent amorphous $Al(OH)_3$ or crystalline gibbsite. The following data are summarized in Stumm and Morgan (1981).

Species	H_f°, kJ mol^{-1}	
Al^{3+}_{aq}	-531	
OH^-_{aq}	-230	
amorphous $Al(OH)_3$	-1278	(Garrels and Christ, 1965)
crystalline gibbsite	-1293	

For amorphous $Al(OH)_3$,

$$\Delta H_r^\circ = \Sigma H_f^\circ{}_{products} - \Sigma H_f^\circ{}_{reactants}$$

$$= -1278 - [-531 + 3(-230)]$$

$$= -57 \text{ kJ mol}^{-1}$$

$$-\Delta H_r^\circ/R = 6856 \,^\circ K$$

For crystalline gibbsite,

$$\Delta H_r^\circ = -1293 - [-531 + 3(-230)]$$

$$= -72 \text{ kJ mol}^{-1}$$

$$-\Delta H_r^\circ/R = 8660 \,^\circ K$$

3. The following relationships are used to develop Figure 9.4. Following the convention of Kittrick (1969), mineral dissolution is expressed as an acid hydrolysis reaction ($K^{\circ'}$).

gibbsite

$$Al(OH)_3 + 3H^+ \rightleftharpoons Al^{3+} + 3H_2O$$

$$pH - 1/3pAl = 1/3pK^{\circ'} = pK_w^\circ - 1/3pK^\circ$$

kaolinite

$$Al_2Si_2O_5(OH)_4 + 6H^+ \rightleftharpoons 2Al^{3+} + 2H_4SiO_4 + H_2O$$

$$pH - 1/3pAl = 1/3pH_4SiO_4 - 1/6pK^{\circ'}$$

montmorillonite

Montmorillonite is incompletely defined in the system Al_2O_3—SiO_2—H_2O because of isomorphous substitution of Mg for Al. Montmorillonite is modeled in this system by use of the pyrophyllite formula, $Al_2(Si_2O_5)_2(OH)_2$.

$$Al_2(Si_2O_5)_2(OH)_2 + 4H_2O + 6H^+ \rightleftharpoons 2Al^{3+} + 4H_4SiO_4$$

$$pH - 1/3pAl = 2/3pH_4SiO_4 - 1/6(pK^\circ + pA_{montmorillonite\ s})$$

4. The relevant mineral relationships are

gibbsite-kaolinite

$$2Al(OH)_3 + 2H_4SiO_4 \rightleftharpoons Al_2Si_2O_5(OH)_4 + 5H_2O$$

$$pK^\circ = -2pH_4SiO_4$$

gibbsite-muscovite

$$2Al(OH)_3 + 3H_4SiO_4 + K^+ \rightleftharpoons KAl_3Si_3O_{10}(OH)_2 + 9H_2O + H^+$$

$$pK^\circ = [pH - pK^+] - 3pH_4SiO_4$$

muscovite-microcline

$$KAl_3Si_3O_{10}(OH)_2 + 6H_4SiO_4 + 2K^+ \rightleftharpoons 3KAlSi_3O_8 + 12H_2O + 2H^+$$

$$pK^\circ = 2[pH - pK^+] - 6pH_4SiO_4$$

kaolinite-muscovite

$$3Al_2Si_2O_5(OH)_4 + 2K^+ + \rightleftharpoons 2KAl_3Si_3O_{10}(OH)_2 + 2H_2O + 2H^+$$

$$pK^\circ = 2[pH - pK^+]$$

Reactions involving montmorillonite model montmorillonite with Al isomorphically substituted into the tetrahedral layer and with K^+ satisfying the charge deficit.

kaolinite-montmorillonite

$$7Al_2Si_2O_5(OH)_4 + 2K^+ \rightleftharpoons 6K_{0.33}[Al_2(Si_{3.66},Al_{0.33})O_{10}(OH)_2] + 9H_2O + 2H^+$$

$$pK^\circ = 2[pH - pK^+] - 8pH_4SiO_4$$

montmorillonite-microcline

$$3K_{0.33}[Al_2(Si_{3.66},Al_{0.33})O_{10}(OH)_2] + 10H_4SiO_4$$
$$+ 6K^+ \rightleftharpoons 7KAlSi_3O_8 + 2H_2O + 6H^+$$

$$pK^\circ = 6[pH - pK^+] - 10pH_4SiO_4$$

kaolinite-microcline

$$Al_2Si_2O_5(OH)_4 + 4H_4SiO_4 + 2K^+ \rightleftharpoons 2KAlSi_3O_8 + 9H_2O + 2H^+$$

$$pK^\circ = 2[pH - pK^+] - 4pH_4SiO_4$$

CHAPTER 10

CHEMICAL AVAILABILITY

The concept of "availability" underlies much of the development and utility of soil solution chemistry. The soil solution chemist applies the term "available" in both biological and geochemical contexts. These contexts differ somewhat when applied to soil environments, but the union of biological and geochemical availability (biogeochemical availability) is largely definable in terms of soil solution composition.

In general, "availability" refers to the fraction of a chemical occurring in the soil environment that participates in biologically or geochemically consequential processes. A more specific meaning of availability is ascribable to biological as compared to geochemical contexts.

10.1 BIOAVAILABILITY

The biologically relevant pool of a chemical in the soil environment is that fraction of the total quantity of the chemical that is or has been available to an organism. *Absolute availability* to an organism is measurable as the quantity of the chemical present in the organism. That is, if a chemical is present in the organism, it has been available to that organism in the environment—perhaps exclusive of some portion that was present in initial genetic reservoir of the organism, as in a seed. The concept of absolute availability of chemicals in the soil environment is unambiguous when plant ion uptake is exclusively from a soil pool. For instance, the source of Ca found in a plant is soil Ca less the initial or reserve of Ca in the seed. Chemicals that are biochemically transformed within the plant or that are obtained from the environment by routes of uptake other than from the soil (that is, foliar uptake) will be less readily interpretable within the context of absolute bioavailability.

Absolute bioavailability is measurable only after the fact as an effect of uptake of a chemical by an organism from the environment. An important thrust of soil fertility and plant nutrition research is the development of measurements of nutrient content in whole plants or plant parts that adequately reflect absolute bioavailability. An alternative approach, for those chemicals accrued by organisms from the soil environment, is the concept soil bioavailability. Expression of bioavailability from the standpoint of the soil pool of a chemical arises from the need to have an antecedent measure for probing the absolute bioavailability of a chemical. Measurements of soil bioavailability must, therefore, strongly correlate with absolute bioavailability. Soil bioavailability is usefully expressed as a dose (the product of concentration with time) or, if sampling intervals are sufficiently short or if concentrations are relatively constant with time, as an intensity (effective concentration) or flux. Soil solution composition proves to be the most directly correlated soil index of absolute bioavailability.

10.1.1 Defining Critical Thresholds

In practice intensity is nearly always used in preference to chronic dose as an index of bioavailability. This is possible for a homogeneous culture medium or for a heterogeneous culture medium that is representatively sampled. Levels of concern or *critical thresholds* are identified for experimental systems through definition of an intensity (dose) response curve where the response is some measure of growth.

There are no consistent means to express plant intensity response or critical thresholds. For plant nutrients, the critical threshold or critical nutrient concentration is definable as the inflection along the intensity response curve leading to the maximum response (Fig. 10-1a). For phytotoxic ions, the critical threshold is frequently identified as the inflection along a downward-declining asymptotic response function (Fig. 10-1b). For xenobiotic molecules, the "dose" resulting in a 10% reduction in growth response (GR_{10}) is frequently used (Fig. 10-1b). Generally, the critical threshold will be more certainly assignable when associated with the region where the observed response is maximized or minimized, rather than at the point of incipient response. Incipient response of a severely nutrient deficient plant will be highly variable until sufficient nutrient is available to allow for normal physiological function of the plant. Similarly, incipient response to a phytotoxic chemical will be highly variable until sufficient amount of the chemical is available to significantly impact the normal physiological function of the plant.

Figure 10-2 illustrates an intensity response curve describing yield of rye seedlings on an initially P-deficient soil in relation to either $H_2PO_4^-$ or total P in soil solution. The growth response is characteristic of many nutrient response curves; the curve from where P is severely growth-limiting to where growth is maximized is so steep that accurate assignment of a critical threshold is difficult. Figure 10-3

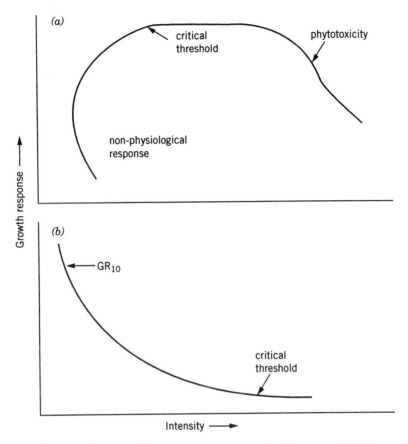

FIGURE 10.1   Idealized plant intensity-response curves for (*a*) an essential plant nutrient and (*b*) a phytotoxic compound.

shows a growth decrement response associated with Al phytotoxicity to cotton. A critical threshold of approximately 2 μmol L^{-1} is assigned on the basis of the approximate point of transition of the intensity response curve from an exponential decline to a linear decline phase.

10.1.2 Soil and Nutrient Culture Comparisons

One of the basic assumptions underlying the utility of soil solution chemistry stated in Chapter 2 was that if soil solution represents the natural medium for plant growth, then soil solution analysis allows for prediction of plant response to chemicals occurring in soil environments. This is another way of hypothesizing the utility of soil solution intensity as a measurement of bioavailability. Furthermore, if chemical intensity in soil solution is indicative of bioavailability in soil culture, it should be possible to relate soil intensity response functions to similarly derived intensity

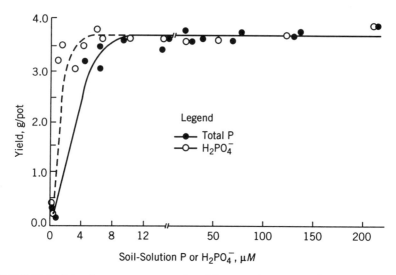

FIGURE 10.2   Intensity response curve describing growth response of rye seedlings on an initially P-deficient soil as a function of $H_2PO_4^-$ or total P intensity in soil solution. [J.F. Adams and J.W. Odom, 1985, "Effects of pH and phosphorus on soil-solution," *Soil Science* 140:202–205.]

response functions for plants grown in nutrient culture. This frequently proves to be the case as exemplified by the data of Figure 10-3 where intensities of Al (Al^{3+} activity) in displaced soil solution and nutrient culture (Al-amended $CaSO_4$ solutions) are shown to follow a similar intensity response function.

The ability to express intensity in terms equally applicable to soil and nutrient culture allows the soil–plant interactions affecting plant nutrition to be probed in nutrient culture systems. Figure 10-4 presents an example illustrating differential uptake of monovalent cations by peanut versus cotton in acid systems. For this case, an *activity ratio* (see section 10.1.4) is used to relate intensity of ions in nutrient solution to absolute bioavailability in seedlings (expressed as an uptake ratio). The monovalent to divalent ion activity ratio shows greater propensity for preferential uptake of monovalent as compared to divalent ions by peanut as compared to cotton. The nutrient culture results provide a bridge (through soil solution analysis) validating field observations regarding K fertilizer response of peanut on acid soils.

The use of nutrient culture systems as models for soil solution requires special design criteria to be utilized so that solution composition is maintained at relatively constant concentrations within the dilute ranges found in typical soil solution (Chapter 7). Dilute flowing culture solutions (Asher and Edwards, 1978; Checkai and Norvell, 1992) utilize large reservoirs of nutrient solution or resin exchange systems in conjunction with rapid flow rate (≈ 10 L hr^{-1}) and monitoring and adjustment of solution composition in order to model soil solution composition properly.

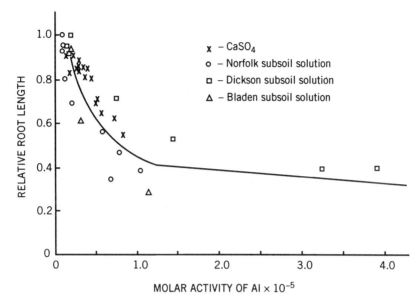

FIGURE 10.3 Intensity response curve describing growth decrement of cotton as affected by Al intensity in soil and nutrient solutions expressed as free Al^{3+} activity. [F. Adams and Z.F. Lund, 1966, "Effect of chemical activity of soil solution aluminum in cotton root penetration of acid subsoils," *Soil Science* 101:193–198.]

10.1.3 Speciation and Complexation Effects on Bioavailability

The means of expression of the intensity parameter has important effects on the predictive utility of intensity response functions. Simple expression of intensity as total chemical concentration in solution is frequently an adequate predictor of critical thresholds when an intensity response curve is developed for a specific plant culture medium. This is seldom the case, however, for intensity response curves developed for diverse plant culture media where each system may exhibit a unique critical threshold when intensity is expressed as a total solution phase concentration. This is shown in Figure 10-5a for cotton root response to soil solution Ca, where critical thresholds vary by about twofold between the Dickson and Norfolk soils.

With respect to inorganic ion availability, soil or nutrient culture solutions may vary widely in the portion of total ion in solution that is effectively bioavailable. Generally, the free ion is considered the effectively bioavailable form of an ion in solution. The free ion concentration in solution may be reduced due to effects associated with other components of the solution. First, the free ion concentrations may be reduced through speciation reactions where nonbioavailable ion pairs or complexes are formed. Second, nonspecific ion effects will alter the effective concentration through effects on ionic strength and, thus, ion activity.

Expression of intensity in a manner that accounts for chemically induced effects of other ions in solution allows for generalization of intensity response across di-

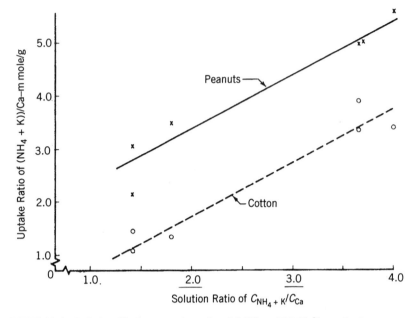

FIGURE 10.4 Relationship between the ratio of $(NH_4^+ + K^+)/Ca^{2+}$ uptake by cotton and peanut seedlings and the ratio of activities of $(NH_4^+ + K^+)/Ca^{2+}$ in nutrient solution. [Adams and Pearson, 1970. Published in *Agron. J.* 62:9–12. 1987. Am. Soc. Agron.]

verse soil and nutrient culture systems resulting in a single diagnostic critical threshold. An example of this is shown in Figure 10-3, where free Al^{3+} activity—that is, total solution Al corrected for ionic strength and for the occurrence of hydrolysis species, ion pairs, and ion complexes—allows for a generalized response expressing Al phytotoxicity to cotton.

The use of activity as an intensity term does not always resolve data from soil and nutrient culture experiments. Figure 10-6 presents the intensity response observed for ammonia phytotoxicity to cotton seedlings where NH_{3aq} concentration (calculated from NH_4^+ activity) was used to express intensity in both soil and nutrient solutions. While the critical threshold is consistent between culture systems, the intensity response is not.

10.1.4 Activity Ratios

Although expression of intensity as an activity may account for chemical effects of ions in solution on observed intensity response, it does not fully account for physiological effects that may restrict availability for plant uptake (as for the case of ammonia phytotoxicity, Fig. 10-6). An additional effect of ions in solution involves specific and nonspecific restriction of the bioavailable fraction through competition or interference with sites of ion uptake by the plant.

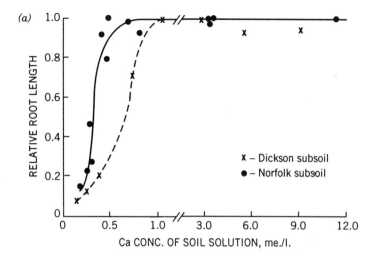

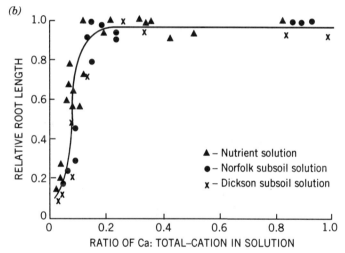

FIGURE 10.5 Intensity response curves describing cotton root response to subsoil Ca as a function of (*a*) soil solution Ca on two soils and (*b*) the ratio of soil or nutrient solution Ca to total cation concentrations in solution. [Howard and Adams, 1965. Published in *Soil Sci. Soc. Am. Proc.* 29:558–562. 1965. Soil Sci. Soc. Am.]

This was demonstrated for cotton root elongation in response to solution Ca by Howard and Adams (1965), who used the ratio of solution Ca concentration to the sum of total cation concentration in soil and nutrient solutions as an intensity term to describe competitive ion effects on bioavailability to cotton (Fig. 10-5*b*). The ratio used,

$$\frac{C_{Ca^{2+}}}{\sum\limits_{i} C_{i^{z+}}}$$

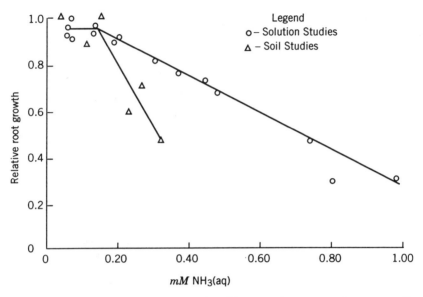

FIGURE 10.6 Intensity response curves describing growth response of cotton seedlings as a function of NH_{3aq} in nutrient and soil solution. [Bennett and Adams, 1970. Published in Soil Sci. Soc. Am. Proc. 34:259–263. 1970. Soil Sci. Soc. Am.]

reduces to

$$\frac{C_{Ca^{2+}}}{(C_{Ca^{2+}} + C_{Mg^{2+}})}$$

since Ca^{2+} and Mg^{2+} were the dominant solution phase cations. This relationship can also be expressed as follows.

$$\frac{A_{Ca^{2+}}/\gamma_{Ca^{2+}}}{(A_{Ca^{2+}}/\gamma_{Ca^{2+}}) + (A_{Mg^{2+}}/\gamma_{Mg^{2+}})} \qquad [10\text{-}1]$$

but $\gamma_{Ca^{2+}} \approx \gamma_{Mg^{2+}}$, so equation 10-1 simplifies to an activity ratio.

$$\frac{A_{Ca^{2+}}}{\sum_i A_i^{z+}} \qquad [10\text{-}2]$$

Adams (1966) subsequently developed a nearly identical Ca intensity response function for cotton on a Lakeland soil, using a computational approach to estimate the activity ratio of equation 10-2.

The activity ratio addresses the competitive effects of soil or nutrient solution components on intensity of a specific ion from both chemical (through expression of intensity in terms of an activity) and physiological (through recognition of competitive ion effects) perspectives. The use of activity ratios may not fully address ion interactions. Lund (1970) and Wolt and Adams (1979a) showed differential intensity response of soybean and peanut, respectively, to Ca^{2+} in soil and nutrient solutions when Mg^{2+} was a dominant competing ion versus K^+. This differential response, leading to higher critical thresholds of Ca^{2+} in high Mg^{2+} systems, was manifested even when an ion activity ratio of Ca^{2+} to total cations was used as the intensity term. This data suggests a physiological effect where divalent Ca^{2+} and Mg^{2+} were competing for specific sites for nutrient uptake or transport.

10.2 GEOCHEMICAL AVAILABILITY

The concept of biogeochemical availability comes into play in soils when processes of phase distribution or transport are considered. Geochemical availability refers to the fraction of a chemical that materially participates in either distribution or transport processes over time and spatial scales of interest. Geochemical availability is less clearly definable than is bioavailability, and its meaning is very much determined by the specific processes considered. In considerations of distribution, for instance, the geochemically available fraction comprises the chemical in soil solution and the solid phase fraction instantaneously interacting with soil solution (the labile pool). When consideration shifts to transport, however, the geochemically available fraction is that chemical in soil solution that is mobile—this may be measured in an absolute sense as the chemical flux through areas of interest in the soil system. For either of these considerations, the pool of a chemical definable as geochemically available may be further modified through consideration of the effective concentration or activity of the chemical.

The concept of geochemical availability is more vague than is bioavailability and, therefore, is less usefully applied. Despite its vagueness, the concept of geochemical availability is important, especially to the recognition of the very limited fraction of a chemical that may be geochemically active in soil at any time. Table 10.1 presents a very general, although useful, comparison between mean concentration of some biogeochemically consequential elements in whole soil and "representative" soil solution composition. If soil solution is taken generally to represent the biologically and geochemically active pool within the soil, it is clear that the geochemically available pool is a very minor part of the total amount of chemical present.

10.2.1 Soil Solution and Solute Leaching

The assignment of soil water into classes of mobile, immobile, or internal catchment water can lead to a misleading appreciation of the conduction of water within soil (see section 2.2.2, Soil Solution and Soil Water). All soil water (exclusive of structural water) is potentially mobile in soils. The various arbitrary classes of soil water occur along a continuous gradient of chemical potential. The greater the water matric potential (that is, volume-weighted chemical potential) associated with a

TABLE 10.1 Representative Concentrations of Some Geochemically Consequential Elements in Soils and Soil Solution

Element	Whole Soil Concentration, mmol kg^{-1}	Soil Solution Concentration		Available Fraction, $\times$ 10^6
		μmol L^{-1}	μmol kg^{-1} at 10% Moisture Content	
Si	11,750	230	23	2.0
Al	2,630	7	1	0.3
Fe	680	2	0.2	0.3
C	1,665	330	33	20
K	360	90	9	25
Ca	340	800	80	230
Na	275	650	65	240
Mg	205	170	17	80
N	70	14	1	20
Mn	15	0.4	0.04	240
S	22	160	16	730
P	21	0.2	0.02	0.8

Adapted from the compilation of Bowen, 1966.

class or fraction of soil water, the more strongly associated with the soil matrix is this class of water and, thus, the less labile it is to move in the bulk phase of water. Despite this, experiments with isotopically labeled water demonstrate that water freely distributes along the water potential continuum from the bulk phase through water adsorbed to soil solid phase surfaces.

Solutes in soil solution, however, will be more freely mobile when in the bulk phase of soil water as contrasted with that portion of water near solid surfaces. Thus, the consideration of a geochemically available fraction of solutes in soil solution is germane for considerations of chemical mobility. Highly weathered sandy soil, for example, may exhibit charge exclusion of Cl$^-$ or Br$^-$ from >20% of the water-filled pore volume of soils near field capacity (Gerriste and Adeney, 1992). Table 10.2 demonstrates this effect for various soils where the pore volume excluded to Cl$^-$ and Br$^-$ ranges from <1 to 17% of the total pore volume. Soil solution composition representative of the bulk phase water composition of soils is, therefore, a useful descriptor of that solute fraction that is geochemically available for transport. However, the extrapolation of bulk phase soil solution composition to total water phase solute content is complicated for those chemicals that, because of charge characteristics, show differential distribution through the soil aqueous phase.

10.3 BIOGEOCHEMICAL AVAILABILITY

Biogeochemical availability is definable as the instantaneous quantity—the intensity—of a chemical that may be taken up by an organism or that may be mobile

TABLE 10.2   Retention Volumes for Isotopically Labeled Water, Cl^-, and Br^- and Anion Exclusion Volumes, Column Saturated Flow

Soil Type	Organic Matter, %	Clay, %	Column Mass, g	Column Volume, mL	Average Retention Volume, mL HDO	Cl^-	Br^-	Exclusion Volume mL cm^{-3}	% of Pore Volume
Karakatta—A1	1	2	14.8	9.1	3.54	3.51	3.50	0.0035	1
Karakatta—A2	1	8	13.7	9.1	3.88	3.73	3.70	0.019	4
Jandakott—A2	0.7	2	15.3	9.1	3.38	3.33	3.33	0.005	1
Gavin—A1	5	5	310.0	216.5	102.0	95.5	95.5	0.027	6
Gavin—A1	5	5	13.0	9.1	4.05	3.80	3.80	0.027	6
Solonchak—A0	50	5	5.0	9.1	7.10	5.90	5.90	0.12	17

Gerritse and Adeney, 1992.

in the soil environment. This intensity is most readily identifiable as the concentration or activity of a chemical in soil solution.

10.3.1   Environmental Chemistry of Anthropogenic Compounds

The concept of biogeochemical availability is typically applied to nutrient or phytotoxic elements in soil systems, but the concept is equally important for consideration of the fate of xenobiotic molecules in soils. The fraction of xenobiotic molecule that participates in biogeochemically consequential reactions of sorption, degradation, and transport depends on the concentration in soil solution. In a soil at 10% moisture, a xenobiotic with a $K_d = 5$ L kg^{-1} has a potentially large fraction of its total mass ($\approx 2\%$) in biogeochemically available form (that is, in soil solution).[1] As K_d increases at constant moisture content, there will be a proportional decrease in the mass associated with the soil solution. As the partitioning of a xenobiotic molecule may vary greatly in time, there is a large degree of variation in the biogeochemically available pool with time as soil moisture fluctuates and as kinetics of the sorption and desorption process shift from adsorptive to desorptive control of the chemical in soil solution. This variation is even further confounded as the molecule is concurrently degraded in the soil environment.

NOTES

1. $K_d = C_{sorbed}/C_{solution}$, where units are mass-based for the sorbed phase and volume-based for the solution phase (for example, units of mol kg^{-1} and mol L^{-1}, respectively). The solution phase concentration can be recast in units of mass through correction for the volumetric water content (Θ_v, L solution per kg soil), where the phase distribution is being considered ($C'_{solution} = C_{solution} \times \Theta_v$). Thus, $K_d = (C_{total} - C'_{solution})/C_{solution}$ and $C_{solution} = C_{total}/(K_d + \Theta_v)$.

CHAPTER 11

SOIL SOLUTION ALUMINUM

Aluminum biogeochemistry is environmentally significant because Al, in its bio-available form, exhibits considerable phytotoxicity as well as aquatic ecotoxicity. Additionally, Al distribution between soil solution and solid phase pools has an important controlling influence over soil pH—the principal master variable governing soil chemical reactivity. The occurrence and distribution of Al in soil solution has long been recognized as being fundamental to describing aluminum reactivity and availability in the environment.

The environmental prevalence of Al, its phytotoxicity, and the amelioration of Al phytotoxicity by dissolved organic acids and certain inorganic ions were all well recognized early in the twentieth century. Magistad (1925) described pH and Al relationships for soil solutions obtained by column displacement and compared these with plant growth responses in solution culture experiments in early investigations of Al toxicity. In the decades since, soil solution chemists have strove to more clearly elucidate these early findings. This chapter presents a comprehensive review of the soil biogeochemistry of Al from the perspective of the soil solution.

Environmental Prevalence of Aluminum. Aluminum comprises from 1 to 30% (mean = 7.1%) of the soil (Bowen, 1966), where it is found predominately as a component of a variety of aluminosilicate, oxyhydroxide, and nonsilicate minerals (Barnhisel and Bertsch 1982). Despite its prevalence in the soil environment, biochemically active Al is principally of concern only in acid environments where hexa-aquo Al [$Al(H_2O)_6^{3+}$], usually designated Al^{3+}, may reach toxic levels. Aluminum in representative soil solutions from temperate regions is typically <0.4 μmol L^{-1} (Bohn et al., 1979), but concentrations ranging from 23 to 410 μmol L^{-1} have been reported for acid, highly weathered surface soils (Kamprath, 1984). Aluminum is reported to range up to 90 μmol L^{-1} in natural waters (Bowen, 1966).

11.1 ALUMINUM DISTRIBUTION IN SOIL SOLUTION

Despite the long-standing recognition of key attributes of aluminum reactivity and distribution in soils and soil solution, the assessment of Al biogeochemical availability has proven difficult. Assessing Al biogeochemical availability is complicated in part by uncertainties regarding the distribution of total Al among a variety of hydrolytic species, ion pairs, and complexes in solution. Burrows (1977) states, "the form and concentration of aluminum in water depends on the pH and nature of substances dissolved in the receiving waters and, to a lesser extent, on the temperature and duration of exposure to the water." Fine-grained crystalline materials and colloidal organo-Al complexes present in soil solution further complicate the determination of biochemically active Al (Beck et al., 1977; Burrows, 1977; Kennedy et al., 1974). The generally recognized toxic form of Al in acid systems is the Al^{3+} ion (Adams, 1984; Foy, 1974), while $Al(OH)_4$ is potentially toxic in strongly alkaline natural waters (Foy, 1974; Freeman and Everhart, 1971).

11.1.1 Aluminum Hydroxy Monomers

The hydrolytic reactions of hexa-aquo Al are important determinates of Al bioavailability (that is, the free ionic activity of Al^{3+}) in soil solutions, but their existence and distribution are by no means certain (Nair and Prenzel, 1978). The important hydrolytic reactions of monomeric Al in an acid system can be represented as (Lindsay, 1979),

$$Al^{3+} + H_2O = AlOH^{2+} \qquad pK = 5.0 \qquad [11\text{-}1]$$

$$Al^{3+} + 2H_2O = AlOH_2^+ \qquad pK = 9.3 \qquad [11\text{-}2]$$

$$Al^{3+} + 3H_2O = AlOH_3^\circ \qquad pK = 15.0 \qquad [11\text{-}3]$$

where pK is the negative logarithm of the stability constant, and hexa-aquo Al $[Al(H_2O)_6^{3+}]$ and hydronium ion (H_3O^+) are represented as Al^{3+} and H^+, respectively. These relationships indicate that Al hydroxy monomers will be the dominant Al species in slightly acid soil solution, while hexa-aquo Al will increasingly dominate at $<pH\ 5$.

11.1.2 Aluminum Hydroxy Polymers

In addition to Al hydroxy monomers, a number of hydrolytic reactions involving the formation of Al-hydroxy polymers have been postulated (Bache and Sharp, 1976; Nair and Prenzel, 1978; Tsai and Hsu, 1984). Although there is no question that these polymeric Al species exist in dilute electrolytic solutions (Bache and Sharp, 1976; Bertsch and Barnhisel, 1985; Smith and Hem, 1972), and that their occurrence in soil solution would prove useful in the explanation of certain soil chemical phenomena (Rich, 1968; Richburg and Adams, 1970), reliable thermodynamic data for stability constants as well as evidence for their presence in soil solution are lacking (Bache and Sharp, 1976; Nair and Prenzel, 1978). The distribution of Al among polymeric forms and the partition of Al between monomers and polymers in solution are kinetically controlled (Bersillon et al., 1980; Burrows, 1977; Tsai and Hsu, 1984). It is unlikely that Al hydroxy species measured in soil extracts are representative of the hydrolysis species present in situ soil solution. Ion speciation models (Adams, 1971; Misra et al., 1974; Nair and Prenzel, 1978) may be more effective for partitioning Al among hydrolytic species; but again, the lack of reliable thermodynamic data is a problem. Bertsch et al. (1986a, 1986b) used ^{27}Al nuclear magnetic resonance (NMR) spectroscopy to identify the "Al_{13}" polymer [$AlO_4Al_{12}(OH)_{12}^{7+}$] as a consequential component of partially neutralized solutions containing >30 mmol L^{-1} Al. Bertsch (1987) subsequently concluded, however, "that Al_{13} polymer formation is an artifact of synthesis procedure . . . it would appear of little significance to secondary Al mineral formation in acidic weathering environments."

11.1.3 Ion Pairs and Complexes

Formation of ion pairs between Al^{3+} and SO_4^{2-} and F^- can reduce Al^{3+} activity in soil solutions substantially. Stability constants for Al–sulfate ion pairs are smaller than those for Al–fluoride ion pairs (Behr and Wendt, 1962; Roberson and Hem, 1969), but the greater abundance of SO_4^{2-} in soil environments may increase the occurrence of Al–sulfate over Al–fluoride pairs (Burrows, 1977). Recent investigations of acid Haplorthod soil solutions and natural waters draining forested watersheds in the northeastern USA indicate F^- is responsible for complexing large fractions of total Al, despite the greater concentrations of SO_4^{2-} relative to F in solution (David and Driscoll, 1984; Driscoll and Newton, 1985; Johnson et al., 1981). Notwithstanding kinetic constraints to Al–fluoride complexation, these ion pairs appear critical to the control of Al^{3+} activity (and therefore Al phytotoxicity) in some watersheds identified as susceptible to acidic precipitation perturbation (Plankey et al., 1986).

11.1.4 Organometallic Complexes

Organometallic complexes are the least understood component of total Al in solution. When present, humic materials appear to complex considerable Al at pH 4.5 to 5.0, but as pH decreases, so does complexation (Krug and Frink, 1983). Organically complexed Al is traditionally considered to be the dominant mechanism for Al mobilization from eluvial (E) to Bs or Bhs horizons in the podzolization process (Farmer et al., 1980; McFee and Cronan, 1982; Nilsson and Bergkvist, 1983), but current evidence suggests dissolved aluminosilicates are involved as well (Childs et al., 1983; Farmer et al., 1980; Wada and Wada, 1980).

High-molecular-weight organic acids (fulvic and humic acids) are frequently identified with complexed Al in soil solution and natural waters, but in most instances the presence of these colloidal organic acids is inferred from measurements of dissolved organic matter (Beck et al., 1974; Nilsson and Bergkvist, 1983), color absorbance (Budd et al., 1981), or other nonspecific measurements (Driscoll, 1984; Krug and Isaacson, 1984). Hay et al. (1985) reported fulvic acids to comprise an average of 46% of the total organic carbon in leachates from the upper 20 cm of a humoferric podzol. Evans (1986) partitioned dissolved organic carbon extracted from a Typic Paleudult Oe horizon into molecular weight classes by ultracentrifugation. Aluminum was complexed dominantly by the 1000 to 2500 molecular weight fraction.

Low-molecular-weight organic acids will also complex Al. The stability of complexes formed is in the approximate order

citric $>$ oxalic $>$ malic $>$ tannic $>$ aspartic $>$ p-hydrobenzoic $>$ acetic

(Hue et al., 1986; Ng Kee Kwong and Huang, 1979a; 1979b; Wang et al., 1983). These complexes will hinder Al oxyhydroxide precipitation in the same relative order (Ng Kee Kwong and Huang, 1979b). The ability of low-molecular-weight organic acids to complex Al appears to be associated with the formation of 5- or 6-bond ring structures with Al, which is dependent on the relative position of OH/ COOH groups along the main C chain (Hue et al., 1986). The kinds and amounts of these organic acids in soil solution are likely to be greatly influenced by microbial activity and will exhibit wide temporal fluctuation. Very little evidence documents the importance of low-molecular-weight organic acids in Al speciation; however, Hue et al. (1986) reported 76 to 93% of total soil solution Al in two acid subsoils was complexed with low-molecular-weight organic acids. Michalas et al. (1992) have shown highly variable amounts of Al complexation by low-molecular-weight organic acids occurring in lysimeter solutions.

11.1.5 Mixed Ligand Systems

Total Al and free Al^{3+} concentrations in soil solution are expected to vary extensively in space and time as a consequence not only of soil pH and degree of Al saturation but also because of variation in the kinds and concentrations of complexing ligands present. Temporal variation in dissolved organic matter is a factor which greatly influences the Al chemistry of soil solutions (Nilsson and Bergkvist, 1983). Seasonal patterns of microbial activity as well as spatially and temporally variable episodes of throughfall, stemflow, and snowmelt will substantially alter the balance of organic acids occurring in soil solution; thus, their importance to complexation of soil solution Al will be highly variable as well. Aluminum–fluoride ion pairs will be more stable than Al–sulfate ion pairs (Behr and Wendt, 1962; Roberson and Hem, 1969); Al complexes with dissolved organic carbon will in general be stronger than with inorganic ligands (Ritchie et al., 1988); and Al complexes with humic or fulvic acid appear stronger than Al complexes with simple carboxylic acids (Plankey and Patterson, 1987; Ritchie et al., 1988; Sikora and McBride, 1989).

Simple models of ligand–ligand complexation for Al^{3+}, however, do not adequately describe the complex equilibria which govern Al speciation in natural waters (Plankey and Patterson, 1988). Residence time of complexing ligands in the soil environment is important for determining which Al complexes will dominate soil solution and will therefore have controlling influence on Al mobility and bioavailability. Mechanistic studies of mixed ligand systems of fluoride and fulvic acid competing for Al have indicated fulvic acid to increase the initial rate of Al complexation with fluoride (Plankey and Patterson, 1988). These studies have shown $AlOH^{2+}$ to be more reactive with all types of complexing ligands than is Al^{3+}. Introduction of complexing ligands to natural waters will shift the ratio $[AlOH^{2+}]/[Al^{3+}]$ to favor $AlOH^{2+}$ and will increase the rate of Al complexation with all types of complexing ligands that are present.

As a consequence of Al speciation, only a relatively small fraction of total soil solution Al may be present as free Al^{3+} ion. Wolt (1981) found free Al^{3+} comprised from 2 to 61% of total Al in soil solutions displaced from acid Hapludults where SO_4^{2-} was the dominant complexing ligand. Free Al^{3+} increased with decreasing solution pH and SO_4^{2-} activity. David and Driscoll (1984) found soil solutions from Haplorthod O and B horizons contained 6 to 7 % and 26 to 28%, respectively, of total inorganic Al as Al^{3+}. Solution Al was associated primarily with organic and fluoride complexes. These examples indicate the inadequacy of total Al as a measure of Al biogeochemical availability (that is, Al^{3+} activity) in soil solution.

11.2   MEASUREMENT OF SOLUTION ALUMINUM

The measurement of Al in soil, nutrient, and dilute electrolytic solutions, in soil extracts, and in natural waters has proven quite problematic as evidenced by the number of analytical procedures employed. Burrows (1977) reviewed various methods that have been used for analysis of Al in water. Methods frequently employed for Al analysis of soil solutions and extracts have been summarized by Barnhisel and Bertsch (1982). Commonly used colorimetric methods utilizing triphenylmethane dyes (aluminon and Eriochrome cyanine R) that react with Al to form deep red colors (Frink and Peech, 1962; McLean 1965) have been largely replaced by procedures that involve Al-hydroxyquinolate complexation, followed by solvent extraction and detection either spectrophotometrically or fluorometrically (Barnes, 1975; Bloom et al., 1978; James et al., 1983; Motojima and Ishiwatari, 1965). Although the latter procedures offer increased sensitivity over colorimetric procedures, the kinetics of the complexation reaction must be considered with respect to the Al species that are determined (Turner 1969, 1971; Turner and Sulaiman, 1971). Kinetic reactions of Al with hydroxyquinolate (James et al., 1983), ferron (Jardine and Zelazny, 1986; Batchelor et al., 1986), fluoride (Ares, 1986), and pyrocatechol violet (Bartlett et al., 1987) have been used in efforts to distinguish Al species in solution.

The measurement of total Al in solution is inadequate for interpretation of Al effects in aqueous systems. Aluminum can form any number of polymeric and monomeric species, pairs, and complexes in solution, and total solution Al is an inadequte indicator of Al biogeochemical availability. Direct measurement of Al species in soil solution is difficult for several reasons. Solution aging (that is, the time between solution sampling and analysis) significantly changes the nature and reactivities of Al species present (Smith and Hem, 1972; Tsai and Hsu, 1984). Shifts in Al speciation will depend on total Al concentration, rate of hydrolysis, anions present, temperature, and pH (Bache and Sharp, 1976). Additionally, fine-grained Al-bearing solids can pass 0.45 and 0.22 μm pore-sized filter membranes leading to erroneous measurements of dissolved Al (Kennedy et al., 1974). Filter membranes can remove substantial quantities of Al from solution due to the presence of Al complexing contaminants or the cation exchange capacities of filter materials (Jardine et al., 1986).

11.3 PARTITIONING SOIL SOLUTION ALUMINUM

11.3.1 Analytical Approaches

Because the Al species present in soil solution or natural waters cannot be precisely ascertained, many researchers dealing with Al chemistry in natural systems determine and express Al speciation in relative terms (Driscoll 1984; Driscoll et al., 1980; James et al., 1983). Frequently solution Al is expressed as nonlabile (monomeric Al-organic complexes), labile (aquo-Al, Al-hydroxy polymers, ion pairs), and acid-soluble Al (polymers, colloids, and strong Al-organic complexes). Even with relative analyses, extraction of labile Al in the field has been necessary to avoid shifts between labile and nonlabile species (Driscoll and Newton, 1985). Uncertainties in direct measurement of Al species in solution make computation of Al speciation a more useful approach in many instances. James et al. (1983) found good agreement between the labile Al measured by 8-hydroxyquinoline extraction and that predicted from ion speciation models.

Effective application of ion speciation models requires reliable thermodynamic data that are frequently lacking in the case of Al-organic complexes and Al polymers. In order to circumvent this problem, reliable measurements of nonorganically complexed monomeric Al are desirable for input into ion speciation models. Table 11.1 compares total solution Al as measured by hydroxyquinolate and pyrocatechol violet with free ionic Al^{3+} as measured by ion chromatography and calculated from known composition of the equilibrating solutions containing approximately equimolar concentrations of Al and organic acids. Hydroxyquinolate and pyrocatechol violet complexed 35 and 97%, respectively, of total Al in the presence of oxalic acid and 78 and 54%, respectively, in the presence of citric acid. Free ionic Al^{3+} measured by ion chromatography or calculated from the solution composition both accounted for 11% of total Al in the presence of oxalic acid; they accounted for 59 and 46% of total Al, respectively, in the presence of citric acid. Ion pair chromatography with sulfonic acid as the ion pair reagent was used by Michalas et al. (1992) to partition soil solution Al among free ions and forms complexed with

TABLE 11.1 Free Ionic Al in Solutions Containing 37.1 μmol L^{-1} Al and 40 μmol L^{-1} Oxalic or Citric Acid at pH 3 as Determined by Various Methods

	Free Ionic Al^{3+}, μmol L^{-1}			
Organic Acid	Hydroxyquinoline[a]	Pyrocatechol Violet[b]	Ion Chromatography	Calculated[c]
Oxalic	13	35	4	4
Citric	29	20	22	17

[a]8-hydroxyquinoline complexed Al corrected for monomeric species.
[b]Pyrocatechol violet complexed Al corrected for monomeric species.
[c]Calculated from solution composition.
After Whitten et al., 1992.

inorganic and organic ligands. Although total Al recovered was comparable to that obtained with electrothermal atomic adsorption spectrophotometry, the distribution among ligands was unreliable because of differential stability of the complexes (Table 11.2).

Hodges (1987) compared the effectiveness of hydroxyquinolate, ferron, and F complexation, ion exchange columns, and chelating resins for the identification of Al species. All methods demonstrated limitations that influenced the distribution of Al between inorganic and organic species:

Hydroxyquinolate, although effective, can degrade Al-organic complexes with low ratios of Al to dissolved organic carbon (James et al., 1983; Adams and Hathcock, 1984)

Ferron is less sensitive than hydroxyquinolate because of a greater tendency to degrade Al-organic complexes

Ion exchange columns and chelating resins both appear to degrade large organic matter complexes leading to poor estimates of Al-organic complexes

Free organic matter reacts with column-sorbed Al on exchangers

Chelation resins underestimate the effects of low-molecular-weight organic matter.

F complexation requires a comprehensive knowledge of the Al-bearing solution being analyzed.

Differing amounts of inorganic Al determined by hydroxyquinolate as compared to Erichrome cyanine R have been attributed to varied degrees of reactivity with dissolved organic carbon (Hathcock and Adams, 1984). The presence of P may form Al-phosphate complexes or fine colloidal precipitates which will be very slowly reactive with hydroxyquinolate (Di Pascale and Violante, 1986). Evans and Zelazny (1986) found crown ether to effectively determine monomeric Al in the presence of Al-organic complexes.

TABLE 11.2 Complexed Al Determined by Ion Pair Chromatography with Fluorescence Detection[a]

Complexing Ligand[b]	Formula	Log K	Retention Time, min	Chromatographic Recovery, %
Monooxalate	$Al(C_2O_4)^+$	7.4	4.48	90.5
Dioxalate	$Al(C_2O_4)_2^-$	12.5	2.48	
Citrate	$Al(C_6H_4O_7)^-$		2.90	74.1
	$Al(C_6H_5O_7)^\circ$	10.7	3.09	
Tartrate	$Al(C_4H_3O_6)^-$		2.80	
	$Al(C_3H_2O_6)^\circ$	6.2	2.96	25.8
Malate	$Al(C_3H_2O_4)^+$	5.7	4.66	40.6
Monofluoro	AlF^{2+}	6.7	8.32	
Difluoro	AlF_2^+	12.0		92.0
Monomeric Al	Al^{3+} species		10.58	

[a]18.5 μmol Al L^{-1}, pH 4.00.
[b]Complexes with formate, lactate, nitrate, and chloride do not chromatograph with the conditions employed.
After Michalas et al., 1992.

11.3.2 Computational Approaches

Computation of Al species entails measurement of total Al and other solution components, and calculation of Al species using an extrathermodynamic modeling routine (section 7.3). The determination of reliable values for the complexation of Al with dissolved organic matter results in ion speciation models that can effectively speciate Al from input values of total Al, thus avoiding the problem of analytical determination of nonorganically complexed monomeric Al. Plankey and Patterson (1987) have determined constants for the complexation of Al with fulvic acid, and this information has been built into an ion speciation model for more effective speciation of total solution Al (Wolt, 1987). Table 11.3 presents typical equilibrium constants for Al complexation in aqueous solution from a number of secondary sources. (Ion speciation and ion speciation modeling of Al are discussed in Chapter 7.)

11.4 ALUMINUM BIOAVAILABILITY

Absolute bioavailability of an element to a plant is measured by the uptake of that element by the plant (that is, if the element is present in plant tissue, it has therefore been available to the plant; section 10.1). Based on this criterion, tissue analysis is frequently used to gauge elemental availability to plants, particularly perennial species such as trees. Unfortunately, this approach to assessment of Al bioavailability has shown little promise because of (1) considerable variation in tree elemental uptake within and between species and across environments, (2) variation in sam-

TABLE 11.3 Equilibrium Constants for Al Species

Reaction Product[a]	Log K^{ob}	Source
$AlOH^{2+}$	−5.02	Lindsay (1979)
$Al(OH)_2^+$	−9.30	Lindsay (1979)
$Al(OH)_3$	−14.99	Lindsay (1979)
$Al(OH)_4^-$	−23.33	Lindsay (1979)
AlF^{2+}	6.98	Lindsay (1979)
AlF_2^+	12.60	Lindsay (1979)
AlF_3	16.65	Lindsay (1979)
HF	3.00	Lindsay (1979)
AlCit	10.18	Motekaitas & Martell (1984)
$AlHCit^+$	13.12	Motekaitas & Martell (1984)
$AlOHCit^-$	6.63	Motekaitas & Martell (1984)
$HCit^{2-}$	6.44	Motekaitas & Martell (1984)
H_2Cit^-	11.32	Motekaitas & Martell (1984)
H_3Cit	14.60	Motekaitas & Martell (1984)
$AlOx^+$	9.15	Martell & Smith (1977)
$AlOx_2^-$	15.16	Martell & Smith (1977)
$AlOx_3^{3-}$	18.17	Martell & Smith (1977)
HOx^-	4.07	Martell & Smith (1977)
H_2Ox	5.44	Martell & Smith (1977)

[a]$nM + mL = M_nL_m$, where M is Al^{3+} or H^+ and L is the appropriate ligand except in the case of AlOH spp, where L is H_2O.
[b]Adjusted where necessary to an ionic strength of zero by the Güntelberg approximation (Table 3.4). Whitten et al., 1992.

pling methodology, and (3) accumulation of Al in plant tops at high concentrations without evidence of toxicity.

Reviews by Foy (1974, 1976, 1983, 1984) have summarized the symptoms and physiology of Al phytotoxicity. Recognizing symptoms of Al phytotoxicity is complicated because several symptoms of acid soil infertility may be manifested concurrently in plants (Adams, 1984; Foy, 1976). Plant yields may be reduced substantially by Al without the occurrence of clearly identifiable symptoms in plant tops. As a consequence, plant growth reduction on acid soils may be frequently associated with some facet of acid soil infertility other than Al toxicity when in fact, Al is frequently the growth-limiting agent (Foy, 1976).

The gross symptoms of Al phytotoxicity in plant tops are frequently similar to symptoms of P or Ca deficiency, or of Fe phytotoxicity (Foy, 1974; 1983) perhaps because of interactive effects of Al with these elements. Aluminum phytotoxicity is more readily characterized by root morphology; Al-injured roots are frequently stunted, root tips are brown, lateral roots are thickened, fine roots are absent, and fungal infection is enhanced (Adams, 1984; Foy, 1974; Reid, 1976). Root length, rate of root elongation, and weight of plant tops are reliable measures of Al phytotoxicity, but root weight is not (Adams, 1984; Adams and Lund, 1966; McCormick and Steiner, 1978; Pavan et al., 1982; Steiner et al., 1980, 1984). According to Foy (1976), "For plants in general, excess Al has been reported to interfere with cell division in root tips and lateral roots, increase cell rigidity by cross linking pectins, reduce DNA replication by increasing the rigidity of the DNA double helix, fix P in less available forms in soils and on root surfaces, decrease root respiration, interfere with enzymes governing sugar phosphorylation and deposition of cell wall polysaccharides, and interfere with the uptake, transport, and use of several essential nutrient elements, including Ca, Mg, K, P, and Fe."

Environmental occurrence of Al phytotoxicity may be confounded by the presence of other factors in acid soil infertility (H and Mn toxicity; Ca, Mg, and Mo deficiency). Therefore, investigators have had relatively greater success in describing Al phytotoxicity in solution culture experiments where greater environmental control is possible or in soil culture studies where occurrence and distribution of Al in soil solution was used to diagnose plant response. Until recently, investigations of Al bioavailability have tended to concentrate on rather gross soil parameters such as soil pH (Hern et al., 1985; Lee et al., 1982), extractable Al (Johnson and Todd, 1984; Lee et al., 1982; Stuanes, 1983), degree of Al saturation (Lee et al., 1982), and total Al in leachates (Budd et al., 1981; Rutherford et al., 1985; Stuanes, 1983). Not surprisingly, most of these investigations have been inconclusive and have not yielded information useful for description of phytotoxicity. Attempts to generalize research regarding Al phytotoxicity have also been largely unsuccessful, which can be attributed to (1) genetic differences in test crops, (2) varied procedures for extracting Al, (3) variation in CEC of soils investigated, and (4) failure to isolate Al as the sole factor adversely affecting plant growth (Adams, 1984). In their effort to define "critical" levels of soil Al responsible for Al phytotoxicity in cotton (*Gossypium hirsutum* L.), Adams and Lund (1966) found soil pH, exchangeable Al, degree of Al saturation, and total soil solution Al to be unsatisfactory indices of Al bioavailability; Al^{3+} activity, however, did satisfactorily describe the observed incidence of Al phytotoxicity. Activity of Al^{3+} represents the fraction of soil solution Al^{3+} concentration that behaves as if in an ideal dilute solution. The computation of Al^{3+} activity accounts for the ionic strength of the solution after correction of total Al in solution for ion complexes and hydrolysis species.

11.4.1 Soil and Nutrient Solution Studies of Aluminum Phytotoxicity

The recognition of the concentration and distribution of Al in solution as a useful index of Al bioavailability has lead to considerable research regarding the most appropriate solution measures for expression of toxicity. In some regards the data are equivocal regarding the appropriateness of Al^{3+} activity as the preferred predictor of Al phytotoxicity. Studies regarding Al phytotoxicity are greatly complicated because of the impossibility of making solution-Al concentration the sole experimental variable. Comprehensive soil solution analysis—coupled with an extrathermodynamic modeling routine to express soil and nutrient solution Al^{3+} activity—remains the most successful technique for generalization of Al effects on plant growth across environments.

Agronomic Crops. Agronomists have attemped since the 1920s to relate plant growth in acid soils to total Al in soil solution (Magistad, 1975; 1984). These attempts failed because the correlation of total Al in soil solution with Al phytotoxicity was inconsistent among soils of varied pH and mineralogy.

The concept of Al^{3+} activity was used by Adams and Lund (1966) to explain cotton root penetration into three acid Ultisols of differing mineralogy as well as into nutrient solutions. They found a common relationship between root penetration and Al^{3+} activity in soil solutions, and this relationship also held for root penetration and Al^{3+} activity in nutrient solutions (Figure 10.3). The threshold activity of Al^{3+} that was toxic to cotton roots was 1.5 μmol L^{-1}.[1] Pavan et al. (1982) studied effects of Al^{3+} activity on root and shoot growth of coffee (*Coffea arabica* L.) growing in acid Ultisols and Oxisols and in nutrient solutions. Both root and shoot growth were inhibited at a toxic threshold of 4.0 μmol L^{-1} Al^{3+} in soil or nutrient solution. Brenes and Pearson (1973) also defined root growth in three *Gramineae* species as a function of Al^{3+} activity in nutrient and soil solutions, where the toxic threshold appeared to be <9 μmol L^{-1}.

Free Al^{3+} activity alone may not be diagnostic of plant growth in acid soils. Other components of acid soil infertility may be involved as well. The toxic effect of Al may sometimes be better expressed in terms of the ratio of soil solution Al^{3+} activity to the sum of the activities of other soil solution cations (for example, Ca^{2+}) that may influence plant growth response (Adams, 1984). Ulrich (1981a) suggested that Al phytotoxicity to forest trees will be manifested with Ca:Al mole ratios of <1 in soil solution. Aluminum antagonism of Ca uptake (Al-induced Ca deficiency) may be the most consequential manifestation of Al phytotoxicity at $<$pH 5.5 (Foy, 1984).

Attempts to define thresholds for Al phytotoxicity better have led to investigation of monomeric Al species, rather than Al^{3+} activity per se as predictors of Al phytotoxicity. Adams and Hathcock (1984) added $Ca(OH)_2$, MgO, or $CaSO_4 \cdot 2H_2O$ to acid Paleudult subsoils sampled from cultivated fields and woodlands in order to isolate the differential effects of Ca and Al on cotton root growth. Aluminum phytotoxicity was observed for some cultivated fields when total soil solution Al was 0.4 μmol L^{-1}, but not for woodlands where up to 11.5 μmol L^{-1} total Al was detected by hydroxyquinolate complexation. Expression of soil solution Al as the sum of Al^{3+}, $AlOH^{2+}$, and $Al(OH)_2^+$ activities was not a diagnostically useful parameter for prediction of Al phytotoxicity to cotton.

Ratios of P to Al were used by Alva et al. (1986) to alter total and monomeric Al in solutions of varied Ca concentration where root elongation of soybean (*Glycine max* L.), subterranean clover (*Trifolium subterraneum* L.), sunflower (*Helianthus annus* L.), and alfalfa (*Medicago sativa* L.) were determined. Using the sum of the activities of monomeric Al species (Al^{3+}, $AlOH^{2+}$, $Al(OH)_2^+$, $Al(OH)_3^\circ$, and $AlSO_4^+$) as a predictor, they observed 50% reductions in root elongation as the sum of monomeric Al activities ranged from 12 to 17, <8 to 16, <7 to 15, and <5 to 10 μmol L^{-1} for soybean, sunflower, subterranean clover, and alfalfa, respectively.

The critical threshold increased as Ca in solution increased, but a sum of monomeric Al activities ≥ 18 μmol L^{-1} was toxic to roots regardless of Ca level.

Wright and Wright (1987) used soil solution composition to evaluate subterranean clover growth on 13 acid surface soils from the southern Appalachians. Reduced root yields occurred when Al^{3+} activity in solution exceeded 3 μmol L^{-1}. The relationship between root growth and soil solution Al was not improved by using the sum of monomeric Al activities. Whitten et al. (1992) evaluated a variety of measures of Al for solutions obtained by centrifugal displacement or 1:5 extraction with 0.01 M $CaCl_2$, for describing Al phytotoxicity of subterranean clover (Table 11.4). Relative dry weight of plants was very strongly regressed on any measure of solution Al determined for uniformly fertilized plants on a single soil where the only experimental variable was soil pH. [For example, on soil 1 (Table 11.4) second-order polynomial fits gave $R^2 \geq 0.97$ for regression of relative yield on Al for any measure of Al shown.] When, however, comparisons were made across soils for uniformly fertilized plants, there was wide variation in the ability of the various measures of Al to predict yield response; free ionic Al^{3+} calculated from comprehensive soil solution composition was the best measure of Al phytotoxicity in this instance.

Baligar et al. (1992) evaluated numerous measures of soil solution and extractable Al to determine phytotoxicity of Al for wheat (*Triticum aestivum* L.) root elongation across diverse soils. Hydroxyquinoline, pyrocatechol violet, and aluminon extracted 75, 57, and 38%, respectively, of total soil solution Al as determined by inductively coupled plasma emission spectroscopy. These measures of Al as well as total soil solution Al and Al^{3+} activity calculated from comprehensive soil solution composition were all significantly correlated with relative root length, but the best predictive measure was 0.01 M $CaCl_2$-extractable Al. The critical threshold Al^{3+} activity was ≈ 20 μmol L^{-1}.

The phytotoxic threshold for barley (*Hordeum vulgare* L.) root elongation in nutrient solution culture was better predicted by Al^{3+} activity (1.5 μmol L^{-1}) than by the sum of total monomeric Al activities (Cameron et al., 1986). Addition of F or SO_4 to solution reduced Al phytotoxicity to the degree of Al complexation into F or SO_4 ion pairs.

In contrast to the preceding studies that support Al^{3+} as the species diagnostic of Al phytotoxicity to plant roots, nutrient culture studies with excised roots of a number of crop species identified Al-hydroxy polymers as the most highly toxic form of Al (Wagatsuma and Kaneto, 1987). In these studies, nutrient solutions of varied composition and pH were used to develop treatments with varied Al species composition, and species distribution was determined by a modified ferron procedure. These results illustrate the difficulty in obtaining unambiguous results from studies of Al phytotoxicity: The conclusions drawn are highly dependent on the ability of the analytical procedure utilized to differentiate Al species on the basis of the kinetics of reaction with Al.

Tree Growth. Several investigators have considered tree growth response in relation to Al in either soil or nutrient solution. Van Praag and Weissen (1985) sampled >700 two-year-old seedlings of Norway spruce [*Picea abies* (L.) Karst.] and

TABLE 11.4 Relative Yield of Subterranean Clover and Al in Displaced Soil Solution and CaCl$_2$ Extracts as Determined by Various Measures

Soil	Relative Yield[a]	Soil Solution							1:5 CaCl$_2$ Extract		
		$[Al^{3+}]_{calc}$[b]	Al_{pvc}	Al_{hq}	$(Al^{3+})_{hq}$	$[Al^{3+}]_{ic}$	$(Al^{3+})_{ic}$	$\Sigma\,Al_{ic}$[c]	Al_{pvc}	Al_{hq}	ΣAl_{ic}[c]
		Concentration or Activity ($\mu mol\ L^{-1}$)									
1	55.1	251.8	516.1	271.8	35.1	187.7	24.2	369.7	74.3	39.2	53.0
	64.3	135.5	312.6	151.0	21.2	77.9	10.9	206.8	59.8	29.6	33.8
	72.3	92.0	225.3	105.1	15.8	44.6	6.7	132.1	52.8	25.9	10.2
	81.4	32.3	127.3	56.7	9.3	17.0	2.8	80.5	43.9	18.9	4.1
	81.8	17.2	86.9	34.4	6.0	7.4	1.3	48.0	36.1	14.8	3.0
	91.8	10.1	49.6	18.0	3.9	2.7	0.6	22.8	27.7	11.2	0.1
2	75.0	120.5	338.9	165.0	22.0	95.1	12.7	231.3	60.8	31.6	26.2
	89.1	67.5	192.8	88.7	12.8	37.5	5.4	120.9	46.9	22.1	4.1
	85.1	34.7	137.4	68.2	10.7	16.8	2.6	82.5	45.1	20.5	0.4
	89.8	42.9	107.4	45.4	7.8	13.1	2.2	65.2	40.3	17.6	1.8
	91.1	26.5	87.7	38.6	7.2	7.1	1.3	48.8	38.4	16.5	1.8
	98.3	17.7	62.9	22.4	4.7	4.6	1.0	31.9	32.4	12.5	0.3
3	73.1	98.7	165.7	114.9	12.8	60.7	6.8	102.1	39.6	25.4	6.2
	74.4		81.9	54.4	6.5	18.2	2.2	48.3	26.8	16.0	0.5
R^{2d}		0.90	0.76	0.82	0.79	0.83	0.83	0.77	0.59	0.73	0.86

[a] Percent of maximum dry weight within soil inoculated treatments.

[b] calc = calculated from soil solution composition; pcv = pyrocatechol violet; hq = 8-hydroxyquinoline; ic = ion chromatography.

[c] Sum of Al monomers detected by ion chromatography.

[d] For second order polynomial.

After Whitten et al., 1992.

associated surface soil horizons (predominately moder and dysmoder humus O horizons) in an attempt to relate seedling dry weight with soil solution Al^{3+} activity. Aluminum phytotoxicity was not evident for soil solutions in which Al^{3+} activity ranged from 7.7 to 64.3 $\mu mol\ L^{-1}$. In related work, total Al of 3.3 mmol L^{-1} in nutrient solution was identified as the toxic threshold for Norway spruce seedlings, and a "somewhat higher" threshold was cited for beech (*Fagus sylvatica* L.) seedlings (van Praag et al., 1985). No effort was made to express nutrient culture results in terms of Al^{3+} activity for comparison against soil solution data. In nutrient solution culture, 0.19, 0.19, 0.37, 1.48, and 2.96 mmol L^{-1} total Al resulted in reduced biomass of red spruce (*Picea rubens* Sarg.), white spruce (*P. glauca*), black spruce (*P. mariana*), jack pine [*Pinus banksiana* (Lamb.)], and white pine (*P. strobus* L.), respectively (Hutchinson et al., 1986).

Scheir (1985) found >1.85 mmol L^{-1} total Al in nutrient solution to inhibit root elongation of red spruce and balsam fir [*Abies balsamea* (L.) Mill.] seedlings. Aluminum was added as $Al_2(SO_4)_3 \cdot 18H_2O$, so it is important to recognize that complexed Al ranged from 17 to 63% of total Al present in solution. When total Al is corrected to Al^{3+} activity using an ion speciation model (Table 11.5), the toxic threshold for Al^{3+} injury appears to be $\approx$0.3 mmol L^{-1}. Thornton et al. (1987) concluded from solution culture experiments that the toxic threshold for root elongation of red spruce seedlings was 0.25 mmol L^{-1} total Al (estimated Al^{3+} activity of 0.05 mmol L^{-1}). The sixfold difference in thresholds for Al^{3+} phytotoxicity to red spruce seedlings as estimated by Scheir (1985) and Thorton et al. (1987) is reasonable when considering differences in experimental protocol and uncertainties in the estimation of Al^{3+} activity from their published results. Red spruce seedlings grown in mixed B and C horizons of a Typic Fragiothod were unaffected by total Al concentrations in soil solution ranging from 0.04 to 0.54 mmol L^{-1} (Ohno et al., 1988).

Honey locust (*Gleditsia triacanthos* L.) root elongation in solution culture was reduced for $\geq$0.15 mmol L^{-1} total Al (estimated Al^{3+} activities of $\geq$0.04 mmol

TABLE 11.5 Relative Root Elongation of Red Spruce and Balsam Fir in Relation to Al^{3+} Activity in Nutrient Solution[a]

[Al] Total mg L^{-1}	[Al] Total mmol L^{-1}	(Al^{3+}) mmol L^{-1}	$[Al^{3+}]$ %	$[AlSO_4^+]$ %	Red Spruce Exp 1 %	Red Spruce Exp 3 %	Balsam Fir Exp 1 %
0	$<10^{-7}$	$<10^{-7}$	83	12	100	100	100
25	.93	.26	64	33	87	—	91
50	1.85	.40	55	43	87	77	76
100	3.70	.57	46	53	47	71	45
200	7.41	.82	37	61	39	61	50

[a]Calculated from the data of Scheir 1985 (Wolt, 1990).
[b]Total Al partitioned into Al^{3+}, $AlSO_4^-$, and Al-hydroxy species, 20% Clark's solution, pH 3.8.

L^{-1}). Secondary root production appeared to be a more sensitive indicator of Al phytotoxicity, however, occurring at ≥ 0.05 mmol L^{-1} total Al in solution (Thornton et al., 1986).

Total Al concentrations up to 3.0 mmol L^{-1} were not detrimental to root biomass production of red oak, American beech (*Fagus gradifolia* Ehrh.), or European beech (*F. sylvatica* L.), although shoot biomass of European beech was decreased as Al in solution increased (Thorton et al., 1989).

McCormick and Steiner (1978) used nutrient solution culture to measure the effect of total Al on root elongation of six genera and 11 species of trees. They found hybrid poplars (*Populus maximowiczii* × *trichocarpa* Schreiner and Stout) to be sensitive to >0.37 mmol L^{-1} total Al in solution. Autumn-olive (*Elaeagnus umbellata* Thumb.) was intermediate in sensitivity, while species of *Quercus, Betula,* and *Pinus* were relatively tolerant of solution Al. When the Al^{3+} activities in these solutions are estimated, the thresholds for Al^{3+} phytotoxicity appear to be ≈0.1 and 0.4 mmol L^{-1} for hybrid poplar and autumn-olive, respectively. Because of the uncertain anion composition of nutrient solutions and the possibility for Al oxyhydroxide precipitation at higher levels of total Al in solution, the threshold tolerance for oak, birch, and pine is less certain (≈0.8 mmol L^{-1}).

The data available concerning forest tree response to Al in solution are restricted to these few experiments, whose objectives were not to define plant growth in terms of Al^{3+} activity per se. More comprehensive studies incorporating soil and nutrient solution approaches are necessary, but forest trees are apparently 10 to 100 times less sensitive to Al^{3+} in solution than are Al-sensitive agronomic crops, which generally exhibit Al phytotoxicity at Al^{3+} thresholds of ≤ 20 μmol L^{-1} (Table 11.6). Organic matter in solution will substantially reduce the toxic effect of Al through Al complexation (Brogan, 1964), so this aspect of Al bioavailability should be considered when defining soil solution Al effects on forest tree growth.

Microbial Toxicity of Aluminum. Soil microflora mutualistically associated with plants may exhibit Al toxicity at lower levels of Al than would be toxic to the host plant itself, but there is little evidence of Al-induced decreases in soil microbial activity under field conditions (Firestone et al., 1983).

Aluminum can inhibit nodulation of legumes. Rhizobium species, in general, are sensitive to Al (Foy, 1984). When 65 Rhizobium strains were screened, most demonstrated slower rates of growth in culture media with >50 μmol L^{-1} total Al; 40% of the strains screened exhibited no growth at this level of Al (Keyser and Munns 1979b). Aluminum toxicity was not ameliorated by increased Ca in the culture medium (Keyser and Munns, 1979a).

Total Al concentrations in leachates greater than 600 μmol L^{-1} inhibited growth of *Aspergillus* flavus spores cultured in soil leachate (Firestone et al. 1983). Ko and Hora (1972) used diluted 1:1 soil–water extracts from a Latisol and artificial media to culture *Neurospora tetrasperma*. They observed inhibition of spore germination with greater than 24 μmol L^{-1} total Al in either soil extracts or artificial media containing various sources of Al. Thompson and Medve (1984) found my-

TABLE 11.6 Threshold of Al Toxicity in Experiments Where Root Elongation Was the Measure of Response and Where Bioavailable Al was Expressed as Al^{3+} Activity in Solution

Plant	(Al^{3+}) at Phytotoxic Threshold	Rooting Medium	Reference
	mmol L^{-1}		
Gramineae spp.	< 0.0009[a]	solution, soil	Brenes and Pearson, 1973
Cotton	0.0015[a]	solution, soil	Adams and Lund, 1966
Barley	0.0015[a]	solution	Cameron et al., 1986
Coffee	0.004[a]	solution, soil	Paven et al., 1982
Cotton	0.006[a]	soil	Adams et al., 1967
Wheat	0.02[a]	soil	Baligar et al., 1992
Honeylocust	0.04[b]	solution	Thorton et al., 1986
Red spruce	0.05[b]	solution	Thorton et al., 1987
Hybrid poplar	0.1[b]	solution	McCormick and Steiner, 1978
Red spruce, balsam fir	0.3[b]	solution	Scheir, 1985
Autumn-olive	0.4[b]	solution	McCormick and Steiner, 1978
Pine, oak, birch	0.8[b]	solution	McCormick and Steiner, 1978

[a](Al^{3+}) computed by authors.
[b](Al^{3+}) estimated from solution composition.

celial growth of ectomycorrhizal fungi commonly associated with trees on acid sites to be sensitive to Al in culture media.

These data suggest toxicity to soil microbes from Al in soil solution at levels lower than would be toxic to forest trees. Data from these few investigations cannot, however, be considered diagnostic of Al toxicity to microbes in natural environments for a number of reasons: (1) growth response of microbes in culture media may not adequately reflect growth in the soil ecosystem, (2) culture media do not realistically model soil solution chemical composition in most cases, and (3) total Al, rather than bioavailable Al (Al^{3+}), is used to describe growth responses.

Studies where ion composition of the culture media would be expected to alter the amount of bioavailable Al present are inconclusive. Firestone et al. (1983) found fluoride addition to culture medium significantly reduced Al toxicity to *Aspergillus flavus,* as would be expected due to formation of Al–fluoride ion pairs and consequent reduction in the fraction of total Al which would be bioavailable. Ko and Hora (1972), however, reported no shift in the toxic threshold of Al for *Neurospora tetraspora* germination when Al was added as AlCl$_3$ · 6H$_2$O versus Al$_2$(SO$_4$)$_3$.

The results of Thompson and Medve (1984) demonstrate the need to study Al toxicity to microflora using better models of soil solution chemistry that consider Al^{3+} activity. Their laboratory screening of fungi for Al toxicity was inconsistent with field observations of fungal occurrence and distribution on acid soils.

11.5   ALUMINUM GEOCHEMISTRY

The principal reactions of Al in soils relate to their association with O^{2-} and OH^- ligands and their precipitation as oxyhydroxides. The solubility products of these oxyhydroxide precipitates have the general form $Al_x(OH)_yO_{(3/2x-1/2y)}$.

11.5.1   Aluminum in the Soil Solid Phase

Chemical reactivity of Al-bearing materials will be largely responsible for soil acidity (Thomas and Hargrove, 1984), as well as for the control of Al concentration in soil solution. Therefore, the potential effects of natural and anthropogenic processes on Al bioavailability involve components of the soil solid phase that control soil solution pH and Al activity.

Aluminosilicate minerals such as feldspars, kaolins, smectites, and mica are the principal Al-bearing material in most soils. Under most conditions in nature they control the concentration of Al in solution (Roberson and Hem, 1969). These crystalline layer silicate minerals are comprised of Si tetrahedral and Al octahedral sheets in varied arrangements and with differing degrees of isomorphous substitution. The aluminosilicate clays are chemically reactive because of their high surface to volume ratios and their permanent charge cation exchange capacities (CEC) resulting from isomorphous substitution.

The aluminosilicate minerals generally exhibit incongruent (incomplete) dissolution (section 9.3.2). When these minerals dissolve, silica is released to solution as silicic acid (H_4SiO_4) while Al is conserved in the solid phase as an oxyhydroxide. Therefore, silica is mobilized more rapidly than Al as soils are weathered and become more acid. If leaching conditions exist in a soil, Si is lost from the system and Al oxyhydroxides accumulate until they dominate the soil solid phase as gibbsite [$Al(OH)_3$]. Gibbsite has been frequently considered with regard to soil pH buffering and the control of Al^{3+} activity in soil solution (Lindsay 1979; Thomas, 1974). The dissolution of gibbsite can be expressed as

$$Al(OH)_{3s} \rightleftharpoons Al^{3+}_{aq} + 3OH^-_{aq}$$

The solubility product is, therefore,

$$K_{sp} = A_{Al^{3+}}A^3_{OH^-} \qquad [11\text{-}4a]$$

which can be written in terms of negative logarithms as

$$pK_{sp} = pAl^{3+} + 3pOH^- \qquad [11\text{-}4b]$$

Substituting $(14 - pH)$ for pOH^-, using the pK_{sp} for gibbsite (34.0; Kittrick, 1966), and rearranging results in the relationship,

$$pAl = 3pH - 8.0 \qquad [11\text{-}4c]$$

which expresses Al^{3+} activity as a function of pH (section 9.2.1). In practice this relationship does not hold for many temperate region soils. There are several conditions that may restrict the ability for equation 11-4 to describe Al^{3+} activity in a given soil.

1. Gibbsite is not present as a stable solid phase component.
2. Solutions have not achieved a steady state with respect to the solid phase (May et al., 1979).
3. The solubility of the Al oxyhydroxide mineral present is other than that of the reference mineral (Lindsay, 1979).
4. Solution or solid phase components have altered the form of Al oxyhydroxide present (Hsu, 1979; Ng Kee Kwong and Huang, 1981; Violante and Jackson 1981; Violante and Violante, 1980).
5. Kinetic considerations preclude gibbsite control over solution Al^{3+} activity (Bloom, 1983; Christophersen and Seip, 1982).

Despite these concerns, gibbsite solubility is still a useful point of departure for consideration of the solid phase control of solution Al.

Other Al oxyhydroxides that have been postulated to control Al^{3+} activity in soil solution are nordstandite or pseudoboehmite which may form in preference to gibbsite in the presence of SO_4^{2-}, Cl^-, or low-molecular-weight organic acids (Violante and Jackson, 1981; Violante and Violante, 1980). Solubility studies suggest boehmite (AlOOH) may control solution Al at pH > 6.7 (May et al., 1979).

Complex hydroxy Al polymers formed as interlayers between the lattices of 2:1 aluminosilicate minerals are another important solid phase pool of Al (Jackson, 1963; Rich, 1968). Hydroxy interlayered Al is strongly fixed and for the most part nonexchangeable. The formation of hydroxy Al polymers and their fixation into intralamellar spaces are strongly influenced by mono- and polyprotic acids (Goh and Huang, 1984; Lind and Hem, 1975; Violante and Jackson, 1981) which may also hinder the normal precipitation reactions of Al oxyhydroxides (Goh and Huang, 1985; Lind and Hem, 1975; Ng Kee Kwong and Huang, 1979b; Wang et al., 1983). Singh and Brydon (1967, 1970) found the stability of interlayered hydroxy Al to be dependent on the anion composition of the system. Sulfate destabilized Al polymers through formation of an Al-hydroxy-sulfate salt in the interlayer. The interlayer material disappeared with the subsequent formation of a crystalline Al-hydroxy-sulfate precipitate (Singh and Miles, 1978).

Sparingly soluble Al-hydroxy-sulfate compounds, although unidentified as discrete minerals in soils, have been frequently associated with the control of solution pH and Al^{3+} activity in acid soils (Adams and Rawajfih, 1977; Eriksson, 1981; Khanna and Beese, 1978; Rhodes and Lindsay, 1978; van Breeman, 1973; Wolt,

1981; Wolt and Adams, 1979b; Wolt et al., 1992). Nordstrom (1982) identified basaluminite [$Al_4SO_4(OH)_{10}$], alunite [$KAl_3(SO_4)_2(OH)_6$], and jurbanite [$AlSO_4OH \cdot 5H_2O$], as Al-hydroxy-sulfate minerals capable of supporting elevated levels of Al in the solutions of acid soils. Control of solution Al by Al-hydroxy-sulfate minerals is possible where large quantities of H_2SO_4 are generated by (1) pyrite oxidation of minespoils, (2) sulfate oxidation following drainage of marine flood-plains (van Breeman, 1973), or (3) anthropogenic acidic deposition (Eriksson, 1981; Khanna and Beese, 1978; Nordstrom, 1982; Wolt, 1981; Wolt and Lietzke, 1982).

Allophane and imogolite are amorphous aluminosilicate gels that are interme-diate products of weathering. They are responsible for large portions of pH-dependent CEC in many soils (Wada, 1977). Allophane may be described as "a hydrous aluminosilicate mineral with short range order and with a predominance of Si-O-Al bonds" (Parfitt and Saigusa, 1985). Imogolite consists of short-range, paracrystalline assemblies with the approximate composition $SiO_2 \cdot Al_2O_3 \cdot 2.5H_2O$ (Wada, 1977). Operationally, amorphous aluminosilicate gels are defined by the methods used in their extraction, which may vary considerably among investigators (Bohn et al., 1979; Wada, 1977). Amorphous aluminosilicates may be important for complexation of Al in humic horizons where Al release by primary mineral weathering exceeds organic matter accumulation (Parfitt and Saigusa, 1985). Metastable cryptocrystalline aluminosilicates are postulated to control solution Al where dissolution kinetics of Al oxyhydroxide or aluminosilicate minerals are limiting (Paces, 1978).

Complexation of Al with organic matter is an important factor controlling soil solution Al, the mobilization of Al in A horizons dominated by organic matter, and the subsequent movement of complexed Al into underlying mineral horizons (van Breeman and Brinkman, 1978). In very acid soils, carboxyl sites of organic matter are largely complexed by Al^{3+} (Bloom et al., 1979b). At pH values < 8.0, Al complexed with organic matter does not behave as an exchangeable cation. There-fore, Al-complexed organic matter contributes little to the pH-dependent CEC of soils (Thomas and Hargrove, 1984).

Bloom et al. (1979a) considered Al binding by organic matter to be an important mechanism for control of Al^{3+} activity in the soil solution below pH 5. In their experiments, addition of 2% leaf humus to soil from the B horizon of an Inceptisol decreased soil solution Al^{3+} by 40%. The degree of Al binding with organic matter will vary among soils depending on the strengths of the carboxylic acid groups present (Bloom and McBride, 1979; Hargrove and Thomas, 1982).

Exchangeable acidity in mineral soil materials (that is, that portion of soil acidity that is extractable with a neutral salt) is due predominately to Al^{3+} (Thomas and Hargrove, 1984). Both the degree of Al saturation of exchange sites and the quan-tity of exchangeable Al present will be important to the Al-buffering capacity of a soil. Unfortunately, neutral salt-extractable Al^{3+} is an inconsistent measure of ex-changeable Al^{3+}, especially in soil surface horizons where organic matter may contribute significantly to CEC (Bloom et al., 1979b). The amount of Al^{3+} extracted by neutral salts will be a function of extraction time, and of cation composition and concentration of the extracting solution (Bloom et al., 1979b; Sivasubramaniam and Talibudeen, 1972). Perhaps because of these factors, the degree of Al saturation has been an inadequate predictor of Al phytotoxicity to plants (Adams, 1984). Soil CEC with a high degree of Al saturation is suggested to be prerequisite to the control of soil solution Al^{3+} by gibbsite (Bohn et al., 1979).

11.5.2 Acid Buffering Capacity in Relation to Aluminum Geochemical Availability

The processes whereby soils are buffered against changes in acidity vary depending on the region of soil pH considered. At >pH 5.0, cation exchange in conjunction with weathering of aluminosilicate minerals buffers pH; from pH 5.0 to 4.0, cation exchange is the predominant buffering process; and in the region of pH 4.0 to 2.8, Al buffers the changes in soil pH (Bloom and Grigal, 1985; Ulrich, 1983). Acidic deposition with a large component of mineral acid will alter soil acidity if it is of sufficient magnitude to shift the acid buffering range of a soil. Soils of low to intermediate acidity are probably the most sensitive to acidic deposition-induced changes to soil acid buffering capacity (van Breeman et al., 1984).

From the standpoint of Al mobilization, soils of >pH 5.0 that are buffered by cation exchange will not be susceptible to increased Al availability until control of soil pH is shifted downward into the region of Al buffering (Bloom and Grigal, 1985; Cosby et al., 1985a, 1985b; Ruess, 1983; Ulrich, 1981a; van Breeman et al., 1984). In acid soils with low base saturation, rates of acidification are slow because of the influence of Al buffering. These soils may produce less acidity internally than is contributed as mineral acidity in acidic deposition; consequently, Al mobility may be enhanced by acidic deposition (van Breeman et al., 1984). Soils in disturbed ecosystems in various stages of vegetative succession may generate much greater quantities of internal acidity than would be contributed by deposition. These ecosystems would not be susceptible to deposition-induced acidification, and Al mobility would be a consequence of natural soil weathering processes (Krug and Frink, 1983).

11.5.3 Podzolization

Intense natural soil weathering, most typically under forest vegetation, results in eluviation of organic acids and metals from soil A horizons in the process of podzolization. Advanced stages of podzolization result in Al (as well as Fe) mobilization from E horizons and subsequent accumulation along with organic matter in Bs and Bhs horizons.

Aluminum transport is a distinct feature of podzolization that is traditionally ascribed to leaching of Al complexed with DOC (the metal-fulvate theory). Recent investigations indicate that amorphous aluminosilicate sols (proto-imogolite) may also be involved in the eluviation of Al from the upper horizons of Spodosols and its subsequent accumulation in Bs and Bhs horizons (Farmer et al., 1980; Parfitt and Saigusa, 1985). Dahlgren and Ugolini (1989) have tested these competing mechanisms for Al transport through analysis and ion speciation modeling of lysimeter solutions collected from a subalpine Spodosol. Using a charge balance approach they concluded cation transport in this soil was dominantly a result of dissociated organic ligands in DOC (Fig. 11-1). The lack of reactivity of soil solution Si with cation exchange resins as well as infrared spectroscopy of freeze-dried soil solution demonstrated the absence of proto-imogolite sol as a form of mobile Al.

11.5.4   Transient Effects of Acid Deposition on Soil Solution Aluminum

During precipitation events, acidic deposition entering the soil is unlikely to reach equilibrium between liquid and solid phases as water rapidly infiltrates and percolates through the profile (Hooper and Shoemaker, 1985). Residence time of water flowing through soil will be the crucial factor when assessing acute effects of acidic deposition on Al bioavailability (Johnson et al., 1981). Rapidly percolating soil water may not achieve a steady state with the soil mineral phase. In such cases, mechanistic rather than equilibrium considerations will determine the availability of Al in leaching waters.

Lateral water flow may further reduce contact time of leaching waters with some soil mineral horizons. When significant lateral flow occurs in shallow soils overlying hardrock, organically complexed Al may be transferred to surface waters (Driscoll and Newton, 1985) rather than undergoing precipitation in soil mineral horizons (McFee and Cronan, 1982). Humic material may act as a solid phase adsorbent, controlling solution Al under conditions of lateral flow through organic surface soil horizons (Cronan et al., 1986).

If anthropogenic SO_4^{2-} is the dominant anion in soil water during transient flow (Cronan, 1980), then Al release to leaching waters will likely involve rapid to moderately rapid events such as exchange reactions, mobilization of organometallic complexes, and possibly dissolution of amorphous Al oxyhydroxides (David and Driscoll, 1984; McFee and Cronan, 1982). Aluminum on exchange sites, in organometallic complexes, and in oxyhydroxide precipitates may represent a labile pool of soil Al released under conditions of transient flow. Because transient snowmelt and storm events may govern the majority of water flow in high elevation watersheds sensitive to acidic deposition (Hooper and Shoemaker, 1985; Johnson et al. 1981), the kinetics of Al release from this labile pool may be of more consequence than the thermodynamics of dissolution–precipitation reactions.

Temporal fluctuations in total and free monomeric Al in soil solutions and surface waters under conditions of high flow will not be adequately explained by models that assume thermodynamic equilibrium between infiltration water and soil mineral phases (Hooper and Shoemaker, 1985). Total and monomeric forms of Al in solution tend to increase during high flow conditions (David and Driscoll, 1984; Hooper and Shoemaker, 1985). This would be the consequence of (1) acidic deposition entering the soil with a corresponding depression in soil water pH, (2) buffering of acidity through release of Al from labile pools, and (3) decreased solubility of nonlabile organoaluminum complexes resulting in a proportionate re-

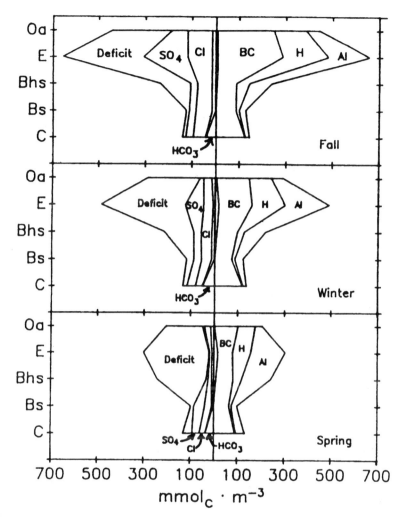

FIGURE 11.1 Charge-balance diagram of Spodosol soil solutions. Organically bound Al is assumed to contribute a 3+ charge while the hydroxyl-bound Al charge is based on summation of species. Basic cations (BC) = ΣCa,Mg,K,Na. Charge deficit is assumed to be the contribution based on dissociated organic ligands. [Dahlgren and Ugolini, 1989. *Soil Sci. Soc. Am. J.*, 1989, 53:559–566.]

duction in the fraction of total solution Al present in organically complexed form (Hooper and Shoemaker, 1985; James and Rhia, 1984). Under high-flow conditions, Al^{3+} activity in soil solution has been reported to range from 7 to 30 μmol L^{-1} (Christophersen and Seip, 1982; Driscoll et al., 1980; Hooper and Shoemaker, 1985). These levels of Al^{3+} activity may pose an environmental hazard to aquatic ecosystems (Cronan and Schofield, 1979; Driscoll et al., 1980), but seem below the toxic Al thresholds for tree species (see Table 11.6). Acute instances of Al phytotoxicity to trees would not appear to be a consequence of acidic deposition if these Al^{3+} activities are representative of waters percolating soils in low-order, acid-sensitive watersheds.

In well-drained soils where percolating soil water makes prolonged contact with the soil mineral phase, soil solution Al activity should be controlled by dissolution–precipitation of discrete mineral phases. Models treating acidic deposition-induced acidification of soils and watersheds assume some crystalline form of gibbsite controls soil solution Al^{3+} (Christophersen and Seip, 1982; Cosby et al., 1985a,b). In some watersheds, gibbsite solubility adequately explains solution Al^{3+} (Budd et al., 1981; Johnson et al., 1981), but for others this does not appear to be the case (David and Driscoll, 1984; Hooper and Shoemaker, 1985; Nilsson and Bergkvist, 1983). The identification of gibbsite as a discrete mineral phase in soils and watersheds being modeled would better justify the association of solution Al^{3+} activity with gibbsite solubility.

11.5.5 Chronic Effects of Acid Deposition on Soil Solution Aluminum

Potential chronic effects of acidic deposition on podzolization and on surface water chemistry relate to considerations of Al complexation, Al^{3+} activity, and the vertical and lateral transport of Al through soil horizons. Throughfall chemistry, soil solution pH, degree of S retention in soil horizons, and the dynamics of Al cycling through the forest canopy will all interact to determine total Al in soil solution, Al speciation, and Al mobility. The variable response of Al to acidification of O horizons indicates that generalizations concerning acidic deposition effects on O horizon Al chemistry will not be possible (James and Rhia, 1984).

For many soils and watersheds considered to be sensitive to acidic deposition, the predominant chronic effect of acidic deposition appears to be a shift in the mineral phase that controls Al^{3+} activity in soil solution. For sulfur-retentive soils, conditions of low pH and high sulfate input (analogous to the hypothesized effect of acidic deposition on soils) may favor the control of soil solution Al^{3+} activity by an Al-hydroxy-sulfate mineral (Nordstom, 1982; van Breeman, 1983; Wolt, 1981; Wolt et al., 1992). The reaction involves gibbsite weathering in an acid, high SO_4^{2-} environment and the formation of an Al-hydroxy-sulfate phase that would support increased levels of Al^{3+} in solution (section 9.2.3). Such a mechanism may explain elevated Al availability in natural waters and soil solutions thought to be adversely affected by acidic deposition (David and Driscoll, 1984; Khanna and Beese, 1978; Johnson et al., 1981; Nilsson and Bergkvist, 1983; Ulrich et al., 1980; Weaver et al., 1985).

A number of Al-hydroxy-sulfate minerals have been postulated to control soil solution Al. However, they represent minor components of the soil solid phase and thus have not been identified mineralogically. Nordstrom (1982) has summarized the evidence for the occurrence of Al-hydroxy-sulfates in soil systems as it relates to acidic deposition.

Investigations of chronic effects of acidic deposition on soils and trees implicate soil solution Al as an important contributing factor to forest decline. Slight declines in soil solution pH in the upper soil profile of a Dystrocrept in the Solling Highlands of Germany during the period 1966 to 1979 resulted in elevated total Al in soil solution (Ulrich et al., 1980) which is related to retention of SO_4^{2-} as an Al-hydroxy-sulfate mineral (Khanna and Beese, 1978). A shift from SO_4^{2-} retention to SO_4^{2-} loss from this site, beginning in 1975, suggests to Khanna et al. (1987) that the Al-hydroxy-sulfate mineral that may have formed has begun to precipitate with continued atmospheric inputs of H^+. Total soil solution Al, which was <27 μmol L^{-1} prior to 1973, remained in the range 37 to 74 μmol L^{-1} during the period 1973 to 1979 (Ulrich et al., 1980). The elevated Al levels were considered sufficient to damage roots of beech *(Fagus silvatica),* although this has not been substantiated in field or greenhouse studies conducted elsewhere (Scheir, 1985; van Praag and Weissen, 1985; see Table 11.6).

Subsequent nutrient culture experiments with beech have served to clarify observations in the Solling Highlands (Hutterman and Ulrich, 1985). Aluminum-induced Ca deficiency was observed as root necrosis and ultrastructural changes in beech seedlings when the Ca:Al mole ratio in solution culture was <1 (Hutterman and Ulrich, 1984; Ulrich 1981a). It is therefore suggested that Ca:Al mole ratios <1 in the presence of free Al^{3+} will damage roots of spruce and beech (Matzner and Ulrich, 1985).

Ulrich and coworkers view acidic deposition as a "predisposing stress" responsible for changes in ion balance in soil solution resulting in sometimes concurrent effects of P, Ca, and Mg deficiency and Al and Mn toxicity (Matzner and Ulrich, 1985; Mayer and Ulrich, 1977; Ulrich et al., 1980). Demonstrating the interrelationship of acidic deposition, ion balance in soil solution, and forest decline is complicated by variation in soil solution composition and fine root turnover both spatially and temporally (Matzner and Ulrich, 1985). For example, beech seedling die-back is observed most often at the base of dominant beech within a stand where atmospheric deposition via stemflow is proportionately greater than for the tree stand in general (Hutterman and Ulrich, 1984).

In contrast to the Solling Highlands, most other forests subject to acidic deposition have much lower levels of atmospheric S deposition and soil solution Al. Comparison of ten forested catchments in western Europe and North America subject to S deposition ranging from 8 to 80 (Solling) kg ha^{-1} yr^{-1}, indicated total monomeric Al [Al^{3+}, $AlOH^{2+}$, $Al(OH)_2^+$] to range from <1 μmol L^{-1} in an Ultisol from the southeastern USA to >240 μmol L^{-1} for an Inceptisol from Solling (Cronan et al., 1987). In Spodosols, the range was 15 to 80 μmol L^{-1}. Differences across locations were associated dominantly with differing mechanisms of soil buffering and anion retention. Fine roots (<2 mm dia) sampled from soil B horizons at sites in West Germany and New York had Ca:Al mole ratios near 0.3 which would suggest limited ability for root regeneration by the criteria of Ulrich et al. (1984). Joslin and Wolfe (1992) associated reduced fine-root biomass of red spruce in high elevation catchments of the southern Appalachians USA with Ca:Al mole ratios <0.05.

In an oak-birch woodland in the Netherlands subject to high inputs of atmospheric $(NH_4)_2SO_4$, soil solutions from Inceptisol and Entisol root zones averaged

620 μmol Al L^{-1} and ranged from 100 to >1300 μmol L^{-1} (Mulder et al., 1987). These levels of solution Al are considerably higher than reported elsewhere and are associated with increased soil solution NO_3, indicating nitrification of NH_4^+ and organic N as the dominant source of acid generation. Although soil solution Al was highest in the summer, Al transport was more consequential in the dormant season when water flux was the highest. Polymeric Al was always negligible, while monomeric Al was dominantly (>80%) Al^{3+}. Organically complexed Al comprised from 30 to 50% of total monomeric Al in shallow depths (10 cm) and comprised <20% of total monomeric Al at greater depths. Theoretically, the Al-hydroxy-sulfate, jurbanite, could control Al^{3+} activities in these soils, but this is not supported by the lack of SO_4 retention by the soil.

Inconsistencies in observations of Al toxicity across experimental locations and over time are attributable to several interacting factors which may be operative in acidic forest soils:

1. Conditions favoring high N availability (Ulrich et al., 1980) or P deficiency (Johnson and Todd, 1984) may increase the incidence of Al-induced damage to roots.
2. Wetting and drying cycles (Nilsson and Bergkvist, 1983; Weaver et al., 1985) may favor release of Al from labile pools (Hooper and Shoemaker, 1985).
3. Ca deficiency may lower the threshold for Al injury of roots (Ulrich, 1981a,b; vanPraag and Weissen 1985).

Other factors in acid soil infertility, such as Mn toxicity, have only been remotely addressed as having a potential role in acidic deposition-induced effects on forest soils.

11.5.6 Aluminum Geochemistry in Agronomic Ecosystems

The fundamental chemistry of Al reactivity in agronomic ecosystems is no different than that described for natural or low-intensity management forest ecosystems. However, the sources, magnitude, and mitigation of acid-generating processes that influence Al geochemical availability in agronomic ecosystems are markedly different from forest ecosystems. A host of anthropogenic processes relating to the management of agronomic ecosystems will alter the Al status of cropped soils, in addition to the natural processes of acid generation occurring as a consequence of soil weathering and the acceleration of these processes by anthropogenic acid deposition. The principal influences on Al availability in agronomic ecosystems are fertilizer, lime, and residue management. Detailed discussions elsewhere of soil acidity in agroecosystems describe how Al availability can be controlled through ecosystem management (Adams, 1984) and how the incidence of phytotoxicity to crops may be reduced by genetic means (Foy, 1976). As for natural systems, soil solution Al is the diagnostic tool whereby the subtleties of Al cycling and avail-ability in managed ecosystems are best described.

NOTES

1. In its proper thermodynamic context, activity is a unitless quantity:

$$A_i = C_i\gamma_i$$

where γ_i is the molar activity coefficient and has units of inverse concentration (see Solution of Electrolytes in section 3.3.5). However, activity is frequently expressed with accompanying units of concentration in soils literature, especially when dealing with soil fertility and plant nutrition, to connote "effective concentration." This convention is used herein; thus, a stated ion activity of 1.5 μmol L^{-1} refers to an ion activity of 1.5×10^{-6} with a corresponding molar activity coefficient with units of L mol^{-1}.

CHAPTER 12

TRACE ELEMENTS IN SOIL SOLUTION

12.1 SOIL BIOGEOCHEMISTRY OF TRACE ELEMENTS

Trace metals and inorganic ligands are of importance to soil and environmental chemistry from the perspectives of plant nutrition, toxicity, and ecotoxicity. Elucidation of the soil environmental chemistry of these trace elements is complicated by several factors.

1. Low natural abundance of trace elements in soils and limited solubilities lead to submicromolar to nanomolar occurrence in soil solution.
2. High degrees of trace element complexation, chelation, and ion pair formation result in uncertain analysis and characterization of the biogeochemically available fractions in soil solution.
3. Speciation is frequently complicated by sensitivity of trace elements to redox transformations.
4. Trace element biological availability and uptake is affected by both antagonistic and synergistic competitive interactions with other trace elements and with solution macrocomponents.

Table 12.1 summarizes the relative natural abundances and soil solution concentrations of some biologically consequential trace elements. The natural abundances may be somewhat instructive, especially with regard to micronutrient elements, but these numbers belie the environmental significance of trace elements. Trace elements are special concerns relative to plant nutrition or ecotoxicity on a localized basis where their incidence in the soil environment results from specialized problems of ecosystem management or waste disposal; therefore, trace element

TABLE 12.1   Representative Natural Abundances of Trace Elements in Soil and Concentrations in Soil Solution

Element	Whole soil Concentration, mmol kg^{-1}	Soil Solution Concentration		
		μmol L^{-1}	μmol kg^{-1} at 10% Moisture Content	Available Fraction, $\times 10^6$
Cr	1.0	0.01	0.001	1
B	0.9	5	0.5	500
Zn	0.8	0.08	0.01	10
Be	0.7	0.1	0.01	17
Ni	0.5	0.17	0.02	730
Cu	0.3	1	0.06	175
Co	0.1	0.08	0.008	63
As	0.1	0.01	0.0013	17
Sn	0.08	0.2	0.02	200
Pb	0.05	0.005	0.0005	10
I	0.04	0.08	0.01	230
Mo	0.02	0.0004	0.00004	80
Cd	0.001	0.04	0.004	8333
Se	0.003	0.06	0.006	2500
Hg	0.0001	0.0005	0.00005	333

Adapted from Bowen, 1966; Swaine, 1969; Schacklette and Boerngen, 1984.

environmental concentrations of concern may be orders of magnitude removed from whole soil natural abundances of Table 12.1.

The topic of biological and geochemical availability of trace elements is very extensive and is shared by a wide variety of disciplines.[1] Even within soil chemistry literature there is considerable information relevant to discussions of soil solution chemistry of trace elements. This chapter is not intended to review this literature in depth, but rather it presents several salient examples of particular importance to considerations of soil solution composition as an indicator of trace element soil biogeochemical availability.

12.2   SPECIATION OF TRACE METALS ARISING FROM WASTE DISPOSAL

Many trace elements occur in soils at elevated concentrations from introduction into the environment through waste disposal or utilization. Waste composition is dependent on the source; Table 12.2 presents examples of trace element content of industrial, municipal, and animal wastes, and projects soil concentrations arising from their application. Comparison with Table 12.1 indicates the degree that waste application to soil can exceed natural abundances. The biogeochemical availability of waste-borne trace elements accumulating in soils are of concern with respect to

TABLE 12.2   Representative Trace Element Concentrations in Three Organic Wastes and the Impact of Soil Application on Soil Trace Element Content

Element	mmol kg^{-1}			Fraction of Natural Abundance When Applied at 10 MT ha^{-1d}		
	Municipal Sludge[a]	Industrial Sludge[b]	Manure[c]	Municipal Sludge[a]	Industrial Sludge[b]	Manure[c]
B	6	10	2	3	5	1
Cd	0.1	0.03		53	16	
Co	0.2	18	0.02	1	88	0.08
Cr	5	122		3	61	
Cu	8	2	0.3	14	3	0.4
Pb	1	0.1		14	1	
Mn	5	57	3	0.2	2	0.09
Mo		0.1	0.02		2	1
Hg	0.01	0.001		47	3	
Ni	1.5	1		1	1	
Sb	0.1	0.6		0.03	0.2	
Zn	18	33	1	11	20	1

[a]Aerobic sludges (Jacobs et al., 1981; Naylor and Loehr, 1982; Sommers, 1977).
[b]Aerobic sludge from textile fiber wastestream.
[c]Dairy manure (USDA, 1979) or farmyard manure (Atkinson et al., 1958).
[d]10 MT per ha, 15-cm depth of incorporation, soil bulk density = 1.33

potential phytotoxicity, plant uptake and bioaccumulation of toxic elements, and transport to ground or surface water of toxic elements.

The biogeochemical availability of a trace element may be influenced to some degree by the waste source in addition to soil environmental conditions; because the waste-borne element must first undergo release to soil solution before it can be environmentally available. The nature of the waste material therefore may control the intensity of a trace element in the soil solution. For example, sludges may harbor trace metals sorbed to organic matter or as soluble precipitates limiting their availability to soil solution, or they may release organic or inorganic ligands to soil solution that effectively complex trace elements to render them nonbiologically active. Thus, the nature of the sludge may determine the predominant form of retained trace metal and the degree to which that form has a controlling influence on soil reactivity. Trace elements are mostly nonlithophillic, so their occurrence and reactivity in soils is not well documented. The fate of waste-borne trace elements in soil environments is incompletely known in terms of the solid phase control of soil solution concentrations and the nature of speciation that occurs within soil solution.

Examples of long-term liquid waste application on trace element concentrations in groundwater are presented in Table 12.3. For these examples, the soil overburden appears to have had a strong ameliorating effect on trace element mobility as there are no clearly discernable differences in groundwater composition from treated and control sites. For the CA location, trace metals in the effluent wastestream were

TABLE 12.3   Effect of Long-Term Wastewater Application on Trace Element Concentrations in Receiving Groundwater

Element	Source	Groundwater Composition μmol L⁻¹ Treated	Control
Rapid Infiltration of Mixed Industrial and Municipal Waste, CA (30 years)			
Cu	0.05	0.06	0.02
Fe	0.7	0.4	0.2
Mn	0.1	2.3	0.0
Ni	0.09	0.17	0.00
Pb	0.03	0.05	0.00
Zn	0.07	0.11	0.01
B	13	11	10
Organic C mmol L⁻¹	21	1	0
Rapid Infiltration of Municipal Waste, WI (20 years)			
Fe	0	38	0
Mn	0	2	0
Zn	34	152	273
As	0	1	0
B	63	51	23
Se	0.15	0.06	0.06
Organic C mmol L⁻¹	3	1	1
Slow Infiltration of Municipal Waste, ND (24 years)			
Cd	0.09	0.00	0.00
Co	0.34	0.00	0.00
Cr	0.39	0.48	0.00
Cu	0	1	0
Fe	13	27	50
Hg	0.005	0	0
Mn	1	16	2
Ni	1	0	0
Pb	0.5	0.0	0.0
Zn	1	5	1
As	0.09	0.18	0.05
B	85	45	61
Se	0.1	0.3	0.1
Organic C mmol L⁻¹	3	7	1

Mullins and Sommers, 1983.

complexed extensively with organic materials, but in groundwater some shifts from organic to inorganic complexes were observed. Generally, distribution of ion species and the proportion of total ion in solution that was in free ionic form was very similar for treated and control groundwater (Mullins and Sommers, 1983).

Emmerich et al. (1982a,b) simulated sewage sludge application to soil in disturbed soil column leaching studies and monitored soil solution composition of saturation extracts in order to better describe the amelioration of waste-borne metals by soil. Although soil solution concentrations of Cd, Cu, Ni, and Zn were elevated with sludge application, the shifts were within the ranges for untreated soils and did not contribute to mobilization of trace metals through the soil columns (Fig. 12.1). Soil solution trace metals were present at $\geq 50\%$ of their total solution phase concentrations in the free ionic form, whereas they were extensively complexed with organic matter at the sludge–soil interface. Downward shifts in solution pH from the sludge–soil interface to the underlying soil resulted in some of the observed changes in free ion activities (Emmerich et al., 1982b). This was particularly true for Cu which occurred exclusively as organic complexes at the sludge–soil interface but was present dominantly as the free ion in the more acid soil solution. Trace metals were present in these soil solutions at activities well below that predicted from solubility relationships for soil minerals.

12.3 COMPLEXATION REACTIONS OF TRACE ELEMENTS

Chapter 7 describes complexation of soil solution components and presents a compilation of principle complexation coefficients for dominant elements of interest in soil solution (see Table 7.6). Trace metals that accrue in soils from waste utilization/ disposal as well as via other anthropogenic processes (such as Cd addition as a trace constituent in fertilizers) exhibit complexation reactions markedly different from common lithophillic metals of interest in soil solution (that is, Na^+, K^+, Mg^{2+}, Ca^{2+}, Fe^{3+}, and Al^{3+}; Table 7.6). Thus, the chemical reactivity, speciation, and bioavailability of certain trace elements is frequently different than that of the more prevalent soil solution components. The soil biogeochemistry of trace elements is related to understanding these differences.

12.3.1 Hard and Soft Acids and Bases

The principle of Hard and Soft Acids and Bases (HSAB) (Pearson, 1973) affords a way for differentiating types of complexation exhibited by trace elements in soils, in contrast to the more commonly occurring lithophillic elements. This principle arises from the theory of hard and soft ions (Ahrland et al., 1958) that characterizes metal ions as class A and B relative to the propensity to form more stable complexes with the initial ligand atom within a periodic group or with a later member of that

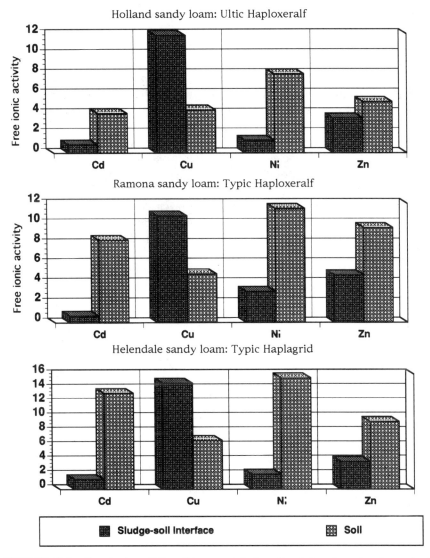

FIGURE 12.1 Free ionic activity of trace metals in soil solution for saturation extracts obtained at the sludge–soil interface and from soil in the 10-cm below the zone of application [from the data of Emmerich et al., 1982]. Free ionic activities expressed as effective concentrations ($\times 10^{-9}$ mol L^{-1} for Cd^{2+}, $\times 10^{-8}$ mol L^{-1} for Cu^{2+}, and 10^{-7} mol L^{-1} for both Ni^{2+} and Zn^{2+}).

group. That is, class A metal ions are those forming stable complexes in the orders:

$F^- \gg Cl^- > Br^- > I^-$ (periodic group VIIB ligands)

O ligands $\gg$ S ligands   (periodic group VIB ligands)

N ligands $\gg$ P ligands   (periodic group VB ligands)

Class B metals exhibit opposite trends and "transition" metals are borderline between these classes. For example, Al^{3+}, a class A metal ion, forms its most stable complex in periodic group VIIB with F^-; Hg^{2+}, a class B metal ion, forms its most stable complex in group VIIB and I^- (Fig. 12.2).

Class A and B distinctions are attributable to the number of outer shell electrons a metal ion possesses, with class A metals generally having d^0 electron configurations and class B generally having ndl^0 and ndl^0 $(n + l)s^2$ configurations. The class A metal ions are visualized as having spherical symmetry with electron sheaths that are not easily deformed by adjacent electric fields, such as those induced by nearby ions. The class B metals have an electron sheath that is nonspherical and which may be deformed by adjacent electric fields; thus, they are more readily polarized.

Class A ligands are F^- and oxygen and nitrogen donor ligands; other halides and sulfur and phosphorus donor ligands are class B ligands. The class A ligands are highly electronegative with low polarization and oxidize with difficulty. Con-

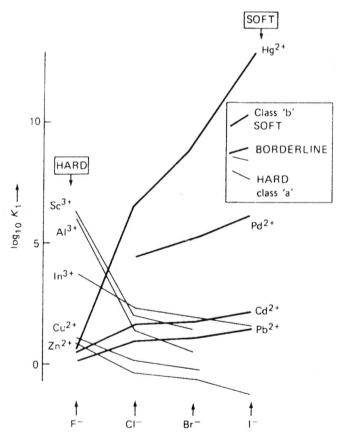

FIGURE 12.2   Stability constant trends and their relation to the Class A/Class B and Hard and Soft Acids and Bases classifications. [Burgess, 1985]

versely, the class B ligands have low electronegativity, are highly polarized, and tend to oxidize.

The HSAB principle postulates that the most stable complexes occur between like combinations of metals and ligands—A with A and B with B; unlike combinations are correspondingly less stable. The terminology popularized by Pearson (1973) defines class A metals and ligands as Hard Lewis Acids and Bases, and class B metals and ligands as Soft Lewis Acids and Bases. Several intermediate cases occur as Borderline Acids and Bases. Classification by HSAB principles is shown in Table 12.4 for important soil solution components.

The HSAB principle is empirical but yields useful generalizations concerning complexation reactions of ions in soil solution. Thermodynamically, hard acid–hard base reactions are entropy-driven whereas soft acid–soft base reactions are enthalpy-driven. Unlike most dominant cations in soil solution, which are hard acids (that is, Na^+, K^+, Mg^{2+}, Ca^{2+}, Fe^{3+}, and Al^{3+}), many trace metals tend to be soft acids. Speciation of trace metals will be more highly sensitive to organic compounds (soft bases, Table 12.4) in solution than will be dominant lithophillic cations. Increased concentrations of Cl^- (a borderline soft base) will significantly affect trace metal speciation while having a limited influence on the speciation of hard acid cations.

TABLE 12.4 Characteristics of Some Metals and Ligands Occurring in Soil Solution Relative to the Principle of Hard and Soft Acids and Bases (HSAB)

	Δ log $K_1{}^a$	Metals	HSAB
	> 4	H^+, Be^{2+}, Al^{3+}, Fe^{3+}, Si^{4+}	Hard
Class A	1 to 4	Na^+, K^+, Mg^{2+}, Ca^{2+}, Mn^{2+}, Sn^{2+}, Cr^{3+}, Co^{3+}	Acid
	0 to 1	Fe^{2+}, Co^{2+}, Ni^{2+}, Cu^{2+}, Zn^{2+}, Ba^{2+}	Borderline
		Pb^{2+}	Acid
Class B	0 to -1	Cd^{2+}	Soft
	< -1	Ag^+, Hg^{2+}	Acid

	Ligands[b]	HSAB
Class A	F^-, NH_3, R—NH_2, H_2O, OH^-, O^{2-}, R—OH, CH_3COO^-, CO_3^{2-}, NO_{3-}, PO_4^{3-}, SO_4^{2-}	Hard Base
	$C_6H_5NH_2$, NO_{2-}	Borderline
	Br^-, Cl^-	Base
Class B	C_6H_6, R_3—As, R_2—S	Soft
	R—SH, I^-	Base

$^a\Delta$ logK_1 = $log_{10}K_1(F^-)$ − $log_{10}K_1(Cl^-)$, where $K_1(L^-)$ represents the stability coefficient for the $1:1$ metal-ligand complex, (ML).
^{b}R is an organic ligand.

12.3.2 Irving–Wallace Order

A further important generalization concerning trends in stability constants for trace metals is *Irving–Wallace order* which applies in particular to divalent first-row transition metals (borderline acids), $Mn^{2+} \rightarrow Zn^{2+}$. Irving–Wallace order shows a general trend in increasing $1:1$ stability of metal–ligand complexes extending along the first-row transition metals from Mn^{2+} through Cu^{2+} but decreasing from Cu^{2+} to Zn^{2+} (Fig. 12.3).

Irving–Wallace order may be explained in terms of *ligand field theory*, which treats metal–ligand reactions variously on the basis of electrostatic crystal theory or from the standpoint of molecular orbital theory.[2] Ligand field theory shows that asymmetries in *d*-shells of the transition metal ions give rise to *ligand field stabilization energies* (LFSE) that account for from 5 to 10% of the total energies of metal–ligand combinations.

The data presented in Figure 12.3 can be generalized in terms of LFSE as follows. First, following from equation 9-1,

$$\log K_1 = \frac{-\Delta F}{2.303RT} \qquad [12\text{-}1a]$$

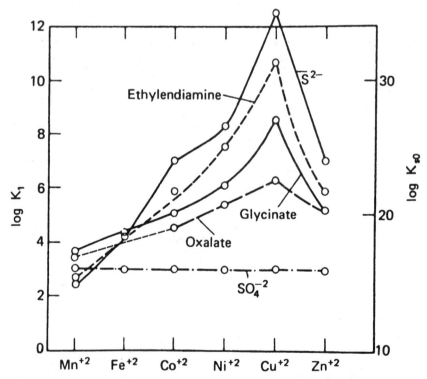

FIGURE 12.3 Stability constants (K_1) of $1:1$ complexes of transition metals and solubility products (K_{SO}) of their sulfides (Irving Williams series). [Stumm and Morgan, 1981]

At standard temperature and pressure (298.15°K and 101.3 kPa),

$$\log K_1^\circ = \frac{-\Delta F^\circ}{5.708} \qquad [12\text{-}1b]$$

Since $T\Delta S^\circ \approx$ constant for metal complexes with a specific ligand along the transition metal series ($Mn^{2+} \rightarrow Zn^{2+}$) changes in $\log K_1^\circ$ are due to changes in ΔH° across the series. Thus, $\log K_1^\circ \propto$ LFSE across the series described by Irving–Wallace order. Ligand field theory shows that LFSE increases within the series as the d-orbital electrons shift from octahedral to tetrahedral coordination; an energy change occurs through stabilization of d-orbital splitting of the aquometal ion as a ligand replaces water in the complex. The Mn^{2+} (d^5) and Zn^{2+} (d^{10}) ions exhibit no LFSE gains, but progressive gains are noted as the series extends from Fe^{2+} through Cu^{2+}.

It is anticipated that LFSE will increase along the transition metal series while all other factors contributing to ΔH° will be relatively constant. Thus, LFSE is the primary cause of the types of stability relationships observed across the Irving–Wallace order. This qualitative interpretation based on ligand field theory has been compared to thermodynamic calculations for tetrahedral to octahedral transformations of the type

$$MCl_4^{2-} + 6H_2O \rightleftharpoons M(H_2O)_6^{2+} + 4Cl^-$$

Figure 12.4 shows the correspondence in change in ΔH° for M^{2+} (where M^{2+} is a member of the series $Mn^{2+} \rightarrow Zn^{2+}$) when the change in ΔH° is thermodynamically based or calculated on the basis of LFSE. The correspondence observed justifies the conclusion that the Irving–Wallace order for $\log K_1$ is explainable on the basis of ligand field stabilization (Cotton and Wilkerson, 1966).

12.4 CADMIUM

Cadmium availability in the environment has implications relative to mammalian toxicity. Animals are more sensitive to Cd than are plants; therefore, plants may accumulate Cd to levels of toxicological concern without evidence of phytotoxicity. [Severe health effects are observed when diets of domestic animals contain 44 μmol Cd L^{-1} (National Research Council, 1980), but plant tissue concentrations will exceed 60 μmol Cd L^{-1} before yield reductions are observed (Sommers, 1980; Bingham et al., 1983).] Cd may be phytotoxic, but yield decrements are not consistently observable until soil Cd reaches concentrations (≈ 0.1 mmol kg^{-1}; Mahler et al., 1980) uncharacteristic of expected environmental exposures (see Tables 12.1 and 12.2). Plant responses are highly variable due to manifold factors of the soil environment and of plant species that will ameliorate potential Cd phytotoxicity.

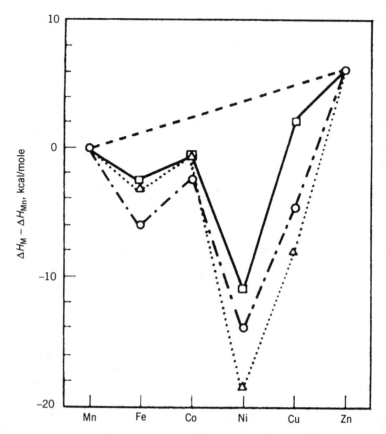

FIGURE 12.4 Enthalpies of reaction for $MCl_4^{2-} + 6H_2O \rightleftharpoons M(H_2O)_6^{2+} + 4Cl^-$ expressed as differences between ΔH for a particular metal, M, and ΔH for the Mn^{2+} compounds. Values represented by □ and ○ were derived in different ways from thermodynamic data, while values represented by △ were calculated from LFSE differences. In each case, the ΔH values are plotted relative to the interpolated value between Mn^{2+} and Zn^{2+} as indicated by the straight line between points for these two ions. [From Blake and Cotton, 1964.]

The soil biogeochemical availability of Cd is related to processes of distribution (exchange and precipitation/dissolution) between soil solid and solution phases and speciation within soil solution.[3] Whole soil parameters such as pH, CEC, organic carbon percent, and clay content are inconsistent measurements of Cd availability, since the biogeochemically relevant form of the molecule is Cd^{2+}. The occurrence of Cd^{2+} in soil environments will largely be a function of soil solution composition and total Cd speciation in soil solution (Bingham et al., 1984; Hirsch et al., 1989; Hirsch and Banin, 1990).

The competitive ion effect between Cd^{2+} and other divalent metals is a further important determinate of Cd soil biogeochemical availability. Divalent ions may compete with Cd^{2+} for binding sites in organic wastes and soil organic matter

(Sikora and Wolt, 1986), ligand complexes in solution phase, and for transporting ligands or sparingly soluble organometallic compounds within plant roots (Chino and Baba, 1981; Girling and Peterson, 1981).

12.4.1 Ion Pairs and Complexes

The biogeochemical availability of total Cd in soil solution is significantly influenced by numerous complexation reactions that reduce the pool of biologically active Cd (that is, Cd^{2+}). Important hydrolysis and complexation equilibria for Cd in soil solution are summarized in Table 12.5. The complexes of greatest importance are those with soft bases, as Cd^{2+} is a soft acid. Thus, complexation equilibria with inorganic ligands such as Cl^- and with certain types of organic ligands may be important (Table 12.4). The relevant Cd–ligand complexes occurring in a particular soil solution, however, will additionally be affected by the concentrations of Cd and ligands present.

Direct determination of Cd speciation in soil solution and other natural waters for the most part leads to uncertain resolution of species due to low concentrations of total Cd (Cd_T) in natural aqueous systems. As a result, computational approaches are widely used to estimate Cd speciation in soil solution.

Mahler et al. (1980) treated soils with sewage sludge amended with $CdSO_4$, measured total metal and ligand concentrations in saturation extracts, and modeled this information to determine the projected distribution of free Cd^{2+} and Cd–ligand complexes. Free Cd^{2+} comprised $\approx 66\%$ of Cd_T in saturation extracts from an acid Redding fine sandy loam (Abruptic Durixeralf, pH 5.1) and a slightly alkaline Holtville clay (Typic Torrifluvent, pH 7.6) where Cd_T ranged from 0.04 to >80 $\mu mol\ L^{-1}$. The dominant complexes occurred with SO_4 and organic carbon in the acid soil. In the alkaline soil, Cd complexes shifted from organic carbon to SO_4^{2-} and CO_3^{2-} as consequential ligands. The formation of Cd–Cl complexes increased in both soils as Cd_T increased.

Bingham et al. (1984) applied combinations of $Cd(NO_3)_2$, $CaCl_2$, and $CaCO_3$ to Redding fine sandy loam, measured total metal and ligand concentrations in saturation extracts, and conducted ion speciation modeling to determine Cd distribution in soil solution. Free Cd^{2+} and the $CdCl^+$ ion pair were the only species of consequence for both limed (pH 7) and unlimed (pH 4) soil. For the limed soil, Cd_T ranged from 0.01 to 0.12 $\mu mol\ L^{-1}$ and Cd^{2+} ranged from 23 to 61% of Cd_T dependent on the rates of Cd and Cl application. The unlimed soil supported considerably higher Cd_T concentrations in soil solution (0.3 to 24.5 $\mu mol\ L^{-1}$) and Cd^{2+} ranged from 24 to 84% of Cd_T, again dependent on rates of Cd and Cl application.

Hirsch and Banin (1990) treated saturated pastes of three calcareous soils (pH 7.5 to 8.5) with $Cd(NO_3)_2$, measured total metal and ligand concentrations in saturated paste extracts, and modeled Cd speciation. Free Cd^{2+} and $CdHCO_3^+$ dominated soil solutions, where they comprised approximately 35 and 45%, respectively, of Cd_T which ranged from 12 to 32 nmol L^{-1}. The computational model for Cd

TABLE 12.5 Hydrolysis Species and Solution Complexes of Cadmium

Equilibrium Reaction	log K°
Hydrolysis Species	
$Cd^{2+} + H_2O \rightleftharpoons CdOH^+ + H^+$	-10.10
$Cd^{2+} + 2H_2O \rightleftharpoons Cd(OH)_2^\circ + 2H^+$	-20.30
$Cd^{2+} + 3H_2O \rightleftharpoons Cd(OH)_3^- + 3H^+$	-33.01
$Cd^{2+} + 4H_2O \rightleftharpoons Cd(OH)_4^{2-} + 4H^+$	-47.29
$Cd^{2+} + 5H_2O \rightleftharpoons Cd(OH)_5^{3-} + 5H^+$	-61.93
$Cd^{2+} + 6H_2O \rightleftharpoons Cd(OH)_6^{4-} + 6H^+$	-76.81
$2Cd^{2+} + H_2O \rightleftharpoons Cd_2OH^{3+} + H^+$	-6.40
$4Cd^{2+} + 4H_2O \rightleftharpoons Cd_4(OH)_4^{4+} + 4H^+$	-27.92
Halide Complexes	
$Cd^{2+} + Br^- \rightleftharpoons CdBr^+$	2.15
$Cd^{2+} + 2Br^- \rightleftharpoons CdBr_2^\circ$	3.00
$Cd^{2+} + 3Br^- \rightleftharpoons CdBr_3^-$	3.00
$Cd^{2+} + 4Br^- \rightleftharpoons CdBr_4^{2-}$	2.90
$Cd^{2+} + Cl^- \rightleftharpoons CdCl^+$	1.98
$Cd^{2+} + 2Cl^- \rightleftharpoons CdCl_2^\circ$	2.60
$Cd^{2+} + 3Cl^- \rightleftharpoons CdCl_3^-$	2.40
$Cd^{2+} + 4Cl^- \rightleftharpoons CdCl_4^{2-}$	2.50
$Cd^{2+} + I^- \rightleftharpoons CdI^+$	2.28
$Cd^{2+} + 2I^- \rightleftharpoons CdI_2^\circ$	3.92
$Cd^{2+} + 3I^- \rightleftharpoons CdI_3^-$	5.00
$Cd^{2+} + 4I^- \rightleftharpoons CdI_4^{2-}$	6.00
Ammonia Complexes	
$Cd^{2+} + NH_4^+ \rightleftharpoons CdNH_3^{2+} + H^+$	-6.73
$Cd^{2+} + 2NH_4^+ \rightleftharpoons Cd(NH_3)_2^{2+} + 2H^+$	-14.00
$Cd^{2+} + 3NH_4^+ \rightleftharpoons Cd(NH_3)_3^{2+} + 3H^+$	-21.95
$Cd^{2+} + 4NH_4^+ \rightleftharpoons Cd(NH_3)_4^{2+} + 4H^+$	-30.39
Other Complexes	
$Cd^{2+} + CO_2(g) + H_2O \rightleftharpoons CdHCO_3^+ + H^+$	-5.73
$Cd^{2+} + CO_2(g) + H_2O \rightleftharpoons CaCO_3^\circ + 2H^+$	-14.06
$Cd^{2+} + NO_3^- \rightleftharpoons CdNO_3^+$	0.31
$Cd^{2+} + 2NO_3^- \rightleftharpoons Cd(NO_3)_2^\circ$	0.00
$Cd^{2+} + H_2PO_4^- \rightleftharpoons CdHPO_4^\circ + H^+$	-4.00
$Cd^{2+} + P_2O_7^{4-} \rightleftharpoons CdP_2O_7^{2-}$	8.70
$Cd^{2+} + SO_4^{2-} \rightleftharpoons CdSO_4^\circ$	2.45

speciation was validated by comparison of modeled Cd^{2+} values with Cd^{2+} measured by ion selective electrode.

Sorption of Cd and Na-montmorillonite exhibits dependence on the dominant complexing ligand in solution (Garcia-Miragaya and Page, 1976; Hirsch et al., 1989) as shown in Figure 12.5 for $Cd_T \leq 1$ μmol L^{-1}. Enhanced sorption in the presence of HCO_3^- may be due to the formation of the $CdHCO_3^+$ ion pair and its subsequent sorption, or from the precipitation of $CdCO_3$. Decreased sorption in the presence of Cl^- may be due to formation of the $CdCl^+$ ion pair, supporting greater total Cd in solution.

From these examples, Cd speciation in soil solution is clearly not easily generalized. The propensity for certain ion pairs to occur from the HSAB principle is complicated by the relative composition and concentrations of soil solution components and the influence of the soil solid phase as a sink for precipitated or adsorbed Cd and Cd complexes.

12.4.2 Bioavailability

As mentioned previously, Cd phytotoxicity resulting in crop yield reduction is uncommon because Cd entering the soil environment is largely immobilized in non-bioavailable form. Page et al. (1972) found the threshold Cd_T concentration for yield reduction of vegetable crops ranged from ≈ 0.4 to 9 μmol L^{-1} depending on plant species. Thus, the critical threshold for the most sensitive plant studied (beet) was about tenfold higher than is the typical Cd_T concentration in soil solution (Table 12.1). In studies where soils have been amended with artificially high levels of Cd, a threshold concentration of ≈ 0.4 μmol L^{-1} has not been observed in soil

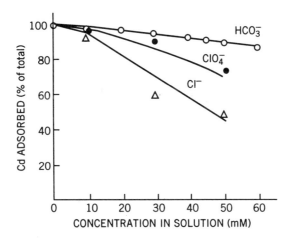

FIGURE 12.5 Effect of type of complexing anion and ionic strength on adsorption of trace concentrations of Cd on Na-montmorillonite. Concentration of adsorbing sites was 2 mmol L^{-1}. (Hirsch et al., 1989)

solution until > 0.001 mmol kg^{-1} total Cd is applied (Mahler et al., 1980; Bingham et al., 1984; Sikora and Wolt, 1986).

Of greater consequence is Cd soil bioavailability for plant uptake of Cd leading to food chain biomagnification. Cadmium that is present in solution will be readily taken up by plants. The ability to predict Cd accumulation by food and feed crops on the basis of waste Cd loadings and waste–soil interactions provides a means to manage Cd biomagnification through the food chain. As described earlier, gross compositional properties of Cd-bearing wastes or soil receiving these wastes are inconsistent measurements of Cd availability for plant uptake. Furthermore, Cd^{2+} is the biologically relevant form of Cd, and its bioavailability is best determined from soil solution composition. Figure 12.6 presents the relationship of leaf Cd concentration to Cd^{2+} activity in soil solution that demonstrates the utility of soil solution composition as a measure of bioavailability.

12.4.3 Competitive Ion Effects

In Chapter 10, various specific and nonspecific ion effects that restrict the biologically consequential pool of an ion are described (section 10.1.4, Activity Ratios).

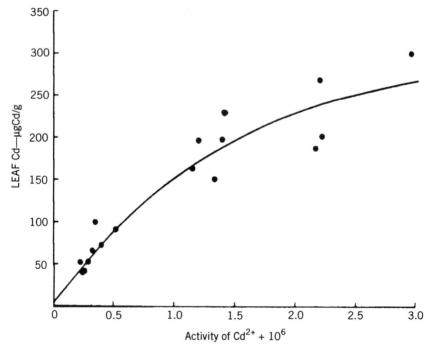

FIGURE 12.6 Leaf Cd concentration of Swiss chard plotted as a function of the activity of Cd^{2+} in the soil saturation extract. Each data point is the mean of five replications. The line represents the equation $y = 310 - 308e^{-0.668x}$. [F. T. Bingham, J. E. Strong, and G. Sposito, "The Influence of Chloride Salinity on Cadmium Uptake by Swiss Chard," *Soil Science,* 135:160–165.]

Cadmium biogeochemical availability will be influenced by the background concentrations of dominant soil solution cations and competitive interactions involving these macrocations as well as trace metals ions in soil solution. Examples of macrocation effects on Cd^{2+} biogeochemistry are ion exchange equilibria and precipitation/dissolution phenomena involving macrocation-containing minerals where Cd may occur as a trace level co-precipitate. An additional important class of competitive ion effects are competitive chelate equilibria described in the following section.

Of particular interest with respect to Cd biogeochemical availability are ion interactions with Zn. As adjacent periodic group IIB metals ions, Cd and Zn are chemically similar and this similarity is manifested in the geological occurrence and soil–plant relationships of Cd and Zn. Geochemical weathering results in Cd depletion relative to Zn (Zn:Cd mole ratios are $\approx$ 600 in primary rock, 700 in the lithosphere, and 800 in soil) largely because approximately equivalent concentrations of Zn^{2+} and Cd^{2+} are supported in soil solution. Certain wastestreams exhibit markedly lower Zn:Cd ratios due to Cd enrichment. Representative municipal sludges, for instance, exhibit a Zn:Cd ratio of 180 (Table 12.2).

Because of chemical similarity, there is an overall interaction of Cd and Zn for release from waste and soil solid phases, speciation in soil solution, and plant root uptake and subsequent translocation (Sikora and Wolt, 1986; Pepper et al., 1983). Sikora and Wolt (1986) noted Cd–Zn interaction on both soil solution composition and root and shoot uptake by corn when sludge-applied Zn and Cd ranged from 0.14 to 11.6 mmol kg^{-1} and from 0.9 to 669 μmol kg^{-1}, respectively. Competition for binding and exchange sites within the waste itself increased Cd extractability from sludge—and the resulting Cd concentrations in soil solution—when Zn content of sludge increased. Cadmium uptake to roots and subsequent accumulation in top growth of corn was influenced by Zn level, probably because of competition for transporting ligands and for sparingly soluble organometallic compounds in root tissue (Chino and Baba, 1981; Girling and Peterson, 1981).

The strong competitive effects of Zn on Cd availability indicate that correlation of plant Cd on soil solution Cd^{2+} such as shown in Figure 12.6 will not be broadly applicable across waste-soil systems where there is a significant variation in concentrations of competing cations. Greater applicability across waste-soil systems will probably require that Cd in soil solution be described in terms of an activity ratio (see section 10.1.4, Activity Ratios). Figure 12.7 shows the Cd/Zn activity ratio in soil solution is a strong predictor of Cd availability to the root (that is, a strong predictor of uptake). This ratio is an ineffective predictor, however, of foliar accumulation of Cd, because of the added effect of plant physiological processes in discriminating against Cd transport from roots to shoots.

12.5 MICRONUTRIENT CATIONS

Fe, Mn, Cu, and Zn—and perhaps Co and Ni as well—are essential plant micronutrients occurring in the environmentally relevant form of Lewis hard acids or

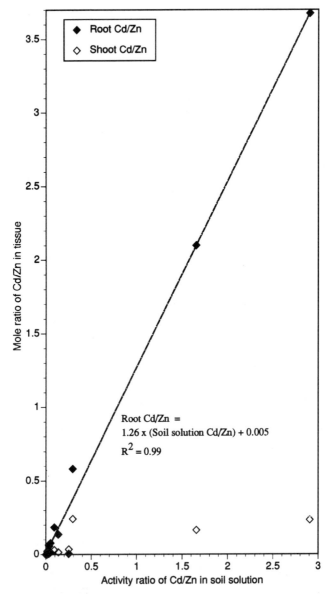

FIGURE 12.7   The relationship of Cd/Zn activity ratio in soil solution to Cd/Zn mole ratios in corn roots and shoots. [After Sikora, 1982.]

borderline hard acids in soil solution. The environmental abundance of these micronutrient cations ranges from macro- (Fe and Mn) through micro-quantities, but they all exist in soil solution at micromolar to submicromolar concentrations (Tables 10.1 and 12.1). Certain of the micronutrient cations (Fe, Mn, and Cu) undergo redox transformations along ranges of pe + pH that are consequential to their distribution between soil solution and solid phases and to their speciation in soil solution. Manganese is described here in some detail to exemplify some aspects of soil chemical reactivity important to micronutrient cation occurrence and distribution in soil solution.

12.5.1 Manganese

The principal reactions of Mn in soils relate to association with O^{2-} and OH^- ligands and precipitation as oxyhydroxides. Manganese oxyhydroxides tend to accumulate in well-oxidized soils, and their solubilities increase under reducing conditions. Manganese may be present at the earth's surface in the Mn(II), Mn(III), and Mn(IV) valence states, but only Mn^{2+} predominates soil solution (Gambrell and Patrick, 1982; Bohn et al., 1979; Lindsay, 1979). Thus, the major transformations of Mn in sumerged soils involve reduction of insoluble forms of Mn^{4+} and Mn^{3+} to Mn^{2+}.

The valence and concentration of soil solution Mn is dependent on pe + pH. A typical Mn concentration in soil solution is 0.4 $\mu mol\ L^{-1}$ (Table 10.1), but levels of 240 to 750 $\mu mol\ L^{-1}$ have been found in soil solutions of Typic Hapludults on Coastal Plains sediments (Adams and Wear, 1957; Morris, 1949). In natural waters, Mn ranges from <0.01 to 2.4 $\mu mol\ L^{-1}$ (Bowen, 1966).

Manganese in Soil Solid and Solution Phases. Soil Mn can take many forms because of the complex chemistry imparted by its many valence states. The solid phases MnO_2, Mn_2O_3, and Mn_3O_4 appear most important for controlling solution Mn in natural waters. At low pe, $Mn(OH)_2$ may be an important solid phase form of Mn, while $MnCO_3$ may be of consequence in high salt systems (Hem, 1972). Pyrolusite (MnO_2) is the most stable Mn mineral under well-oxidized conditions, while Mn^{2+} is the most common solution species across a wide range of pe + pH.

The dissolution of pyrolusite leading to Mn^{2+} in solution is expressed

$$MnO_{2s} + 4H^+ + 2e^- \rightleftharpoons Mn^{2+} + 2H_2O$$

which yields the relationship, $pMn^{2+} = pK° + 4pH + 2pe$, where $pK° = -41.89$ (Lindsay, 1979). Bohn (1970) found that this relationship did not explain variation in solution Mn^{2+} in 1:10 soil/0.01 M $CaCl_2$ suspensions as a function of pe and pH, when E_H (mV, = 59.2 pe at 298.15 °K) was measured with a platinum (Pt) electrode or calculated from the O_2—H_2O redox couple. Failure for these measurements to be thermodynamically predicted and the observed internal disequilibrium among redox couples in natural systems makes measured E_H an unsatisfactory parameter for describing redox equilibria in soils (Bohn, 1968, 1969; Lindberg and Runnels, 1984; see section 4.3.3, Interpretation of Measured E_H). The equilibrium between MnO_2 and $MnCO_3$ expresses Mn^{2+} activity independent of pe as pH $-$ 1/2 pMn = pK° $-$ 1/2 pCO_{2g} (Bohn, 1970).[4] Sadana and Takkar (1988) observed that Mn^{2+} in submerged sodic soils was well-described by this relationship when $pK^\circ = 4.4$. Olomu et al. (1973) monitored changes in soil solution pH, E_H, and Mn in saturated soils over a six-week period, during which E_H fluctuated from reducing to oxidizing conditions. Dissolution of Mn_2O_3 best explained the results by the reaction

$$Mn_2O_{3s} + 6H^+ + 2e^- \rightleftharpoons 2\ Mn^{2+} + 3H_2O$$

In an anoxic environment, Fe^{2+} will begin to precipitate from soil solution at pe + pH values lower than necessary for Mn^{2+} precipitation; Mn^{2+} in solution is decreased and tends to precipitate as a ferromanganous material that does not re-solubilize with short-term shifts in pe + pH (Collins and Buol, 1970).

Equilibrium relationships for some of the more consequential Mn species in soil solution are summarized in Table 12.6. Manganese forms a number of hydrolytic species and ion pairs in aqueous solutions. In acidic soil systems, the dominant hydrolytic reactions of Mn are

	pK°
$Mn^{2+} + H_2O \rightleftharpoons MnOH^+ + H^+$	10.9
$Mn^{2+} + 3H_2O \rightleftharpoons Mn(OH)_{3^-} + 3H^+$	34.0

And $MnCl^+$ and $MnSO_4^\circ$ are important ion pairs (Hem, 1972; Lindsay, 1979). The dominant solution species of Mn shift with changing redox of the system, depending on the solid phase controlling Mn solubility. Sanders (1983) found that at pH < 7.0, from 70 to 90% of soil solution Mn was present as Mn^{2+}. Small amounts of Mn were complexed as $MnSO_4^\circ$ at low pH. As pH decreased, free Mn^{2+} decreased, probably because it complexed with dissolved organic matter. This contrasts with the work of Geering et al. (1969) who found Mn^{2+} in displaced soil solutions of varied pH represented from 1 to 16% of total Mn in solution. Organic matter complexation with Mn^{2+} was suspected.

Rate Dependence of Manganese Transformations. Consideration of redox equilibria is useful for gaining insight concerning control of soil solution Mn^{2+} (section 9.2.3). Predictions of soil solution Mn content based solely on redox equilibria are frequently inadequate, however, because of slow rates of redox transformation of Mn species.

Rates of Mn transformation in soils leading to Mn^{2+} release to soil solution may be strongly influenced by reduction/chelation reactions involving low-molecular-weight polyphenolic compounds. The general reaction of a polyphenolic (Ph) with soil Mn can be expressed

$$soil\text{-}Mn + 2H^+ + Ph_{red} \rightleftharpoons Mn^{2+} + Ph_{ox} + 2H_2O$$

where a reactive polyphenolic is oxidized as Mn is reduced from Mn(IV) to Mn(II).

TABLE 12.6 Redox Equilibria, Hydrolysis Species and Solution Complexes of Manganese

Chemical Reaction	log K°
Redox Reactions	
$Mn^{4+} + e^- \rightleftharpoons Mn^{3+}$	25.51
$Mn^{3+} + e^- \rightleftharpoons Mn^{2+}$	25.55
$Mn^{4+} + 2e^- \rightleftharpoons Mn^{2+}$	51.06
$Mn^{2+} + 2e^- \rightleftharpoons \alpha\text{-}Mn(c)$	−40.40
$MnO_4^{2-} + 8H^+ + 4e^- \rightleftharpoons Mn^{2+} + 4H_2O$	118.31
$MnO_{4-} + 8H^+ + 5e^- \rightleftharpoons Mn^{2+} + 4H_2O$	127.71
Hydrolysis Reactions	
$Mn^{2+} + H_2O \rightleftharpoons MnOH^+ + H^+$	−10.95
$Mn^{2+} + 2H_2O \rightleftharpoons Mn(OH)_2^o + 2H^+$	
$Mn^{2+} + 3H_2O \rightleftharpoons Mn(OH)_3^+ + 3H^+$	−34.00
$Mn^{2+} + 4H_2O \rightleftharpoons Mn(OH)_4^{2-} + 4H^+$	48.29
$2Mn^{2+} + H_2O \rightleftharpoons Mn_2OH^{3+} + H^+$	10.60
$2Mn^{2+} + 3H_2O \rightleftharpoons Mn_2(OH)_3^+ + 3H^+$	23.89
$Mn^{3+} + 3H_2O \rightleftharpoons MnOH^{2+} + 3H^+$	0.40
Mn^{2+} Complexes	
$Mn^{2+} + Cl^- \rightleftharpoons MnCl^+$	0.61
$Mn^{2+} + 2Cl^- \rightleftharpoons MnCl_2^o$	0.04
$Mn^{2+} + CO_2(g) + H_2O \rightleftharpoons MnHCO_3^+ + 2H^+$	−6.02
$Mn^{2+} + CO_2(g) + H_2O \rightleftharpoons MnCO_3^o + 2H^+$	−18.87
$Mn^{2+} + SO_4^{2-} \rightleftharpoons MnSO_4^o$	2.26

The phenolic molecule in either the reduced or oxidized form may additionally act as a chelating ligand increasing Mn solubilization and potential mobilization through the soil profile (Pohlman and McColl, 1989).

Manganese solubilization in the presence of hydroxyquinone and naturally occurring polyhydroxyphenolic acids is kinetically well described as an overall second-order reaction that is first order with respect to both unreacted (that is, reduced) phenolic and solid phase Mn(IV) (Stone and Morgan, 1984; Pohlman and McColl, 1989).

$$\frac{dC\ Ph_{ox}}{dt} = kC\ Ph_{red}C_{\text{soil-Mn}} \qquad [12\text{-}2]$$

Rate constants (k) ranging from 0.16 to 0.39 L mol^{-1} sec^{-1} have been reported for soil or birnessite ($MnO_{1.8s}$) in suspension with sodium acetate buffer (pH 4.5) where the initial concentration of solid phase Mn ranged from 18 to 36 mmol L^{-1} and the initial concentration of Ph_{red} was 25 mmol L^{-1} as gallic acid (Pohlman and McColl, 1989).

The suite and concentration of reactive polyphenolics present in soil solution will constantly vary due to production and transformation (Chapter 13), so redox transformation of Mn will dynamically vary in a way not predicted by static redox equilibria in those environments rich in dissolved organics.

Bioavailable Forms of Manganese. Sequential extraction techniques have been employed in an effort to better understand Mn distribution among soil fractions (Goldberg and Smith, 1984; Jarvis, 1984; Tokashiki et al., 1986). Estimation of the labile Mn pool using ^{54}Mn in conjunction with sequential extraction indicates labile (that is, potentially bioavailable) Mn is present in water-soluble, exchangeable, organically bound, and easily reducible fractions (Goldberg and Smith, 1984). The establishment of equilibrium between ^{54}Mn and soil Mn required from 30 min to 80 h for various soils, indicating the importance of kinetic considerations when assessing Mn bioavailability. Easily reduced Mn decreased as a result of soil drying, and other soil Mn fractions—including more resistant mineral phases—increased.

Up to 38 and 54% of total Mn in surface and periodically flooded subsoils, respectively, has been associated with reducible oxyhydroxides, while exchangeable Mn^{2+} is of minor occurrence (Jarvis, 1984). Acid, sandy lateritic soils under forest cover have been found to contain from 20 to 70% of total Mn in an easily reducible form (Sannigrahi et al., 1983). It appears, therefore, that transient changes in soil pe + pH may result in acute Mn toxicity through shifts in the labile pool of soil Mn.

Both quantity and composition of organic matter must be considered relative to the solid and solution phase complexation of Mn by organic matter. Organic matter reportedly has a great capacity for complexation of Mn (Adams, 1984), but organic Mn is probably important only in soils with appreciable organic matter content (Gambrell and Patrick, 1982). In comparison with other trace metals, Mn is either weakly complexed or not complexed by organic matter (Bloom and McBride, 1979; Gambrell and Patrick, 1982; Olomu et al., 1973). Organic (O) horizons containing biologically active acid mull humic materials accumulate Mn; O horizons containing moder and mor humus do not (Duchaufour and Rousseau, 1960; Rousseau, 1960).

Water-soluble Mn may be diagnostic of Mn availability, especially in acid soils or under reducing conditions. Care must be taken, however, to maintain soil conditions so that Mn status is not altered in the interval between sampling and analysis. Soils should not be dried or crushed prior to analysis, and determinations should be made as soon as possible after sampling (Gambrell and Patrick, 1982). Measurement of soil solution Mn has also been diagnostically useful in the assessment of Mn toxicity (Adams and Wear, 1957; Morris, 1949), but as with water-soluble Mn, alteration of pe + pH can invalidate the values of solution Mn obtained.

Toxicity of Manganese to Plants and Soil Microflora. Nutrient and soil solution Mn appear to exhibit a toxic threshold for sensitive agronomic crops at concentrations of 0.2 mmol L^{-1} (Adams, 1984; Adams and Wear, 1957; Morris, 1949). Nutrient solution experiments with birch (*Betula verrucosa* Ehrh.) indicated Mn toxicity occurred with >9 µmol L^{-1} in solution (Ingestad, 1964). European fir seedlings in nutrient solution culture exhibited increased mortality when Mn exceeded 0.1 mmol L^{-1} (Rousseau, 1960). Based on these few findings it is difficult to suggest levels of soil solution Mn that may be phytotoxic.

Soil microflora mutualistically associated with plants may exhibit Mn toxicity at lower levels of soil Mn than would be toxic to the host plant itself. Mn can inhibit nodulation of legumes. Rhizobium species, in general, are more sensitive to Al toxicity than to Mn (Foy, 1984). Hepper (1979) observed severe depression of spore germination and growth of *Glomus caledonium* (a vesicular-arbuscular mycorrhizal endophyte) with 2.6 µmol L^{-1} Mn in culture media; germination was nil with 25.5 µmol L^{-1} Mn present. Growth of *Aspergillus flavus* spores cultured in soil leachate was not inhibited by 3 to 470 µmol L^{-1} Mn.

12.5.2 Competitive Chelate Equilibria

Chelates of natural or anthropogenic origin are important to trace metal biogeochemistry because they affect increases in metal ion solubility in soil solution. In managed ecosystems, chelates are used as sources of plant-available Fe, Zn, Mn, and Cu; while naturally occurring chelates in unmanaged ecosystems may be important for trace metal transport. Additionally, chelates are useful as extractants for evaluating the availability of micronutrients and other trace metals in soils. Understanding metal–chelate stability relationships yields information relative to (1) micronutrient availability, (2) complexation reactions involving trace metals and natural organic ligands, (section 13.2), and (3) the use of competitive-ion equilibria in soil solution (see section 5.7, Complexation and Exchange Techniques).

Consider the example of a chelate (L) interacting with Ca as the dominant soil solution cation as well as with the micronutrient cations Fe and Mn. The distribution of L in solution is described.

$$L_T = L + \Sigma H_n L + \Sigma CaXL + \Sigma Fe(III)XL + \Sigma MnXL + \Sigma MXL \quad [12\text{-}3]$$

where the total concentration of L in solution (L_T) is distributed among free L, protonated species ($\Sigma H_n L$), species containing Ca^{2+}, Fe^{3+}, and Mn^{2+} ($\Sigma CaXL$, $\Sigma Fe(III)XL$, and $\Sigma MnXL$, respectively); X signifies certain chelates containing H^+ or OH^-, and ΣMXL refers to any other metal-chelate species that may occur. Note that if system redox favors the occurrence of Fe(II), a term $\Sigma Fe(II)XL$ needs to be included into equation 12-3. Similarly, if some other metal M occurring in the term ΣMXL forms significant concentrations of complexes with L relative to other metals, it would need to be broken out separately in equation 12-3 was well. The mole fraction of any metal–chelate species occurring in solution can be calculated from equation 12-3. For example, if those species occurring from Fe(III) chelation with

L are considered,

$$\Sigma Fe(III)XL = K^m_{Fe(III)L}[Fe^{3+}]_T[L^{4-}] + K^m_{Fe(III)HL}[Fe^{3+}]_T(H^+)[L^{4-}]$$

$$+ K^m_{Fe(III)OHL}[Fe^{3+}]_T(H^+)^{-1}[L^{4-}] \qquad [12\text{-}4]$$

where $K^m_{Fe(III)XL}$ is a conditional stability constant for an ionic strength of 0.01 mol L^{-1} (Sommers and Lindsay, 1979). When $[L^{4-}]$, (H^+), and the total ion concentrations of consequential chelating metals are known (ΣMXL is negligible), equation 12-3 becomes

$$1 = \frac{L}{L_T} + \frac{L\Sigma K^m_{CaXL}[Ca^{2+}][X]}{L_T} + \frac{L\Sigma K^m_{FeXL}[Fe^{3+}][X]}{L_T} + \frac{L\Sigma K^m_{MnXL}[Ma^{2+}][X]}{L_T}$$

$$[12\text{-}5]$$

From equation 12-5, the mole fraction of L_T complexing Fe(III) is

$$MF_{Fe(III)} = \frac{L\Sigma K^m_{FeXL}[Fe^{3+}][X]}{L_T} \qquad [12\text{-}6]$$

Sommers and Lindsay (1979) used stability constants for metal–ligand reactions for various common chelates, along with predicted solution concentrations of ions based on mineral solubilities, to create stability relationships describing competitive chelate equilibria. Examples of these are shown in Figure 12.8 for chelation with DTPA when pH varies at fixed pe + pH and when pe + pH varies at fixed pH. Relationships such as these can provide useful experimental approaches for discerning trace metal availability in soil solutions (such as described in section 5.7), but care must be taken in design and interpretation of such experiments because of the consequential effect of acidity and redox in shifting the competitive chelate equilibria observed.

12.6 TRACE INORGANIC LIGANDS

Several multivalent ions [Si(IV), C(IV), N(V), S(IV), P(V), B(III), Se(VI), Mo(VI) and As(V)] occur in soil dominantly in the form of weak acid oxyanions. The anionic forms of these weak acids occur in soil solution— SiO_4^{4-}, CO_3^{2-}, NO_3^-, SO_4^{2-}, PO_4^{3-}, BO_3^{3-}, SeO_3^{-2}, MoO_4^{2-}, AsO_4^{3-}—and frequently act as bridging ligands. Silicon, N, C, S, and P occur in the lithosphere in relatively high concentrations and tend to be key components of soil solution both because of the relative concentrations at which they occur and because of their biogeochemical importance (Chapter 7). The remaining weak acid anions occur in soil solution at trace concentrations but are of importance because they are essential plant micronutrients

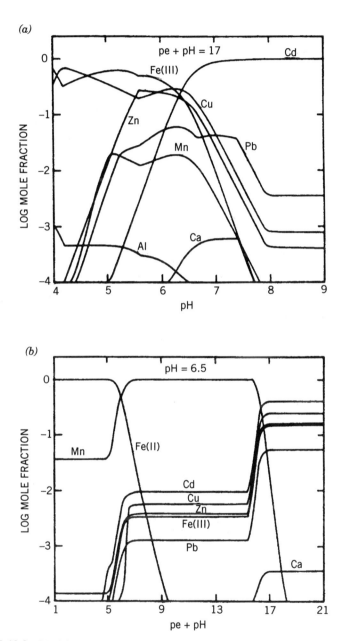

FIGURE 12.8 Modeled competitive complexation equilibria for trace metals and DPTA. (*a*) Mole fraction diagram for complexation by DPTA in soils at pe + pH = 17. (*b*) Mole fraction diagram for complexation by DPTA in soils at pH 6.5 with precipitation of Cu_2S, PbS, and CdS. [Sommers and Lindsay, 1979. *Soil Sci. Soc. Am. J.,* 1979, 43:39–47.]

(B and Mo), or they are of nutritional/toxicological concern in the food chain (As and Se).

12.6.1 Natural Abundance and Anthropogenic Occurrence

Natural abundances of the weak acid oxyanions of B, Mo, As, and Mo are quite low on a whole soil basis and are in the micromolar to nanomolar range in "typical" soil solutons (Table 12.1). The typical soil solution concentrations of these elements can be misleading, however, because complicated environmental reactivity leads to widely varied availability. Molybdenum, for example, occurs at submicromolar concentrations in acid soil solutions and may be deficient for adequate plant N nutrition (Mengel and Kirkby, 1979); but in anaerobic, highly saline agricultural drainage, Mo may occur at $\approx$ 50 μmol L^{-1} and pose concerns relative to toxicity and food chain bioaccumulation (Presser and Barnes, 1985).

The occurrence of trace weak acid oxyanions at concentrations of environmental concern may arise from athropogenically induced alterations of ecosystems. Molybdenum may occur at elevated concentrations from coal combustion and the disposal/utilization of municipal wastes and fly ash (Jarrell et al., 1980); arsenical pesticide use may cause localized instances of elevated soil As (Johnson and Hiltbold, 1969; Jacobs et al., 1970); and both Mo and Se have been found in elevated concentrations in saline agricultural drainage waters (Presser and Barnes, 1985). The solution chemistry of Se in relation to food chain Se toxicity is a recent interesting example of the soil environmental chemistry of trace inorganic ligands and will be described in further detail here.

12.6.2 Selenium

Biogeochemical Importance. Selenium availability is very low in acid to neutral soils, where it may be specifically adsorbed or precipitated; but in arid-region soils, available Se may accumulate to higher levels. In seleniferous soils, soil solution Se concentrations are typically $\approx$ 10 μmol L^{-1} (about a hundredfold higher than the desired environmental concentration from the standpoint of avian toxicity; Tokunga and Benson, 1992).

The selenate oxyanion, SeO_4^{2-}, appears to be the biologically relevant form of Se in terms of both phytotoxic effect and plant uptake (Zhang et al., 1988; Khattak et al., 1989). Ion speciation of model soil solutions indicates $HSeO_3^-$ and $CaSeO_4^\circ$ are dominate ion pairs in acid solutions, while these ion pairs as well as $CaSeO_3^\circ$ dominate neutral solutions. In these model systems, SeO_4^{2-} was the principal free ionic form in either acid or neutral solutions where it comprised $\approx$ 40% of total solution Se (Mikkelsen et al., 1987).

Selenate is moderately mobile in neutral to alkaline soils. Reduction of Se(VI) to the selenite oxyanion (SeO_3^{2-} or $HSeO_3^-$), Se_s°, or organic Se will reduce Se availability in arid-region soils. Reductive immobilization of Se is an important process for mitigating the toxicity hazard of Se in those environmental compartments where it may accumulate (Oremland et al., 1989; Tokunga and Benson, 1992).

Oxidation–Reduction Relationships. Both Se and S are periodic group VIA elements; therefore, they share common chemical and biological properties. Both may be specifically adsorbed to soil solids with release governed by hydroxyl ion exchange. Significantly, however, Se exhibits a differing pathway of microbially mediated reduction than does S (Oremland et al., 1989).

The biogeochemical pathway whereby Se(VI) is reduced to Se(IV) and where Se(IV) in turn is reduced to Se_s° is important for understanding the mitigation of Se toxicity arising from Se occurrence in waters draining seleniferous soils. The stepwise reduction of Se(VI) to Se_s° is expressed

$$SeO_4^{2-} + 2H^+ + 2e^- \rightleftharpoons SeO_3^{2-} + H_2O$$

$$SeO_3^{2-} + 6H^+ + 4e^- \rightleftharpoons Se_s^{\circ} + 3H_2O$$

Oremland et al. (1989) have shown complete reduction of Se(VI) to Se_s° in anaerobic sediment–slurry systems. The rate of reduction is enhanced in the presence of H^+ and inhibited in the presence of oxygen or when the systems are autoclaved. Providing acetate as a source of reducing power further stimulates Se reduction; the reaction

$$4CH_3COO^- + 3SeO_4^{2-} \longrightarrow 3Se_s^{\circ} + 8CO_2 + 4H_2O + 4H^+$$

is highly exergonic ($\Delta F_r = -326$ kJ mol^{-1}).

Competitive Ion Effects. Weak acid oxyanions exhibit a high degree of interaction with respect to sorption/precipitation, complex formation in soil solution, and for uptake/transport by organisms. For example, SeO_4^{2-} plant uptake may be antagonized by SO_4^{2-} (Zhang et al., 1988); SeO_4^{2-} reduction is inhibited in the presence of SO_4^{2-} (Oremland et al., 1989); and uptake and biotic effects of SeO_4^{2-} are influenced by interactions with MoO_4^{2-} and AsO_4^{3-} (Khattak et al., 1989).

Khattak et al. (1989) investigated the interactive effects of Se(VI), Mo(VI), and As(V) on plant growth and uptake using solution culture studies with alfalfa. At approximately equimolar concentrations in solution, uptake was in the order Se > Mo ≫ As. Selenium and As were mutually antagonistic for uptake, while Se enhanced Mo uptake without a countering effect of Mo on Se uptake.

12.7 TECHNETIUM

The longest lived isotope of Tc is ^{98}Tc ($t_{1/2} = 4.2 \times 10^6$ years); therefore, no primordial Tc exists and environmental occurrence of Tc is entirely of anthropogenic origin—aside from traces occurring in uranium ore. Thermal fission of ^{235}U and ^{239}Pu in nuclear reactors produces ^{99}Tc ($t_{1/2} = 2.1 \times 10^5$ years). Technetium-99 is a weak β emitter with low specific activity (0.63 GBq g^{-1}). Environmental release of Tc occurs from liquid effluents of uranium enrichment plants processing recycled nuclear fuel where it will occur as TcO_4^- (Henrot, 1988).

12.7.1 Soil Biogeochemistry of Technetium

Environmental behavior of Tc is somewhat similar to that of Mn (section 12.5.1) since both are periodic group VIIIB elements. Technetium exhibits several oxidation states, but Tc(VII) and Tc(IV) are the most environmentally relevant forms. The most stable form of Tc in soil will be TcO_4^-. The reduction of Tc(VII) to Tc(IV) results in precipitation as $TcO_2 \cdot H_2O$ or complexation of Tc(IV) with organic ligands (Carsen et al., 1984; Stalmans et al., 1986). The reduction of Tc(VII) leading to precipitation is rate-limited however, so TcO_4^- occurs under conditions where TcO_2 formation is thermodynamically favored.

Oxidation–Reduction Relationships. Reduction of Tc(VII), although thermodynamically favored as soils become reducing (Fig. 12.9), appears to be rate-limited such that TcO_4^- is frequently observed at pe + pH, where its reduction and subsequent precipitation as TcO_2 or Tc-organic complexes would be anticipated. Mixed culture studies with aerobic and anaerobic microbial consortia indicate microbially mediated reduction of Tc occurs in the presence of anaerobes but not in the presence of aerobes (Henrot, 1988). It is likely that in anoxic systems TcO_4^- is used as an electron acceptor for anaerobic respiration—the standard electrode po-

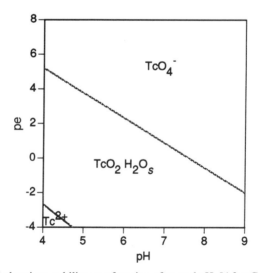

FIGURE 12.9 Technetium stability as a function of pe and pH. [After Carlsen et al., 1984.]

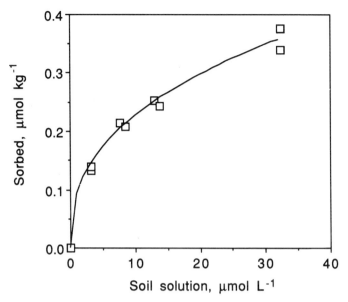

FIGURE 12.10   Effect of soil solution technetium on Tc sorption by an Aquic Hapludult Ap horizon at field moisture content ($\theta_v = 0.33$) (Redrawn from the data of Hernot, 1988).

tential of the TcO_4^-/TcO_2 couple ($E_H^\circ = 738$ mV) is similar to that of common soil metal oxide couples.

Chemisorption.   Since distribution of Tc between soil solution and solid phases is redox-controlled (Fig. 12.9), differential removal of Tc from soil solution is anticipated as soil pe + pH shifts. Differential chemisorption of Tc in Ap (1.4% OC) and Bt (0.2% OC) horizon soil samples maintained at field capacity or flooded was attributed to (1) the activity of anaerobes to reduce Tc, (2) the importance of organic carbon as a sink for reduced Tc, and (3) rate effects on the degree of chemisorption (Henrot, 1988). Displacement of soil solution from soils incubated at field moisture content proved a useful approach to monitoring the concentration-dependence of Tc reduction and subsequent immobilization in the soil solid phase (Fig. 12.10). Technetium sorption in field-moist soils was attributed to microbial activity at anaerobic microsites. Soil sterilization by γ-irradiation altered the kinetics of Tc reduction to reduce Tc chemisorption in surface soil. The kinetics of chemisorption were well-described by the Elovich equation.

NOTES

1. See for example the twenty-plus volumes of *Metal Ions in Biological Systems,* H. Sigel (ed.), Marcel Dekker, New York.

2. An excellent description of ligand field theory can be found in Chapter 26 of Cotton and Wilkinson, 1966.

3. Although Cd undergoes reduction to form a crystalline phase,

$$Cd^{2+} + 2e^- \rightleftharpoons Cd_{crystalline} \quad \log K^\circ = -13.64$$

Cd occurrence in soils is limited to Cd(II) (Lindsay, 1979). This is because for the redox equilibria ($\log Cd^{2+} = \log K^\circ + 2pe = 13.64 + 2pe$), a Cd^{2+} activity typical of soil solution (≈ 0.04 μmol L^{-1}; Table 12.1) will require pe < -10.5 ($E_H = -622$ mV). For the pH range of soil solution and natural waters, this pe falls below the stability field for H_2O (Figure 4.2).

4. Soil solution Mn^{2+} in equilibrium with solid phase MnO_2 and $MnCO_3$ is described by the following reactions (values of log K° are from Lindsay, 1979).

		log K°
$MnCO_{3s} + 2H^+$	$\rightleftharpoons Mn^{2+} + CO_{2g} + H_2O_l$	8.08
$Mn^{2+} + 2H_2O_l$	$\rightleftharpoons 2e^- + 4H^+ + MnO_{2s}$	-41.89
$Mn^{2+} + 2e^-$	$\rightleftharpoons Mn_c$	40.40

Net: $MnCO_{3s} + Mn^{2+} + H_2O_l \rightleftharpoons MnO_{2s} + CO_{2g} + 2H^+$ 6.59

Thus,

$$pK^\circ = -\log CO_{2g} + 2pH + pMn^{2+}$$

Rearranging results in,

$$pH - 1/2\, pMn^{2+} = pK^\circ - 1/2\, pCO_{2g}$$

CHAPTER 13

DISSOLVED AND COLLOIDAL ORGANICS

This chapter deals with the nature, distribution, and reactivity of dissolved organic molecules in soil solution.[1] In section 4.4 suspended organic carbon was discussed as one of several master variables important for the control of chemical composition, speciation, and activity of ion components in soil solution. Other sections describe additional specific instances of how reactivity of organic molecules influences the biogeochemical availability of soil solution constituents.

13.1 REACTIVE NATURAL ORGANIC CHEMICALS IN SOIL SOLUTION

The pool of organic carbon in soil solution is typically described on the basis of water extraction as "dissolved organic carbon" (DOC) when in fact it comprises a variety of low-molecular-weight molecules that are dissolved into solution and higher molecular weight molecules that are colloidally suspended in solution. The two dominant classes of reactive natural organic chemicals in soil solution are biochemical components of living organisms and the products of microbial decomposition of organic matter (fulvic acids).

13.1.1 Biochemicals

Soil solution contains a transitory mixture of biochemicals of natural origin reflecting the nature and activity of soil biota. The kinds and concentrations of these biochemicals fluctuate widely both spatially and temporally reflecting the balance of metabolic and catabolic activities occurring as these chemicals are microbially cycled in soil ecosystems. A variety of low-molecular-weight organic acids (simple aliphatic acids, amino acids, and sugar acids), hydroxamate siderophores, and polyphenols are important constituents of soil solution due to reactivity with other soil solution components.

Low Molecular Weight Organic Acids. Low-molecular-weight organic acids are important dissolved constituents of natural waters, but they have been little characterized and are frequently lumped with higher molecular weight organic molecules (that is, fulvic acids) in the water soluble pool of organic carbon (DOC). Low-molecular-weight organic acids, however, are distinctly different from higher molecular weight organic molecules in terms of their occurrence, distribution, and reactivity. The relative abundance of low-molecular-weight organic acids in soil solutions appears to be

volatile aliphatic acids > nonvolatile aliphatic acids > aromatic acids.

Few studies of soil solution composition detail organic carbon distribution sufficiently to determine the contributions of low-molecular-weight organic acids to the pool of soil solution DOC.

Low-molecular-weight organic acids are the most transitory component of the soil solution pool of organic carbon. In the rhizosphere, low-molecular-weight organic acid occurrence in soil solution is largely related to biological activity. Elsewhere in the soil, adsorption may be a more consequential mechanism for controlling low-molecular-weight organic acids in soil solution, especially for those organic acids that possess higher chemical reactivity (McKeague et al., 1986; Qualls and Haines, 1992).

Table 13.1 presents examples of aliphatic and phenolic organic acids known to occur in soils. Soil solution is dominated by the more soluble aliphatic acids. Phenolic acids exhibit greater reactivity with soil solids than do aliphatic acids due to the presence of both hydroxyl and carboxyl groups. Phenolic acids tend to show limited solubility as evidenced by concentrations of < 1 μmol L^{-1} occurring in water extracts from soils (McKeague et al., 1986).

Typically, the concentrations of organic acids in rhizosphere soil solution range manyfold higher than in ambient soil solution because the rhizosphere represents the prevalent region of intense biotic activity within the soil. A limited database of organic acid occurrence in "soil solution" (Huang and Violante, 1986, and Table 13.2) indicates millimolar concentrations of simpler aliphatic acids occur in soil solution, while the more strongly sorbed aromatic acids are usually present in sub-millimolar concentrations.

Variability in soil solution organic acids is expected, because of their occurrence as products arising from biochemical cycling of carbon. Description of this variability, however, is complicated by the sampling, storage, extraction, and analytical techniques employed in organic acid characterization. Biological alteration, volatilization, limited sensitivity to detection, and incomplete chromatographic resolu-

TABLE 13.1 Formulae of Some Low-Molecular-Weight Organic Compounds That Occur in Soils

HCOOH	COOH | CH$_3$	COOH | CH$_2$ | CH$_3$	COOH | (CH$_2$)$_2$ | CH$_3$	COOH | (CH$_2$)$_3$ | CH$_3$	COOH | (CH$_2$)$_4$ | CH$_3$
formic acid	acetic acid	propionic acid	butyric acid	valeric acid	hexanoic acid
COOH | CHOH | CH$_3$	COOH | COOH	COOH | CH$_2$ | CH$_2$ | COOH	COOH | CHOH | CH$_2$ | COOH	COOH | CH$_2$ | HO-C-COOH | CH$_2$ | COOH	COOH | C = O | (CHOH)$_3$ | CH$_2$OH
lactic acid	oxalic acid	succinic acid	malic acid	citric acid	2-ketogluconic acid

benzoic acid	salicylic acid	p-hydroxybenzoic acid	orcinol	3,4 dihydroxybenzoic acid

vanillic acid	syringic acid	orsellinic acid	gallic acid	phenylacetic acid

4-hydroxy- phenyl- propionic acid	3,4-dihydroxy- phenylpropionic acid	cinnamic acid	ferulic acid	p-coumaric acid	caffeic acid	3,5-dimethoxy- 4-hydroxy cinnamic acid

TABLE 13.1   *Continued*

| quercetin | catechin | chrysotalunin |

McKeaque et al., 1986. Published in *Interaction of Soil Minerals with Natural Organics and Microbes.* Soil Sci. Soc. Amer.

tion all limit the ability for refined measurements of organic acids in soil solution.

Low-molecular-weight-organic acids are viewed as important complexing agents affecting the complexation, availability, and transport of metals in soils and consequently influencing the mineral weathering of soils (see section 13.2). However, evidence for the direct influence of low-molecular-weight organic acids in this regard is limited, as most studies have treated DOC as an uncharacterized pool. Laboratory studies point strongly to complexation of low-molecular-weight organic acids with metals as well as reactivity with soil minerals (see, for example, the discussion of organometallic complexes of Al, section 11.1.4), but these studies are conducted at concentrations and under conditions unlike those of ambient soil solution.

Amino Acids. Amino acids are prevalent products of soil microbial activity occurring in free, complexed, and polymeric forms. Micromolar concentrations of amino acids have been identified in soil solution (Table 13.2). Amino acids are unlikely to remain at high concentrations in soil solution for extended periods because they are readily biodegraded or immobilized through reaction with soil solids. Amino acids react with soil solids through (1) sorption in neutral and basic forms, (2) polymerization, perhaps with clay acting as a catalyst, and (3) nucleophilic addition to quinone groups of humic acid by way of amine and sulfhydryl groups. Amino acids stabilized by complexation with humic acids have an approximately fourfold slower rate of soil decomposition than do free amino acids in soil solution (Martin and Haider, 1986).

Although concentrations of amino acids present in soil solution are expected to be low, when present they may form relatively stable complexes with metals such as Cu, Zn, and Al (McKeague et al., 1986). Cysteine appears to form the most stable amino acid complexes with metals.

Products of amino acid metabolism may be of consequence in solutions of certain soils. For example, uric acid—the chief form of amino group excretion from α-amino acid metabolism in uricotelic organisms (terrestrial reptiles and birds) as well as the end product of purine metabolism in primates, birds, and terrestrial

TABLE 13.2   Ranges of Organic Acids Observed in "Soil Solution"

Organic Acid	Concentration Range, mmol L^{-1}	Source
Acetic	2.65–5.7	Stevenson (1967); Rao and Mikkelson (1977)
Citric	0.014–0.21	Stevenson (1967); Bruckert (1970); Forstner (1981); Hue (1992)
Formic	0.25–4.35	Stevenson (1967); Fox and Comerford (1990)
Oxalic	0.004–1.04	Stevenson (1967); Hue et al. (1986); Bruckert (1970); Fox and Comerford (1990)
Malic, tartaric, malonic	1.0–4.0	Stevenson and Ardakani, (1972)
Benzoic	<0.075	Shorey (1913); Whitehead (1964); Wang et al., (1967); Davies (1971)
Amino	0.08–0.60	Stevenson and Ardakani, (1972)
Uric	0.0004–0.0005	Hue (1992)
Aromatic	0.05–0.30	Stevenson and Ardakani (1972)
Phenolic	0.05–0.30	Coulson et al. (1960); Davies (1971); Ladd and Buttler (1975); Stevenson (1982); Hue (1992)
Glycolic	0.0004–0.0015	Hue (1992)
Phthalic	0.0001–0.011	Hue (1992)

reptiles (Lehninger, 1970)—was found in the soil solution of manure- and sludge-treated soils in μmol L^{-1} concentrations where it acted to complex Al (Hue, 1992).

Uric acid (keto form)

Sugar Acids. Sugar acids are released to soil solution through plant and microbial exudates and through cleavage of polysaccharides. Sugar acids are highly transient components of soil solution subject to rapid microbial assimilation and immobilization in the soil solid phase through cation bridging reactions. Uronic acids are the most biologically consequential family of sugar acids. Both glucuronic and galacturonic acid have been identified as soil microbial products, in addition to 2-ketogluconic acid, an aldonic acid (Stevenson, 1982).

```
      COOH              HC=O              COOH
       |                 |                 |
      HCOH              HCOH              C=O
       |                 |                 |
      HOCH              HOCH              HOCH
       |                 |                 |
      HCOH              HCOH              HCOH
       |                 |                 |
      HCOH              HCOH              HCOH
       |                 |                 |
     CH₂OH              COOH             CH₂OH
```

D-Gluconic acid	D-Glucouronic acid	2-ketogluconic acid

Hydroxyamate Siderophores. Hydroxyamate siderophores are a class of naturally occurring red-brown iron transporting biochemicals with a characteristic adsorption band of 420 to 440 nm and having complexation constants with iron of $\approx 10^{30}$ (Stevenson and Fitch, 1986). Siderophores are peptide derivatives containing an acyl hydroxyamine functionality that effectively complexes Fe^{3+}.

Powell et al. (1980, 1982) found hydroxyamate siderophore concentrations in the range 0.1 to 0.01 μmol L^{-1} in soil solutions of soils at field moisture content; concentrations were 10 to 50% higher in rhizosphere soils.

Polyphenols. Phenolic acids occur in soil solution as degradation products of polyphenolics arising from plant residues and as products from the oxidative decomposition of humic and fluvic acids. Polyphenolics occur as somewhat recalcitrant residues from the degradation of plant cell walls. Phenolic acids are hypothesized to comprise 20 and 30%, respectively, of model fulvic and humic acids (Schnitzer, 1986).

13.1.2 Fulvic Acid

The process of humification—humus formation—occurs as plant and microbial biomass senesces and dies. Soil fauna and microbes affect the decomposition and partial biodegradation of plant materials. Autolytic degradation by enzymes and the action of heterotrophs release low-molecular-weight biochemicals to soil solution. Chemical oxidation and enzyme catalysis transform these biochemicals to humus-like substances (protohumus). Further degradation results in the formation of humic materials.

Fulvic acids are the most consequential form of humic material in soil solution. The fulvic acid fraction classically defines a pool of organic matter that is alkaline-extractable but remains soluble in acid solutions. The classically defined fulvic acid fraction, therefore, includes the fulvic acids, a wide variety of low-molecular-weight organic molecules, and artifacts formed from acid and base hydrolysis of organic matter. Fulvic acid is characterized as yellow- to black-colored polyelectrolytes with molecular weights ranging from 100 to several thousand. In comparison with humic acids, fulvic acids are (1) lower molecular weight; (2) contain relatively similar amounts of carbon and hydrogen but higher amounts of oxygen and nitrogen; and (3) are somewhat more reactive due to size and greater total concentrations of total and carboxyl acidity.

The structure of fulvic acid is not known with certainty because the methodologies involved in the isolation and characterization of fulvic acid lead to uncertain interpretations—the creation of artifacts is a distinct possibility. Schnitzer (1986) and Stevenson (1982) describe isolation and characterization of fulvic acid and present several hypothesized structures. Figure 13.1 presents the partial structure of fulvic acid developed by Schnitzer (1978) where the molecule is not a single distinct entity but is an assemblage of biochemicals associated primarily through hydrogen bonding. These assemblages are dominantly aromatic structures heavily substituted with carboxyl and hydroxyl groups containing adsorbed alkanes, fatty acids, carbohydrates, and nitrogenous substituents.

Fulvic acids are more important as metal-complexing ligands in soil solution in comparison to biochemicals due to greater prevalence and the stability of complexes formed.

13.2 METAL–HUMIC MATTER INTERACTIONS

Dissolved organic molecules are critically important with regard to trace metal biogeochemical availability in soil solution. Chelation of trace metals by soil solution organics can enhance bioavailability of some metals (Fe^{3+}), decrease phytotoxicity of others (Al^{3+}), and increase mobility of environmental toxicants. The understanding of trace element chemistry in soil solution (Chapter 12) is hampered by (1) analytical difficulties associated with trace metals present in soil solution at typical concentrations of 10^{-8} to 10^{-9} mol L^{-1}, (2) the complexities of trace element speciation in aqueous solutions, and (3) difficulties in obtaining soil solutions

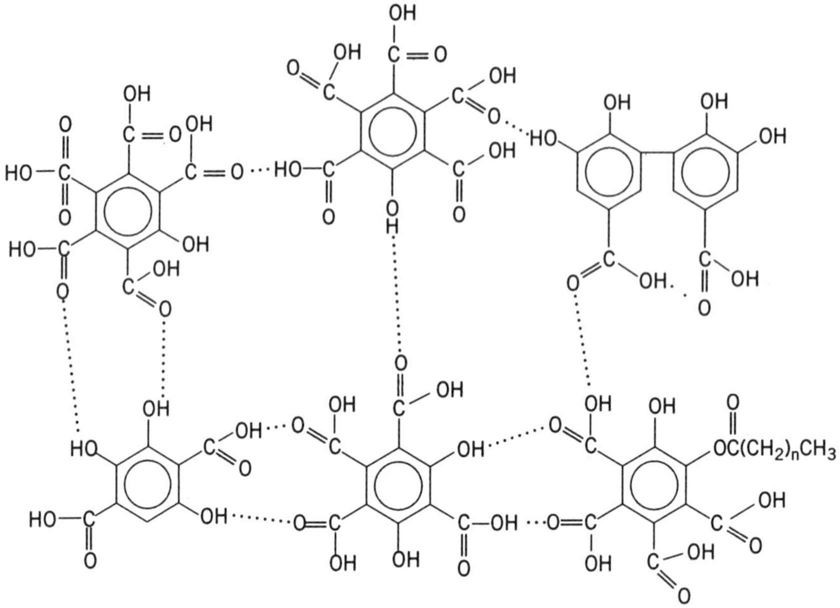

FIGURE 13.1   Partial structure for a hypothetical fulvic acid. [From M. Schnitzer, 1978 "Humic Substances: Chemistry and Reactions," in Schnitzer and Khan (eds.), *Soil Organic Matter* (Elsevier Science), pp. 1–64.]

that are representative of the trace element concentration and distribution of true soil solution. These concerns are overridden, however, by the strong controlling influence of soil solution organics on free trace metal distribution in soil solution and the uncertain nature of these complexing ligands.

13.2.1   Measurement of Organically Complexed Forms of Metals

Various measurement techniques have been employed for estimating the free and complexed forms of metals in soil solution. Direct measurement of total, free, or complexed forms of trace metals occurring in soil solution may be used in conjunction with ion speciation modeling to estimate the distribution of trace metals. Complexation and solvent extraction or separation on ion exchange resins have been used to partition the total concentration of a trace metal occurring in soil solution into pools of similarly reacting forms of the metal. One-step analytical methodologies involving in situ analysis for the metal, ligand, or the metal–ligand complex are preferred to two-step analytical techniques involving isolation and subsequent analysis because of the possibility of artifact formation as trace metals are redistributed among various forms during the analytical workup. (For example, these problems are addressed relative to the analytical determination of total, free, and complexed Al in sections 11.2 and 11.3). Prevalent side reactions leading to redistribution of trace metals are protonation of the ligand or strong ion pairing reactions of the metal with other ligands in solution, such as Cl^-.

Methods for measuring metal–humic acid interactions seek to determine one or all of the concentrations of free metal (M), complexed metal (M_a), free ligand (L, the organic moiety), and complexed ligand (L_b) occurring in solution, where the parameters measured refer to the reaction

$$aM + bL \rightleftharpoons M_aL_b$$

Which is described by the stability constant, K, as,

$$K = \frac{A_{M_aL_b}}{A_M^a\, A_L^b} \qquad\qquad [13\text{-}1]$$

Total metal in solution ($M + M_aL_b$) can be accurately determined. A number of techniques—dialysis, ion exchange, gel permeation chromatography (GPC), bio-assay, UV and fluoresence spectrophotometry, ion selective electrode (ISE), anodic stripping voltametry (ASV)—have been developed whereby reliable measurements of free metal (M) can be obtained. Reliable measurements of total, free, or complexed ligand are less reliably obtained; therefore, measurements of ligand concentration are frequently indirect, leading to large uncertainties as to the accuracy of measured stability constants. In most studies L is a fulvic acid; therefore, concentration of L is not measured per se. Instead, the *reactive site concentration* (j[L]$_t$, the molar concentration of reactive sites) is determined using potentiometric titration, spectrophotometry, spectrofluorometry, dialysis, or resin exchange (Stevenson and Fitch, 1986).

For certain applications, concurrent measurement of M and M_aL_b, M and L, or L and M_aL_b may be possible, leading to greater certainty as to the interpretation of metal–ligand reactivity (Stevenson and Fitch, 1986). Examples are use of

- ASV to measure both M and L
- GPC to identify M and M_aL_b
- electron spin resonance (ESR) to measure M and M_aL_b
- kinetic studies using UV or fluoresence spectroscopy, or UV spectroscopy with scanning or photodiode array detection to determine M or L and M_aL_b
- competitive equilibration studies involving chelation or ion exchange that measure M and M_aL_b.

None of these approaches is without difficulties in application, and the method of choice may be restricted by the particular metal–ligand system investigated. For example, ASV determination of M and L requires that the regions of reduction do not overlap, and kinetic studies may be applicable only to 1:1 metal–ligand complexes. Many methods for measurement of the metal actually measure "labile" metal (free M and M that is present as ion pairs and complexes with inorganic ligands, or as hydrolysis species). Thus, analytical or computational corrections are needed to estimate the portion of the labile pool that is free. Other methods may involve uncertain assumptions concerning kinetics of metal–ligand reactivity, as when competitive equilibria are used to assign total metal among free and complexed pools.

13.2.2   Modeling Reactions of Metals with Organic Macromolecules

Measurements of metal reactivity with organic macromolecules lead to conditional stability constants (K^c). Conditional stability constants, unlike thermodynamic stability constants, measure concentrations (as opposed to activities) of products and reactants under specified conditions of ionic strength, background solution composition, and pH.

The analytical complexity of K^c measurements depends on the nature of the binding reaction involved. There are three general types of binding reaction, in order of increasing analytical and interpretive complexity.

1. Mono–mono binding complexes, M_aL_b where a = b = 1.
2. Mononuclear complexes with ≥ 2 binding substrates where the central molecule of the complex is either the ligand (M_aL, where a ≥ 2) or the metal (ML_b where b ≥ 2).
3. Polynuclear binding where for M_aL_b, a > 1 and b > 1. Polynuclear binding includes both homopolynuclear (involving a single type of metal) and heteropolynuclear (involving more than one type of metal) binding reactions.

A common complexation model of the second type described above considers a central ligand, L, forming n complexes with a metal, M, each exhibiting a unique binding constant. For the n complexes formed (ML, M_2L, M_3L ... M_nL), there are n unique binding constants, K $\{K_1 = [ML]/([M][L]), K_2 = [M_2L]/([M][ML]), K_3 = [M_3L]/([M][M_2L]) \ldots K_n = [M_nL]/([M][M_{(n-1)}L])\}$. The degree of saturation of possible binding sites is described by the formation function v.

$$v = \frac{[M_aL]}{[L]_t} = \frac{[ML] + 2[M_2L] + 3[M_3L] + \ldots n[M_nL]}{[L] + [ML] + [M_2L] + \ldots [M_{(n)}L]} \qquad [13\text{-}2]$$

When all binding sites behave identically and independently at the molecular level, this equation resolves to

$$v = \frac{jK_0[M]}{1 + K_0[M]} \qquad [13\text{-}3a]$$

where K_0 is the overall binding constant and j is the number of reactive sites.[2]

Equation 13-3 can be linearized as,

$$\frac{\nu}{[\text{M}]} = j\text{K}_0 - \nu\text{K}_0 \qquad [13\text{-}3b]$$

A plot of $\nu/[\text{M}]$ against ν (a Scatchard plot) yields a line of slope K_0, the overall binding constant.

The formation function is frequently expressed relative to the reactive site concentration ($j[\text{L}]_t$)—for example, as moles of reactive sites per liter—when the mole weight of the ligand is unknown,

$$\theta = \frac{[\text{M}_a\text{L}]}{j[\text{L}]_t} = \frac{\nu}{j} \qquad [13\text{-}4]$$

If the complexing ligand contains i populations of binding classes each with j reactive sites, equation 13-3 is rewritten as

$$\nu = \frac{j_1\text{K}_1[\text{M}]}{1 + \text{K}_1[\text{M}]} + \frac{j_2\text{K}_2[\text{M}]}{1 + \text{K}_2[\text{M}]} + \cdots \frac{j_i\text{K}_i[\text{M}]}{1 + \text{K}_i[\text{M}]} \qquad [13\text{-}5]$$

where K_i represents the binding constant for the ith binding class. A Scatchard plot for a metal–ligand complexation reaction involving > 1 classes of binding sites will be nonlinear with different regions representing each binding class (Fig. 13-2).

Little agreement is found amongst published metal–ligand binding constants in part because of the wide variety of measurement and modeling approaches that have been used. Numerous techniques other than the Scatchard plot approach have been used for determining metal–ligand binding constants (Stevenson, 1982); these include ligand exchange, chelation, partition, base titration, and spectrophotometry. Additionally, (1) the ligands investigated may vary significantly dependent on source and method of preparation, (2) the composition of ligands may be inadequately described so as to distinguish features affecting differences in reactivity, and (3) the conditional nature of the test systems results in differences in ionic strength, ion composition, and pH of the background solutions.

13.3 QUANTIFYING DISSOLVED ORGANIC CARBON EFFECTS ON SOIL SOLUTION COMPOSITION

Thermodynamic approaches for quantifying organic colloid effects on the distribution of chemicals in soil solution suffer from inconsistencies and uncertainties in the measurement of binding constants. Therefore, equilibrium modeling of ion species and complexes in soil solution (Chapter 8) can only estimate the relative importance of dissolved and colloidal organics. Coupled with uncertainties in the binding constants selected for use in equilibrium models are further uncertainties regarding the true chemical natue of DOC in soil solutions. Consideration of the effect of DOC on biogeochemical availability of metals and other complexed chemicals in soil solution is perhaps of greater importance than is variability of the thermodynamic data used, especially when DOC is present in millimolar concentrations (as in soils treated with sludges, manures, green manures, or other concentrated sources of readily degraded organic material). Idealized approaches for describing the composition and reactivity of soil solution DOC represent a useful approach to the description of metal–ligand complexation in soil solution when the limitations in source thermodynamic and compositional data are considered.

The mixture model approach for idealized description of metal–ligand complexation (Dudley and McNeal, 1987; Mahler et al., 1980; Mattigod and Sposito, 1979) considers fulvic acid as a mixture of organic acids selected proportionally to represent the proton titration curve for a sludge-derived fulvic acid (Fig. 13-2). For a solution of unknown fulvic acid composition, the ratio of DOC measured to that of mixture model reference fulvic acid is used to project a model distribution of organic acids representative of the reactivity of fulvic acid.

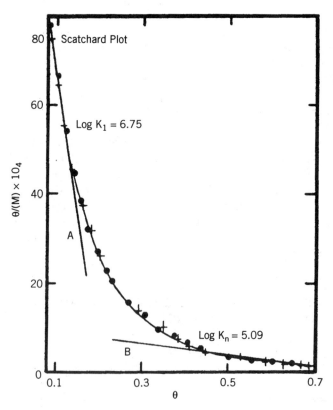

FIGURE 13.2   Scatchard plot for binding of Cu(II) by soil humic material. [Fitch and Stevenson, 1984. *Soil Sci. Am. J.* 1984. 48:1044–1049.]

$$\sum_i [OA_i]_{estimated} = \sum_i \frac{[DOC]_{measured}}{[DOC]_{reference}} [OA_i]_{reference} \qquad [13\text{-}6]$$

Conditional binding constants for individual metal–organic acid complexation reactions are then used to estimate the metal–fulvic acid complexation in an ion speciation model.

The mixture model assumes that free energies of metal binding for the fulvic acid are describable as the sum of partial free energies of binding for the organic acids comprising the mixture model

$$dF = \sum_i \mu_i dn_i = \sum_i RT\ln C_i f_i \qquad [13\text{-}7]$$

where f_i is the extended Debye–Hückel activity coefficient. Table 13.3 presents

TABLE 13.3 Mixture Models of Fulvic Acid in Saturation Extracts from Sludge-Amended Soils (Normalized to 10 mmol DOC L^{-1})

	Mixture Model		
	I^a	II^a	III^b
		$\mu mol\ L^{-1}$	
Acetic acid		438	439
Arginine	17		
Benzenesulfonic	29		
Benzoic		100	84
Citric	38		
Diethanolamine		83	158
Lactic acid		90	133
Lysine	23		
Maleic	57		
2-Methoxybenzoic acid		100	84
Ornithine	23		
Phenol		326	211
Phthalic acid	57		
Pyruvic acid		326	269
Salicylic acid	29		
Valeric acid		113	110
Valine	23		
Mole fraction of DOC comprised of reactive organic acids	0.18	0.65	0.58
Reference	Sposito et al., 1982	Dudley and McNeal, 1987	Dudley and McNeal, 1987

[a]Electrostatic interactions are assumed to be adequately described by Debye–Hückel theory.
[b]Incorporates a correction for electrostatic interaction between mixture model components.

mixture models for fulvic acids obtained from saturation extracts of sludge-amended soils. Models I and II compare model organic acid distributions for two different fulvic acid sources. These models assume that partial free energies of binding are adequately described on the basis of chemical potential as treated by extended Debye-Hückel theory (equation 3-7)—that is, electrostatic interactions occurring among organic acids comprising the mixture are considered insignificant. Differences between mixture models I and II reflect differing reactivities of the source fulvic acids; but more importantly, they reflect very different assumptions as to the organic acids comprising the hypothesized mixture. Mixture model I is comprised of organic acids exhibiting a higher density of binding sites per mole of carbon than does model II; thus, the proton titration curve for the model I fulvic acid is represented by organic acids comprising 18% of the DOC as compared to 65% for model II fulvic acid.

For a polyion where electrostatic interactions are significant, such as that described by the mixture model, free energy of binding involves both chemical and electrostatic components. Although it is not possible to isolate electrochemical potential into discrete chemical and electrostatic terms, it is useful conceptually to describe the free energy of binding of a mixed model polyion as

$$dF = \sum_i \mu_i^{el} dn_i = \sum_i RT \ln C_i f_i + \sum_i RT \ln \phi_i \qquad [13\text{-}8]$$

where free energy is described in terms of electrochemical potential (μ_i^{el}). The "electrostatic activity coefficient," ϕ, describes departure from an idealized standard state condition where electrostatic interactions among organic acid components of the model are zero at infinite dilution. Dudley and McNeal (1987) have employed a statistical mechanical approach to estimate ϕ for mixture model polyions. Models II and III (Figure 13.3) compare the hypothetical effect of electrostatic interactions among organic acids for the same fulvic acid when model III incorporates an electrostatic activity coefficient into model II. Inclusion of a correction term (ϕ) for electrostatic effects among organic acids within the mixture model resulted in a proportional redistribution of organic acids in the mixture model slightly favoring organic acids with a higher density of binding sites per mole of carbon in model III, as compared to model II. Thus, the proton titration curve for the model III (corrected for electrostatic effects) is represented by organic acids comprising 58% of the DOC as compared to 65% for model II (not corrected for electrostatic effects). This result indicates that improved mixture model development should include considerations of the physical confirmation of the postulated mixture model polyion as this relates to the effect of interparticle electrostatic interactions on the chemical reactivity that is modeled.

The mixed model approach has been compared with an alternative approach whereby metal–fulvic acid complexation in saturation extracts from sludge-amended soils is estimated based on a quasiparticle model that projects two classes of binding sites (Sposito et al., 1982). Roughly similar results were achieved in model comparisons for Cd^{2+} speciation in saturation extracts, but not for Cu^{2+}

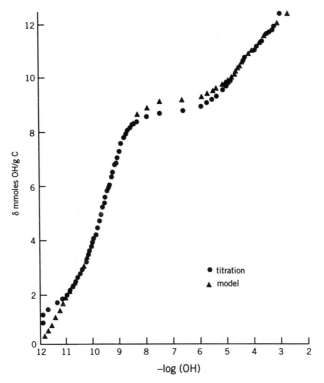

FIGURE 13.3   Proton titration curve for fulvic acid from a saturation extract of a sludge-amended soil and a corresponding fit to the data from a mixture model. [L. M. Dudley and B. L. McNeal, 1987 "A Model for Electrostatic Interactions among Charged Sites of Water-Soluble, Organic Polyions: I. Description and Sensitivity," *Soil Science*, 143:329–340.]

speciation. Generalization of metal–ligand complexation based only on measurement of soil solution DOC certainly will not provide refined estimates of organic ligand reactivity with trace metals in soil solution; but this approach is useful for considering the relative magnitude of influence organic ligands may have on ion speciation within a soil solution.

13.4 DISSOLVED AND COLLOIDAL ORGANIC CARBON EFFECTS ON BIOAVAILABILITY AND TRANSPORT

A considerable body of research points to important potential influences of organic materials on bioavailability and transport of chemicals. The influences of organic material on soil chemical reactivity are manifold and involve both direct and indirect processes occurring in the soil solid and solution phases. The more consequential and more readily observable effects of organic materials on soil chemical reactivity relate to organic material degradation and incorporation into the solid phase pool of colloidal soil organic matter. It is in this pool that large quantities of highly reactive organic carbon act as a solid phase sink for chemicals.

Several investigations, however, point to the importance of organic ligands, introduced to soil solution from concentrated external sources of organic carbon, on the composition of soil solution and the consequent bioavailability and transport of nutrients and phytotoxic chemicals. The presence of dissolved and colloidal organic matter in soil solution can affect ionic strength and charge balance in soil solution as well as total concentrations of chemicals supported in soil solution and the degree to which these chemicals are available in the environment. The relative importance of organic ligand occurrence in soil solution to bioavailability and transport is expected to vary markedly due to temporal and spatial variation in organic carbon flux through the solution phase of soils. Localized occurrences can deliver high concentrations of transient organic carbon to soil solution. Natural events such as stem flow in forest landscapes can cause uneven distribution of organic carbon within the landscape. Management practices in agroecosystems such as incorporation of green manures, animal wastes, or sludges can lead to short-term flushes of organic carbon through the soil solution. Shifts in localized quasi-equilibrium distributions of chemical species within soil solution are an important aspect of organic carbon flux through soil solution that have been inadequately described.

Extensive laboratory research documents the ability of organic acids to affect mineral dissolution (Chapter 9) but there is considerable uncertainty regarding in situ organic carbon occurrence in soil solution relative to the experimental conditions typically employed. Therefore, the application of this information to the description of processes occurring in natural environments is tenuous (McKeague et al., 1986) because

1. Concentrations of organic acids used in closed system laboratory studies may be excessive in comparison to natural concentrations.
2. Minerals used in laboratory studies may be freshly prepared and thus offer freshly exposed surfaces for reaction when in nature mineral surfaces are coated with weathering products and organic matter.
3. Open system studies of leaching may model unrealistic conditions of organic ligand concentration, water flow and volume, and temperature in comparison to natural systems.

Misleading interpretations may result because of the uncertain nature of the chemical and physical mechanisms involved in weathering reactions.

Despite the foregoing uncertainties, there appear to be significant effects of organic carbon on mineral dissolution at concentrations of organic carbon that can be considered reasonable, at least for short-term localized conditions. For example, Pohlman and McColl (1986) have shown rapid Mg release from forest soils when leached with naturally occurring organic acids at 0.2 to 5.0 mmol L^{-1} concentrations; rates of Mg release by organic acids were greater than by mineral acids. Thus, under conditions of rapid organic carbon flux in localized environments (for example, at the base of a tree receiving stem flow), mineral weathering may be more rapid than "typical" conditions within the landscape might suggest. Other processes in the environment that can cause concentrated localized flushes of organic carbon through soil solution, and, thus, localized effects on the chemistry of soils, are decomposition of carcasses (Vass et al., 1992) or urination by animals in heavily grazed paddocks (Haynes and Williams, 1992).

The ability of organic materials to affect changes in soil solution composition have been used as a tool for managing soil fertility. An important example is the use of green manures, animal wastes, or sludges to ameliorate the causes of acid soil infertility. High amounts of organic material incorporation to soil may act to neutralize soil acidity, shift the ratio of soil solution Ca^{2+} to Al^{3+} to favor establishment of calcicole plants, or alter the availability of trace nutrients and phytotoxic elements. While many of these beneficial effects of organic materials involve solid phase reactivity, altered organic carbon composition of soil solution has been shown as a causal mechanism as well (Hue et al., 1986; Hue, 1992; Besho and Bell, 1992). Detoxification of soil solution Al can occur when organic material amendment causes altered total concentration and composition of the dissolved and colloidal pool of soil solution carbon. Hue (1992), for example, has shown the occurrence in soil solution of urate arising from chicken manure amendment, the presence of citrate and tartrate arising from sludge amendment, and the occurrence of phthalate, glycolate, and perhaps oxalate as well from sludge or manure amendment. These organic acids had a favorable effect on establishment of *Desmodium intorum*, an Al intolerant calicole tropical forage legume, on an acid tropical Ultisol in part because they complexed soil solution Al to reduce levels of phytotoxic monomeric Al^{3+} in soil solution during the initial period of crop establishment.

Studies such as these, as well as investigations of dissolved and colloidal organic matter effects on podzolization (section 11.5.3) and xenobiotic availability and mobility (section 14.1), serve to indicate why DOC is a key master variable for characterization of soil solution. Quantification of the extent of DOC complexation as a parameter controlling ion distribution in soil solution remains a critical area of investigation for soil solution chemists.

NOTES

1. The discussion herein focuses exclusively on organics in soil solution. More detailed consideration of soil organic chemistry and humus chemistry may be found in several recent texts such as those of Huang and Schnitzer (1986), Stevenson (1982), Aiken et al. (1985).

2. When all sites behave identically and independently, the concentration of bound metal ($[M_aL]$—the numerator of equation 13-2) can be expressed as

$$[ML]_0 + 2[ML]_0 + 3[ML]_0 + \ldots n[ML]_0 = [ML]_0 \Sigma n = [ML]_0 n(n + 1)/2.$$

In this expression, $[ML]_0$ represents the overall concentration of metal–ligand complexes

($= K_0[M][L]$, where K_0 is the overall binding constant). The number of reactive sites, j, is defined by the number of complexes formed with M [$j = \Sigma n = n(n + 1)/2$]. The total ligand concentration ($[L]_t$—the denominator of equation 13-2) is

$$[L] + \Sigma[M_iL] = [L] + K_0[M][L] = [L](1 + K_0[M])$$

Thus, equation 13-2 can be expressed

$$\nu = \frac{jK_0[M][L]}{[L] + K_0[M][L]} = \frac{jK_0[M]}{1 + K_0[M]}$$

CHAPTER 14

XENOBIOTICS IN SOIL SOLUTION

Xenobiotic substances—from the Greek word *xenos,* meaning strange or foreign—are synthetic organic chemicals that are not natural to biological organisms; or, if they are of natural biological origin, they occur in environments and/or at concentrations that are biologically unnatural. The occurrence of xenobiotics in nature is therefore a consequence of anthropogenic processes. Xenobiotics occur in soil either purposefully (as when they are used for their beneficial effects in controlling unwanted plants or plant pathogens, or when soil is used as a medium for organic waste disposal or utilization) or they occur inadvertently (through contamination of soils with organic wastes).

Applying soil solution chemistry to interpret synthetic organic chemical availability in soils is a recent development for which there is relatively little published research. For certain aspects of xenobiotic fate and availability assessment, soil solution chemistry shows great promise. Both biological availability (efficacy/toxicity) and geochemical availability (environmental fate) are functions of chemical intensity in the soil solution for xenobiotic molecules that are biologically active in the soil environment.

This chapter reviews the use of soil solution chemistry for diagnosing xenobiotic fate in the soil environment. The relatively limited body of research describing xenobiotic behavior in soil solution has had three goals.

1. Develop and validate methodology for sampling and analysis of unaltered soil solution.
2. Demonstrate the diagnostic utility of soil solution composition as an indicator of xenobiotic bioavailability.

3. Predict soil environmental fate (that is, partitioning, degradation, and mobility) in situ.

Traditional Approaches to Understanding Soil Environmental Chemistry of Xenobiotics. Given the complexity of soil environments and the diverse nature of xenobiotics occurring in the unsaturated zone, it is not surprising that predicting xenobiotic bioavailability and mobility is often uncertain.

Current knowledge of xenobiotic bioavailability is mostly derived from case studies of a specific molecule under narrowly defined environmental conditions. Generally, xenobiotic bioavailability is inferred from field and greenhouse studies directed toward evaluation of efficacy or phytotoxicity to target and nontarget organisms. Studies of this type use semiquantitative measurements of biotic response (for example, relative scores for plant injury) and relate this response to an initial gross level of xenobiotic applied. Such approaches suffer from several limitations—principally, the inability to define the biologically active pool and the inability to discriminate between genetic and environmental contributions to the observed response.

The mobility of xenobiotics is frequently addressed from the standpoint of adsorption on soil or isolated soil components (Bailey and White, 1964; Bouchard and Lavy, 1985; Madhun et al., 1986; Majka and Lavy 1977; Scott and Weber, 1967). Although such studies provide valuable insights into the processes governing solid phase retention of xenobiotics, they do not adequately describe the relationship of water solubility to leaching potential (Bailey and White, 1970). For moderately to highly water soluble chemicals batch adsorption studies mostly measure the effect of solubility. Therefore, traditional sorptivity studies based on batch systems give little insight concerning vadose zone mobility of many molecules.

Most research regarding xenobiotic fate and transport has investigated either behavior in isolated compartments of the soil (most often distribution between the solid and liquid phases) or the losses from the solum (that is, dissipation). Such approaches ignore or overly simplify the processes of xenobiotic transfer and degradation. A more comprehensive approach integrates considerations of transfer and degradation among and within isolated soil compartments with description of solute flow paths in the unsaturated zone. Soil solution composition is integral to any consideration of xenobiotic bioavailability and mobility, so it is the logical focus for monitoring and prediction of xenobiotic fate and transport in soil.

14.1 SPECIATION AND COMPLEXATION IN SOIL SOLUTION

Speciation and complexation reactions modify the total concentration of a xenobiotic molecule occurring in the solution phase. As for inorganic chemicals in soil solution (Chapter 7), speciation and complexation influence both bioavailability and mobility of xenobiotics.

Weak acid dissociation is the principal speciation reaction of importance for synthetic organic molecules having a negative logarithm of the acid dissociation constant (pK_a) near the typical pH range for agricultural soils (pH 5 to 9). Dissociated versus undissociated forms of a molecule have markedly different reactivity with both soil solution and solid phase components. For example, flumetsulam, a triazolopyrimidine sulfonanalide herbicide ($pK_a = 4.6$) exhibits increased sorptivity with decreasing soil pH. This is a function of the differing sorptivities of dissociated ($K_{OC} = 12$ L kg^{-1}) versus undissociated ($K_{OC} = 450$ L kg^{-1}) forms of the molecule; the relative proportion of the dissociated versus undissociated forms present in soil solution varies with pH (Fontaine et al., 1991). Thus, pH-dependent sorptivity can be described as a function of soil pH as[1]

$$K_d = K_{OC}^n \ OC \left[\frac{10^{-pH} + \dfrac{K_{OC}^{an}}{K_{OC}^n} 10^{-pK_a}}{(10^{-pH} + 10^{-pK_a})} \right] \qquad [14\text{-}1c]$$

where K_{OC^n} is for the neutral form of the molecule and $K_{OC^{an}}$ is for the anionic form.

Additionally, complexation of xenobiotics may occur between both organic molecules and inorganic ions. Complexation with dissolved organic carbon (DOC) has been hypothesized to alter xenobiotic distribution in soil solution. However, there are inconsistent research findings to support this hypothesis (Stevenson, 1972). Pennington et al. (1991) isolated soil DOC by water extraction of low organic carbon content mineral soils and measured its complexation with the herbicides bromacil, metrabuzin, alachlor, diquat, and paraquat; these herbicides were weakly sorbed by DOC. Conversely, Madhhun et al. (1986) reported substantial binding of bromacil, diuron, chlorotoluron, simizine, glyphosate, and diquat with DOC isolated by water extraction of a peat soil. A major difference between these studies was the source of DOC, which may have influenced the chemical reactivity observed. In any event, both authors agreed that actual DOC present in ambient soil solution would be too limited to have a controlling influence on the partitioning and transport of the herbicides studied.

An example of complexation with inorganic ions is that of the glyphosate. Glyphosate, a nonspecific herbicide active when applied directly to plants, is rendered nontoxic in the soil by strong complexation with Ca^{2+}.

14.2 SAMPLING AND ANALYTICAL APPROACHES

Displaced soil solutions, lysimeter solutions, and soil extracts are all used to measure available pools of xenobiotics. The adequacy of any of these measurements depends on the ability to distinguish the readily available pool (that is, concentrations occurring in soil solution) from the labile or potentially available pool. Where adequate means to measure soil solution pools are available, it is possible to describe in dynamic terms the distribution between soil compartments concurrent with degradation of the molecule.

14.2.1 Soil Solution Displacement

Soil solution displacement procedures (Chapter 5) provide a means whereby xenobiotic availability in the soil environment can be rapidly and effectively evaluated. The displacement and analysis of soil solution provides

1. refined measurements of the bioavailability of soil active molecules,
2. static measurements of phase partitioning under conditions that closely model soil moisture regimes in field environments, and
3. dynamic measurements of availability as a function of residence time in the soil.

Soil solution displacement appears a logical approach for monitoring bioavailability and mobility of xenobiotics in soil because it affords greater control and certainty as to the meaning of analytical results than does field-based soil water sampling or extraction techniques. Soil sampling and displacement under laboratory conditions have proven useful for obtaining unaltered soil solution for determining toxic organic (Ononye et al., 1987) and pesticide (Goetz et al., 1986; Patterson et al., 1982) bioavailability and mobility in the soil liquid phase.

The modification of traditional aerobic soil metabolism experiments to incorporate measurement of soil solution composition can be used to investigate dynamic concurrent processes of degradation and partition occurring in soil environments (Wolt, 1993). This approach allows for generation of parameters relevant to environmental fate (soil metabolic half-life and apparent distribution coefficients) and phytotoxicity from the same test system under conditions of controlled moisture and temperature. An example of this approach is illustrated in Figure 14-1. Field-moist soils treated with a xenobiotic of interest are incubated and displaced after various times of incubation using a vacuum displacement technique (section 5.3.2). The soil solution recovered is analyzed for the xenobiotic of interest as well as for pH and ionic strength (as estimated from electrical conductivity). Subsequently, aliquots of soil solution may be used for plant bioassay. The soil residue following displacement (or, alternatively a separate aliquot of treated soil) may be extracted with an organic solvent system and analyzed to determine the labile pool.

14.2.2 Use of Extracts to Mimic Soil Solution Composition

Conventional batch studies (Majka and Lavy, 1977) utilize soil in contact with an excess of equilibrating solution, typically 0.01 M Ca^{2+} as $CaCl_2$ or $CaNO_3$, to estimate phase partitioning of xenobiotics. Distribution coefficients derived from batch studies attempt to describe chemical partitioning in the absence of physical constraints to partitioning. Thus, the intent of equilibrium batch studies is to measure chemisorption under conditions where physical constraints to the sorption process (for example, diffusion) are minimized. Equilibration batch extraction may, however, exhibit interpretive limitations or inadequate sensitivity for assessing sorptive partition of xenobiotics. Whereas this method may lead to appropriate estimates

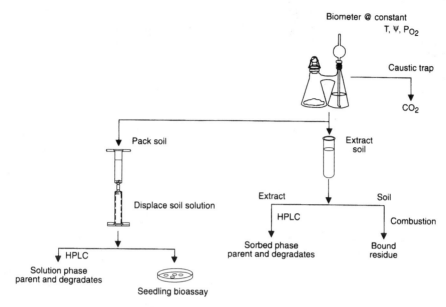

FIGURE 14.1   Flow sheet of partition and analysis of soils treated with a xenobiotic molecule and incubated aerobically in biometer systems. Replicate biometers sampled and analyzed after varied times of treatment yield soil metabolic half-life, time-dependent sorptivity, and bioactivities as well as quantity and nature of parent molecule and metabolites. [Wolt, 1993]

of static partitioning for nonpolar molecules of low water solubility, it can vastly underestimate the sorptive partition for highly water soluble, weak organic acids or bases (Table 14.1).

The adequacy of an equilibrium batch system to model sorption xenobiotic relates to the sorption mechanism involved. These mechanisms may be viewed as "enthalpy-driven" or "entropy-driven" (Sims et al., 1992). Enthalpy-driven mechanisms involve surface reactions between soil particles and charged organic molecules. Ionic bonding, ligand exchange, charge transfer, and short-range electrostatic dipole interactions are types of mechanisms that may result in sorption to surfaces; these mechanisms may involve physical and chemical barriers to the sorption process that are not realistically reflected in batch equilibration studies. Entropy-driven mechanisms, however, involve thermodynamically favorable entropy gains when an organic solute is eliminated from the aqueous phase coupled with preferential solvation of lipophilic nonpolar organic molecules into soil organic matter. While entropy-driven mechanisms may be well represented in batch equilibration studies with freshly added nonpolar molecules, there can be distinct hysterisis in the sorptive versus desorptive behavior of these molecules. Thus, for aged residues of both polar and nonpolar xenobiotics, the dynamics of desorption from the soil solid phase may not be well represented in batch systems.

Weakly acidic organic molecules that predominate soil solution in dissociated forms in typical soil solutions (pH 5 to 9) tend to sorb by enthalpy-driven mech-

TABLE 14.1 Herbicide Sorptivity Relations as Determined by Batch Equilibration, Soil Solution Displacement, and Simulation Modeling of Soil Solution Data

Soil	Batch K_d, L kg^{-1}	Soil Solution k_{app}, L kg^{-1}					Soil Solution AR_{eq}, L kg^{-1}
		Herbicide W					
		0.2 d	*2 d*	*5 d*	*12 d*	*18 d*	
Tama sil[a]	0.26	0.32	0.41	1.15	1.65	1.24	2.36
Barnes l	0.35	0.16	0.34	0.60	0.83	1.12	1.99
Cecil sl	0.06	0.16	0.22	0.20	0.31	0.53	0.18
		Herbicide X					
		0.2 d	*2 d*	*5 d*	*12 d*	*18 d*	
Tama sil	0.07	0.25	0.57	0.60	0.52	1.33	0.32
Barnes l	0.00	0.10	0.29	0.30	0.30	0.41	0.18
Cecil sl	0.04	0.14	0.16	0.16	0.19	0.23	0.12
		0.5 d	*1.5 d*	*3.5 d*	*4.5 d*	*9.5 d*	
Etowah l	0.17	0.20	0.30	0.43	0.31	0.37	0.25
		Herbicide Y					
		0.2 d	*2 d*	*4 d*	*8 d*	*42 d*	
Hastings sil	0.64	1.43	1.25	1.91	5.84	5.21	9.91[b]
Catlin sil	0.16	0.40	0.21	0.29	0.43		0.57[b]
Barnes l	0.10	0.13	0.25	0.53	0.50		0.63[b]
		Herbicide Z					
		0.2 d	*2 d*	*5 d*	*12 d*	*18 d*	
Tama sil	0.25	0.27	0.59	0.81	1.32	1.22	1.20
Barnes l	0.39	0.22	0.16	0.53	0.46	0.71	0.61
Cecil sl	0.06	0.15	0.13	0.20	0.18	0.26	0.29

[a]sil = silt loam, l = loam; sl = sandy loam.
[b]Parent half-life is < 2 days; calculated soil solution AR_{eq} is for the primary degradate.
Wolt, 1993. Copyright ASTM. Reprinted with permission.

anisms. The initial ability to sorb may be hindered by exclusion from sorptive surfaces (for instance, anion exclusion), and this tendency will be exaggerated for batch systems where the excess solution phase encourages distribution into the aqueous phase simply on the basis of high water solubility of the molecule. In situ sorptive partition of such molecules is anticipated to be very different at field soil moisture contents—low initial sorption, arising from diffusive limitations and charge exclusion near surfaces of soil particles, will be followed by progressive sorption as molecules approach proton swarms near particle surfaces by simple mass action, become neutral, and subsequently sorb as a nonpolar molecule (enthalpy-driven charge transfer reactions followed by entropy-driven sorption). Such dynamic reactivities can result in distinct hysterisis in sorption versus desorption and a pronounced "aged soil effect"—the commonly observed increase in sorptive K_d with time of reaction in soil at field moisture contents—and cannot be evaluated with traditional equilibrium batch systems. This phenomenon is illustrated in Figure 14-2 for the triazine herbicide, simazine. A significant discrepancy in the seasonal distribution of simazine into soil solution is shown by comparing the measured concentration of simazine occurring in soil solution (obtained from centrifugal displacement of field-moist soil) to the predicted aqueous phase concentration (determined from equilibrium batch K_{OC}, soil moisture, and total simazine residues in soil). Equilibrium batch sorption adequately predicted simazine sorption by field soil only in the initial hours following application; at all other times measured soil solution simazine concentrations were significantly below predicted values.

A unique approach to assessing the dynamic sorption of xenobiotics is the use of instantaneous batch desorption or miscible displacement to estimate the available pool through extraction of the kinetically labile pool of molecule after various times of soil incubation at field moisture contents. This approach was discussed in Chapter 3 (Table 3.9 and Fig. 3-3) for describing the dynamics of simultaneous sorption, desorption, and degradation of picloram in soil. McCall and Agin (1985) generated apparent K_d for desorption of picloram from soils aged at field moisture content by restricting extraction ratio (2:1 water to soil) and time (2 minutes) to sample

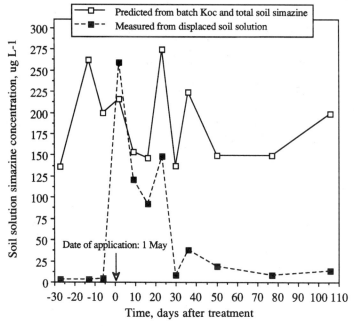

FIGURE 14.2 Comparison of simazine measured in displaced soil solution with that predicted from batch K_{OC} and total soil simazine residues. [From the data of Scribner et al., 1992.]

the readily available soil pool (that is, soil solution). An alternative approach to instantaneous batch desorption may be the use of miscible displacement (Carski and Sparks, 1985) to model the soil available pool of soil residues.

14.2.3 Lysimetry

Field Lysimetry. As detailed in Chapter 6, tension lysimeter sampling techniques have restricted usefulness because of nonuniform sampling and artifact effects that make the interpretation of lysimeter water composition uncertain. The primary utility of tension lysimetry for field monitoring of solution phase xenobiotics is to determine if movement occurs, rather than to obtain diagnostic measurement of soil solution intensities for interpretation of chemical reactivity and availability in the soil environment. An example of this approach is the use of transient tension point lysimeters to monitor picloram mobility following aerial application to a forested watershed as influenced by soil type and landscape position (Fig. 14-3; Michael et al., 1989). The concentrations measured in the lysimeter solutions have limited interpretative meaning per se. However, the occurrence of the xenobiotic in lysimeter solutions with time—as related to rainfall distribution and intensity, landscape position, and physical properties of the soils within a landscape—provides useful information for the generalized description of mobility within a specific landscape.

Column Leaching and Breakthrough Studies. Plug flow studies utilizing packed columns of disturbed soil are frequently employed to determine the potential mobility of xenobiotics and their degradates in soils. The solution composition obtained for leachates from these column leaching systems cannot be properly associated with soil solution (systems are saturated and contact times are too brief to adequately model soil solution) nor with soil water (flow velocities are too high and flow paths too uniform).

A closer approach toward systems representative of soil water and water-borne solutes interacting diffusively with soil solids is achieved when uniformly packed columns of disturbed soil are used for breakthrough studies. Unlike column leaching, saturated flow column breakthrough studies can be designed with varying degrees of sophistication to assure that flow velocities are uniform and represent natural systems. If the chemical composition of the eluting solution adequately models that of the bulk soil solution, the results of column breakthrough studies with disturbed soils can be used to interpret the relative controlling influence of physical versus chemical processes for sorption of xenobiotics by soil.

Column breakthrough studies utilizing undisturbed soil columns and conditions of unsaturated flow allow for assessment of flow-restrictive control over leaching of xenobiotics (that is, the influence of channelized flow). When coupled with soil solution measurements of the diffuse transport of xenobiotics, undisturbed columns studies allow for determination of the relative influence of flow paths and mechanisms on the mobility and sorptivity of chemicals by soil (sections 6.2.2 and 14.4.1).

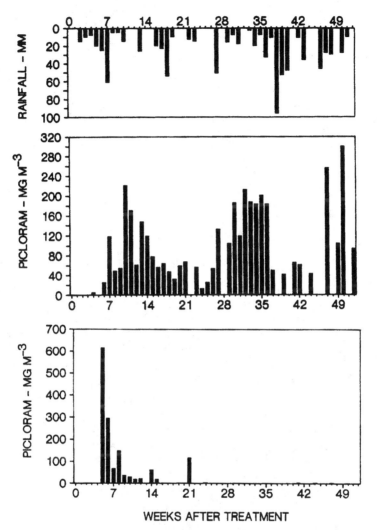

FIGURE 14.3 Mean concentrations of picloram in soil solution sampled with porous cup lysimeters. (*a*) Midslope. (*b*) From under an ephemeral channel in a Coastal Plain forest. [Michael et al., 1989. *J. Environ. Qual.,* 18:89–95. Am. Soc. of Agronomy, Crop Sci. Soc. Am., Soil Sci. Soc. Am.]

14.2.4 Expression of Soil Solution Composition

Measuring xenobiotic concentrations occurring in soil solution may inadequately describe bioactivity in the vadose zone. Researchers interested in the fate and behavior of inorganic chemicals have frequently demonstrated that the thermodynamic activity, as opposed to the total concentration, of these chemicals in soil solution is often the more diagnostically meaningful parameter (Chapter 8). This is in part because of the tendency for solution ionic strength and complex formation to obscure the actual fraction of a dissolved species in soil solution that may be of chemical or biological consequence (Sparks, 1984). The occurrence of colloidally bound materials in the soil liquid phase further complicates the interpretation of soil solution composition. Approaches that may be used to determine the activity of a dissolved organic species are the use of thermodynamic modeling routines to correct measured concentration (Arbuckle, 1986) or the use of thermodynamic computations based on the measured partial pressure of the dissolved organic exerted on a contacting gas phase (Yin and Hassett, 1986).

Ionized Molecules. Simple thermodynamic models based on molecule pK_a and mole weight and soil solution pH and ionic strength have been used to correct soil solution herbicide concentrations (intensities) to quasi-thermodynamic activities (effective concentrations). For example, Wolt et al. (1989) estimated the chemical activity of soil solution imazaquin from the relationship

$$(A^-) = [HA]_i \frac{10^{(pH-pK_a)}10^{(-0.509)(0.013)(EC)}}{1 + 10^{(pH-pK_a)}} \qquad [14\text{-}2]$$

where activity of the ionized molecule (A^-) is related to total herbicide concentration in soil solution (the herbicide intensity, $[HA]_i$), the negative logarithm of the acid dissociation constant (pK_a), and soil solution pH and electrical conductivity (EC). Such approximations may provide a first estimate of the bioavailable portion of the molecule as influenced by various environmental variables for xenobiotics that predominate soil solution in ionized form.

Un-ionized Molecules. The activity coefficient of a nonelectrolyte in an electrolyte solution (γ_S) can be developed from water solubility (M_O), solubility in the electrolyte solution (M_S), and the respective equilibrium partial pressures $(P_O$ and $P_S)$.

$$\gamma_S = \frac{M_O}{M_S} \frac{P_S}{P_O} \qquad [14\text{-}3a]$$

where gas phase behavior is assumed to approach ideal behavior (Garrels and Christ, 1965). If P_S is constant, equation 14-3a can also be expressed as

$$\gamma_S = \frac{K'}{M_S} \qquad [14\text{-}3b]$$

providing an experimental basis for the determination of γ_S. Experimental results indicate that, in general, M_S decreases at constant P_S as ionic strength (I) of the electrolyte solution increases—the *salting-out effect*. The empirical relationship,

$$\log y_S = kI \qquad [14\text{-}4]$$

where k is the *salting coefficient* and y_S is the empirical estimate of γ_S, provides a useful first approximation for determining activities of undissociated, neutral molecules.

Static equilibrium partitioning experiments can be used to directly measure γ_S for volatile, nonelectrolytes (Gossett, 1987). Table 14.2 compares γ_S measured using equilibrium partitioning in a closed system with y_S estimated by equation 14-4 for the *cis* and *trans* forms of 1,3-dichloropropene (1,3-D) in displaced soil solutions. The salting coefficient was determined for *cis-* and *trans*-1,3-D equilibrated with KCl solutions of varied ionic strength; the calculated k and estimated ionic strengths of displaced soil solution (I = 0.13 × EC) were used to estimate y_S. The y_S compares favorably with measured γ_S.

14.3 AVAILABILITY

14.3.1 Soil Solution Composition as an Indicator of Xenobiotic Availability

The interest in using soil solution composition as an indicator of xenobiotic bioavailability stems from the success of soil scientists in defining critical thresholds for metal toxicity and nutrient deficiency to plants and microbes (Chapter 10). These critical levels frequently correspond to similar determinations made in hydroponic culture studies, especially when solutes are expressed in terms of their chemical thermodynamic activity rather than in terms of concentrations.

In recent years, soil solution chemistry has been employed to better ascertain herbicide behavior in soils. Patterson et al. (1982) related fluometuron efficacy in soils of differing mineralogy with measurement of fluometuron in displaced soil solutions. Goetz et al. (1986) used a similar approach to characterize imazaquin sorptivity and availability in soils of differing organic matter content. These authors found analysis of displaced soil solution offered greater sensitivity than a batch equilibration technique. In contrast, Wehtje et al. (1987) considered batch and soil

TABLE 14.2 Estimated and Measured Activity Coefficients for *cis-* and *trans*-1,3-dichlorpropene in Displaced Wahiawa Soil Solutions[a]

	cis-1,3-dichloropropene		*trans*-1,3-dichloropropene	
	Estimated (y_S)	Measured (γ_S)[b]	Estimated (y_S)	Measured (γ_S)[b]
Surface soil[c]	0.99	1.15	1.01	1.07
Subsoil[d]	1.00	1.30	1.00	1.18

[a]Tropeptic Eutrostox.
[b]Relatively large standard deviations are associated with these measurements and are characteristic of error associated with application of the modified equilibrium partitioning in closed system procedure to low to moderately volatile molecules (Gossett, 1987).
[c]0 to 20 cm; pH 6.35, DOC 2.39 mmol L^{-1}, EC 0.535 mmhos cm^{-1}.
[d]60 to 80 cm; pH 6.05, DOC 1.93 mmol L^{-1}, EC 0.305 mmhos cm^{-1}.

solution techniques to exhibit equivalent sensitivities for determination of sulfo-meturon and imazapyr sorptivities and, thus, bioavailabilities. Goetz et al. (1989) considered soil solution apparent K_d measurements to overestimate chlorimuron sorptivities, probably as a consequence of how they used their soil solution com-positional analyses for expression of apparent K_d.

Measurements of imazaquin concentration in displaced soil solution have been used to interpret the influence of crop management on persistence (Wolt et al., 1989). Figure 14-4 illustrates the experimentally modeled effect of residue man-agement on plant biotic response (relative root elongation in displaced soil solution) as a function of readily available imazaquin (herbicide in soil solution expressed as chemical activity of the molecule) where the chemical activity of imazaquin was estimated from equation 14-2,

$$(\text{imazaquin}^-) = [\text{imazaquin}]_i \frac{10^{(\text{pH}-3.8)}10^{(-0.509)(0.013)(\text{EC})}}{1 + 10^{(\text{pH}-3.8)}} \qquad [14\text{-}5]$$

Soil solution monitoring of soil-active herbicides is a more direct measure of bioavailability than are measurements based on total quantity of applied herbicide or sorbed phase concentration. For example, Ononye et al. (1987) have observed that the concentrations of benzidine supported in soil solution as a function of time after introduction to the soil environment are environmentally consequential, despite

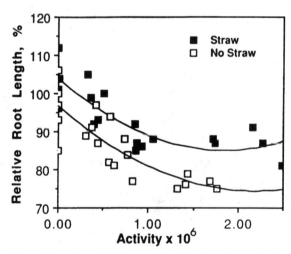

FIGURE 14.4 The relationship of relative root length of sunflower (at 0 activity = 100%) germinated in soil solution to ion activity of imazaquin in soil solution for straw-amended and unamended soils. (For straw-amended soils, $y = 103 - 20.54x + 5.59x^2$, $R^2 = 0.70$; for unamended soils, $y = 97 - 20.45x + 4.67x^2$, $R^2 = 0.69$). [J. D. Wolt, G. N. Rhodes, Jr., J. G. Graveel, E. M. Glosauer, M. K. Amin, and P. L. Church, 1989, "Activity of Imazaquin in Soil Solution as Affected by Incorporated Wheat *(Triicum aestivum)* Straw," *Weed Science,* 37:254–258.]

the fact that $> 99\%$ of added benzidine is sorbed. Additionally, the sensitivity of the soil solution displacement procedure when coupled with analysis via high-performance liquid chromatography and mass spectroscopy allowed for rapid quantitation of benzidine without the possibility of side reactions which could alter benzidine when obtained by solvent extraction.

Figure 14-5 illustrates the relationship between quantity of applied herbicide, the intensity and effective concentration in soil solution, and the sorbed phase concentration. Varied appreciations are gained regarding the relationships of soil herbicide concentration to biotic response depending on the way soil herbicide concentration is expressed. Quantity of herbicide applied bears only a gross relationship to biotic response because the processes of partition and degradation interact to modify the soil herbicide pool with time. This information is conveyed to some extent through consideration of the sorption coefficient and the soil metabolic half-life, but a mix of static (K_d) and dynamic (half-life) terms are used to describe the dynamic interactive processes of sorption, desorption, and degradation. Herbicide intensity in soil solution measured at any given time reflects this dynamic interaction and can be further refined to express the interaction of soil properties (pH and ionic strength) and physical properties of the molecule (pK_a) on the bioavailability of molecule (for example, as in equation 14-2). Herbicide intensity or effective concentration represents an integrated measurement of dynamic processes, so the relationship of these measurements to biotic response will be less variable

a

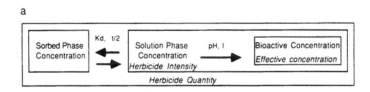

b Barnes silt loam: 8-day incubation at 25C and 16.12% moisture

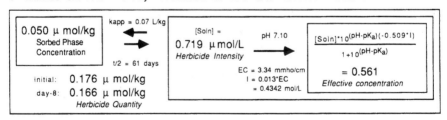

FIGURE 14.5 (*a*) Relationship between herbicide quantity, intensity, and effective concentration. (*b*) Example of herbicide phase partition where soil solution composition (herbicide intensity, pH, and electrical conductivity) is used to determine half-life, the apparent distribution coefficient (k_{app}), sorbed phase concentration, and effective concentration. [Wolt, 1993]

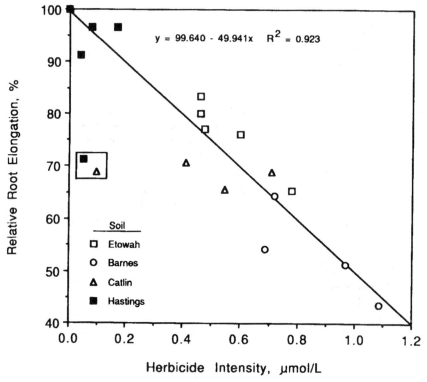

FIGURE 14.6 Herbicide phytotoxicity, expressed as seedling bioassay (relative root length) in displaced soil solution, as functionally related to soil solution herbicide concentration (intensity). (Outliers [box] are excluded from the regression equation.) [Wolt, 1993]

across environments than will be grosser measurements such as herbicide quantity or sorbed phase concentration.

Soil solution pesticide concentration (herbicide intensity) or quasi-thermodynamic estimates of activity (effective concentration) can be used in conjunction with bioassays to determine phytotoxic thresholds for growth or to identify the biologically active pool of a xenobiotic molecule in soil (Wolt et al., 1989; Wolt, 1993). Figure 14-6 presents the correlation of biotic response [relative root elongation of sunflower (*Helianthus annus* L.) in displaced soil solution] on flumetsulam intensity in soil solution. Use of the herbicide intensity to reflect the integrated effect of soil environment on bioavailability of a pesticide allows for generalization of response across environments.

14.4 ENVIRONMENTAL CHEMISTRY

Even though there can be little question as to the value of determining activities of xenobiotics in soil solution toward the elucidation of mechanisms whereby toxic organics are retained or are mobilized in the unsaturated zone, one must be cautious as to how this information is applied towards the evaluation of xenobiotic transport under conditions of unsaturated flow. Soil solution composition is representative of soil water in psuedo-equilibrium with the soil solid phase and subject to diffuse flow. As such, for a given concentration of synthetic organic introduced to a soil, solution composition represents a condition favoring maximum reactivity with the solid phase. The assumption of psuedo-equilibrium in soil environments has not been adequately detailed especially for those cases in which biotransformation dominates the rate of xenobiotic transfer. Additionally, for soils of well-defined structure, macropore flow may be an important conduit for the unsaturated flow of water. In these soils, short contact times of solutes with the soil solid phase may influence xenobiotic mobility and may further limit the applicability of static models.

The applicability of static models to description of xenobiotic fate and transport cannot be discounted out-of-hand because such models find nearly universal use in current models of unsaturated flow; but little has been done to validate the typical research approach (equilibrium batch experiments) as a reliable measure of unsaturated conditions. As a refinement of static models, dynamic studies of xenobiotic behavior can provide useful information for describing the rates as reactions move toward equilibrium and the mechanisms involved, but the necessary theoretical and experimental approaches are difficult to apply to heterogeneous systems (Sparks, 1986).

14.4.1 Static versus Dynamic Measures of Sorptivity

The retardation of xenobiotics as they leach through the unsaturated zone is a key consideration of xenobiotic transport. Retardation is frequently expressed relative to the soil–water partition constant or distribution coefficient (K_d, section 8.3). This approach is limited because equilibrium batch K_d determinations at high water-to-soil ratios may be inappropriate models of soil environments and their use implies solution–solid phase equilibrium exists for xenobiotic transport through the vadose zone.

Traditional equilibrium batch K_d are compared with soil solution apparent distribution coefficients (k_{app}) in Table 14.1 for a suite of water-soluble, anionic herbicides. These molecules exhibit lower sorptivities in batch systems than in soil solution. Soil solution k_{app} are termed "apparent" for several reasons. First, the parent molecule and degradates are assumed to partition similarly between soil solid and solution phases. Second, k_{app} are measured for dynamic systems and therefore would be anticipated to be time-dependent measures, unlike the more commonly reported K_d that are generated for systems near equilibrium. Third, k_{app} reflects both the physical process of diffusive movement of a molecule from bulk soil solution to a sorptive surface and the chemical process of sorption into the soil solid phase; whereas, K_d, as measured by typical equilibrium batch methods, attempts to minimize physical effects on sorption. The k_{app} is a dynamic measure of chemical partitioning in a system that has not achieved equilibrium, and it more realistically represents the phase partitioning of herbicides that is likely to be manifested in the field environment. Values of k_{app} (Table 14.1) tend to increase with time for all molecule $\times$ soil combinations, which is consistent with diffusion from bulk soil solution into interaggregate pores and subsequent chemisorption.

14.4.1 Equilibrium versus Nonequilibrium Xenobiotic Behavior

On a time-averaged basis, most components of soil environments may be at steady state with respect to each other. However, highly dynamic events may dominate the actual mobilization of xenobiotics and drive their migration through the soil profile; xenobiotic distributions during these events may be far from equilibrium. The driving force in xenobiotic transport is water flow, the dynamics of which are complicated by storm events and heterogeneities in hydrologic flow paths. Large quantities of water (and possibly mobile organic colloids containing significant amounts of bound residues) are flushed through soil during rain events. Mesopore and macropore flow through root channels and other subsurface irregularities may account for a significant portion of the water flux through a soil system. Diffuse flow of water through the small interstices between soil particles is sufficiently slow to permit equilibration of solutes among different soil compartments, but the rapid flushing through large (relative to the size of solute molecules) channels may limit steady state approaches.

Flow paths of water, and thus the contact time of soil water with the soil solid phase, will influence interpretations of physical, chemical, and biological reactions governing both the fate and transport of organic solutes in the vadose zone. A more certain understanding of water flow in unsaturated soils and its linkage to transfer and degradation processes is required for assessment of xenobiotic transport (that is, leachability). A comparison of xenobiotic in soil solution (diffuse soil water) with xenobiotic in leachates from undisturbed soil columns (diffuse plus channelized soil water) is one approach whereby a clearer understanding of leaching potential and the relationship of water flow paths to transfer and degradation processes in the unsaturated zone may be ascertained. This approach was used by O'Dell et al. (1992) to describe imazethapyr transport through undisturbed soil columns. Time-dependent sorptivity of imazethapyr was measured as a function of residence time in field-moist soil using displacement and analysis of soil solution. The kinetics of partitioning were well described by a biphasic general-order reaction mechanism (Jardine and Zelazny, 1986), and this information was incorporated into a two-site nonequilibrium convective-dispersive transport model (Parker and van Genuchten, 1984). Comparison of breakthrough curves experimentally derived for imazethapyr leaching through intact soil columns with modeled results indicated the occurrence of preferential solute flow. Preferential solute flow resulted in short contact times of imazethapyr with the soil solid phase and limited exposure to sorptive sites; thus, sorptivities were restricted to low initial values (Figure 14-7).

The determination of bioavailability and mobility of environmentally consequential chemicals in vadose environments requires a more comprehensive under-

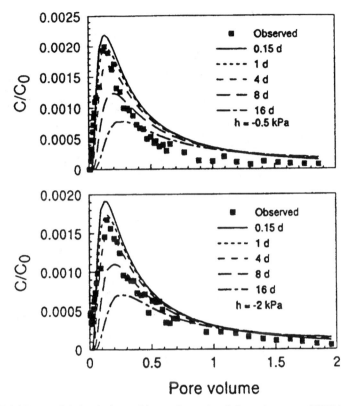

FIGURE 14.7 Model simulations of imazethapyr breakthrough curves (BTCs) using parameters determined from Br^- BTC and imazethapyr decay curves; h = potential. Decay curves developed from 0.15 to 16 days after treatment reflect increased imazethapyr sorption with time of reaction in soil. [O'Dell et al., 1992] The observed BTC are best modeled for decay curves reflecting limited reaction with soil solids. [Published in *Soil Sci. Soc. Am. J.,* 56:1711–1715. 1992. Soil Sci. Soc. Am.]

standing of the partitioning and rates of transfer of xenobiotics between soil compartments. Therefore, development of transport models that treat partitioning as a rate-dependent process are needed for refined modeling of xenobiotic fate in soil.

14.5 SUMMARY

The methodology for displacement, analysis, and expression of soil solution components is well established and has been widely utilized for evaluation of the integrated effect of soil processes on the availability of nutrients, trace metals, and ligands in the unsaturated zone. This same methodology shows considerable promise for elucidation of xenobiotic bioavailability and mobility in vadose environments, although it has not been widely used to date. A focus on soil solution chemistry is especially appropriate for considerations of the fate and behavior of many complex organic molecules that are moderately to highly water-soluble, ionic, and weakly sorbed in typical soil environments. Soil solution approaches to elucidating the environmental fate and behavior of such molecules have provided greater insights as to bioavailability, sorptivity, and mobility. The utility of these approaches will be further enhanced through refining methods for obtaining and analyzing soil solutions, describing soil solution composition effects on xenobiotic chemical activity, and defining the relative importance of static versus dynamic processes in the control of xenobiotic fate in soils.

NOTES

1. Fontaine et al. (1991) describe the effect of soil pH on flumetsulam sorption as

$$K_{OC} = \frac{K_{OC}^n + K_{OC}^{an} \, x}{(1 + x)} \qquad [14\text{-}1a]$$

where the net sorptivity, expressed as K_{OC}, is a function of the K_{OC} for neutral (undissociated) and anionic (dissociated) species (K_{OC}^n and K_{OC}^{an}, respectively). [$K_{OC} = K_d/OC$, where OC = the organic carbon fraction.] The modeled fitting parameter x can be defined as

$$x = \frac{Ka \, M_{an}}{[H^+] \, M_n}$$

where Ka is the acid dissociation constant and M_n and M_{an} represent the mole weight of neutral and anionic forms of the molecule. Thus, $x \approx K/[H^+]$. Substitution into equation 14-1a gives

$$K_{OC} = K_{OC}^n \left[\frac{10^{-pH} + \dfrac{K_{OC}^{an}}{K_{OC}^n} 10^{-pKa}}{(10^{-pH} + 10^{-pKa})} \right] \qquad [14\text{-}1b]$$

or,

$$K_d = K_{OC}^n \, OC \left[\frac{10^{-pH} + \dfrac{K_{OC}^{an}}{K_{OC}^n} 10^{-pKa}}{(10^{-pH} + 10^{-pKa})} \right] \qquad [14\text{-}1c]$$

REFERENCES

Abrams, M. M. and W. M. Jarrell. 1992. Bioavailability index for phosphorus using ion exchange resin impregnated membranes. *Soil Sci. Soc. Am. J.* (56):1532–1537.

Adams, F. 1966. Calcium deficiency as a causal agent for ammonium phosphate injury to cotton seedlings. *Soil Sci. Soc. Am. Proc.* (30):485–488.

Adams, F. 1971. Ionic concentrations and activities in soil solution. *Soil Sci. Soc. Am. Proc.* (35):420–426.

Adams, F. 1974. Soil solution. In E. W. Carson (Ed.), *The Plant Root and Its Environment* (pp. 441–481). Charlottesville: University Press of Virginia.

Adams, F. 1984. Crop response to lime in the southern United States. In F. Adams (Ed.), *Soil Acidity and Liming,* 2nd ed. (pp. 211–265). Agronomy Monograph 12. Madison, WI: American Society of Agronomy.

Adams, F. 1984. *Soil Acidity and Liming* (2nd ed.). Madison, WI: American Society of Agronomy.

Adams, F., C. Burmester, N. V. Hue, and F. L. Long. 1980. Comparison of column-displacement and centrifuge methods for obtaining soil solution. *Soil Sci. Soc. Am. J.* (44):733–735.

Adams, F. and P. J. Hathcock. 1984. Aluminum toxicity and calcium deficiency in acid soil subhorizons of two coastal plains soil series. *Soil Sci. Soc. Am. J.* (48):1305–1309.

Adams, F. and Z. F. Lund. 1966. Effect of chemical activity of soil solution aluminum in cotton root penetration of acid subsoils. *Soil Sci.* (101):193–198.

Adams, F. and R. W. Pearson. 1970. Differential response of cotton and peanuts to subsoil acidity. *Agron. J.* (62):9–12.

Adams, F., R. W. Pearson, and B. D. Doss. 1967. Relative effects of acid subsoils on cotton yields in field experiments and on cotton roots in growth-chamber experiments. *Agron. J.* (59):453–456.

Adams, F. and Z. Rawajfih. 1977. Basaluminite and alunite: A possible cause of sulfate retention by acid soils. *Soil Sci. Soc. Am. J.* (41):686–692.

Adams, F. and J. I. Wear. 1957. Manganese toxicity and soil acidity in relation to crinkle leaf of cotton. *Soil Sci. Soc. Am. Proc.* (21):305–308.

Adams, J. F., F. Adams, and J. W. Odom. 1982. Interaction of phosphorus rates and soil pH on soybean yield and soil solution composition of two phosphorus-sufficient Ultisols. *Soil Sci. Soc. Am. J.* (46):323–328.

Adams, J. F. and J. W. Odom. 1985. Effects of pH and phosphorus on soil-solution phosphorus and phosphorus availability. *Soil Sci.* (140):202–205.

Ahrland, S., J. Chatt, and N. R. Davies. 1958. The relative affinities of ligand atoms for acceptor molecules and ions. *Quart. Rev.* (12):265–276.

Aiken, G. R., D. M. McKnight, R. L. Wershaw, and P. MacCarthy. 1985. *Humic Substances in Soil, Sediment, and Water.* New York: Wiley.

Aitken, R. L. and R. J. Outhwaite. 1987. A modified centrifuge apparatus for extracting soil solution. *Comm. Soil Sci. Plant. Anal.* (18):1041–1047.

Alexander, M. and K. Scow. 1989. Kinetics of biodegradation in soil. In B. L. Sawheny and K. Brown (Eds.), *Reactions and Movement of Organic Chemicals in Soils* (pp. 243–269). Madison, WI: Soil Science Society of America.

Alva, A. K., D. G. Edwards, C. J. Asher, and F. P. C. Blamey. 1986. Effects of phosphorus/aluminum molar ratio and calcium concentration on plant response to aluminum toxicity. *Soil Sci. Soc. Am. J.* (50):133–137.

Alva, A. K., M. E. Sumner, and W. P. Miller. 1991. Relationship between ionic strength and electrical conductivity of soil solutions. *Soil Sci.* (152):239–242.

Amacher, M. C. 1984. Determination of ionic activities in soil solutions and suspensions: Principal limitations. *Soil Sci. Soc. Am. J.* (48):519–524.

Arbuckle, W. B. 1986. Using UNIFAC to calculate aqueous solubilities. *Environ. Sci. Technol.* (20):1060–1064.

Ares, J. 1986. Identification of aluminum species in acid forest soil solutions on the basis of Al:F reaction kinetics: 1. Reaction paths in pure solutions. *Soil Sci.* (141):399–407.

Asher. C. J. and D. G. Edwards. 1978. Relevance of dilute solution culture studies to problems of low fertility tropical soils. In C. S. Andrew and E. J. Kampreth (Eds.) *Mineral Nutrition of Legumes in Tropical and Subtropical Soils* (pp. 131–152). Melborne, Australia: Commonwealth Scientific and Industrial Organization.

Atkinson, H. J., G. R. Giles, and J. G. Desjardins. 1958. Effect of farmyard manure on the trace element content of soils and of plants grown thereon. *Plant Soil* (10):32–36.

Bache, B. W. and G. S. Sharp, 1976. Soluble polymeric hydroxy-aluminum ions in acid soils. *J. Soil Sci.* (27):167–174.

Baham, J. 1984. Prediction of ion activities in soil solutions: Computer equilibrium modeling. *Soil Sci. Soc. Am. J.* (48):525–531.

Bailey, G. W. and J. L. White. 1964. Review of adsorption and desorption of organic pesticides by soil colloids, with implications concerning pesticide bioavailability. *J. Agric. Food Chem.* (13):324–332.

Bailey, G. W. and J. L. White. 1970. Factors influencing the adsorption, desorption, and movement of pesticides in soil. *Residue Rev.* (32):29–92.

Baker, D. E. 1973. A new approach to soil testing: II. Ionic equilibria involving H, K, Ca, Mg, Mn, Fe, Cu, Zn, Na, P, and S. *Soil Sci. Soc. Am. Proc.* (37):537–541.

Baker, D. E. and M. C. Amacher. 1981. The development and interpretation of a diagnostic soil-testing program. *Penn. St. Univ. Agric. Exp. Sta. Bull.* 826.

Baligar, V. C., R. J. Wright, and K. D. Ritchey. 1992. Soil acidity effects on wheat seedling root growth. *J. Plant. Nutr.* (15):845–856.

Ball, J. W., E. A. Jenne, and D. K. Nordstrom. 1978. WATEQ2—A computerized chemical model for trace and major element speciation and mineral equilibrium of natural water. In E. A. Jenne (Ed.). *Chemical Modeling in Aqueous Systems* (pp. 815–836) ACS Symposium Series No. 93. Washington, DC: American Chemical Society.

Barnes, R. B. 1975. The determination of specific forms of aluminum in natural water. *Chemical Geol.* (15):177–191.

Barnhisel, R. and P. M. Bertsch. 1982. Aluminum. In A. L. Page et al. (Eds.), *Methods of Soil Analysis, Part 2. Chemical and Microbiological Properties.* 2nd ed. (pp. 275–300) Agronomy Monograph 9. Madison, WI: American Society of Agronomy.

Barrow, G. M. 1966. *Physical Chemistry* (2nd Ed.). New York: McGraw-Hill.

Barry, D. A. and G. Sposito. 1988. Application of the convection-dispersion model to solute transport in finite soil columns. *Soil Sci. Soc. Am. J.* (52):3–9.

Batchelor, B., J. B. McEwen, and R. Perry. 1986. Kinetics of aluminum hydrolysis: Measurement and characterization of reaction products. *Environ. Sci. Technol.* (20):891–894.

Bartlett, R. and B. James. 1980. Studying dried, stored soil samples—Some pitfalls. *Soil Sci. Soc. Am. J.* (44):721–724.

Bartlett, R. J., D. S. Ross, and F. R. Magdoff. 1987. Simple kinetic fractionation of reactive aluminum in soil "solutions," *Soil Sci. Soc. Am. J.* (51):1479–1482.

Bates, D. M. and D. G. Watts. 1988. *Nonlinear Regression Analysis and Its Applications.* New York: Wiley.

Beck, K. C., R. H. Reuter, and E. M. Perdue. 1977. Organic and inorganic geochemistry of some Coastal Plain rivers of the southeastern United States. *Geochim Cosmochim Acta* (38):341–364.

Behr, B. and H. Wendt. 1962. Schnelle Ionenreaktionen in Losugen. I. Die Bildung des aluminiumsulfatokomplexes. *Z. Electrochem.* (66):223–228.

Beier, C. and K. Hansen. 1992. Evaluation of porous cup soil-water samplers under controlled field conditions: Comparison of ceramic and PTFE cups. *J. Soil Sci.* (43):261–271.

Bennett, A. C. and F. Adams. 1970. Concentration of $NH_3(aq)$ required for incipient NH_3 toxicity to seedlings. *Soil Sci. Soc. Am. Proc.* (34):259–263.

Bergstrom, L. 1990. Use of lysimeters to estimate leaching of pesticides in agricultural soils. *Environmental Pollution* (67):325–347.

Bersillon, J. L., P. H. Hsu, and F. Fiessinger. 1980. Characterization of hydroxy-aluminum solutions. *Soil Sci. Soc. Am. J.* (44):630–634.

Bertsch, P. M. 1987. Conditions for Al_{13} polymer formation in partially neutralized aluminum solutions. *Soil Sci. Soc. Am. J.* (51):825–828.

Bertsch, P. M. and R. I. Barnhisel. 1985. Speciation of hydroxy-Al solutions by chemical and Al NMR method. *Agron. Abstr.* (77):145.

Bertsch, P. M., W. J. Layton, and R. I. Barnhisel. 1986a. Speciation of hydroxy-aluminum solutions by wet chemical and aluminum-27 NMR methods. *Soil Sci. Soc. Am. J.* (50): 1449–1454.

Bertsch, P. M., G. W. Thomas, and R. I. Barnhisel. 1986b. Characterization of hydroxy-Al solutions by ^{27}Al NMR spectroscopy. *Soil Sci. Soc. Am. J.* (50):825–830.

Bessho, T. and L. C. Bell. 1992. Soil solid and solution phase changes and ming bean response during amelioration of aluminum toxicity with organic matter. *Plant Soil* (140): 183–196.

Bingham, F. T., G. Sposito, and J. E. Strong. 1984. The effect of chloride on the availability of cadmium. *Soil Sci. Soc. Am. J.* (13):71–74.

Bingham, F. T., J. E. Strong, and G. Sposito. 1983. Influence of chloride salinity on cadmium uptake by Swiss chard. *Soil Sci.* (135):160–165.

Birkeland, P. W. 1974. *Pedology, Weathering, and Geomorphological Research.* London: Oxford Univ. Press.

Bitton, G. and R. A. Boylan. 1985. Effect of acid precipitation on soil microbial activity: I. Soil core studies. *J. Environ. Qual.* (14):66–69.

Bjerrum, N. 1926. Ionic association. I. Influence of ionic association on the activity of ions at moderate degrees of association. *K. Dan. Vidensk. Selsk.* (7):1.

Black, S. A. and A. S. Campbell. 1982. Ionic strength of soil solution and its effects on charge properties of some New Zealand soils. *J. Soil Sci.* (33):249–262.

Blake, A. B. and F. A. Cotton. 1964. Relative enthalpies of formation of some tetrachlorometallate ions. *Inorg. Chem.* (1):5–10.

Blancher, R. W. and G. K. Stearman. 1984. Ion products and solid-phase activity to describe phosphate sorption by soils. *Soil Sci. Soc. Am. J.* (48):1253–1258.

Blancher, R. W. and G. K. Stearman. 1985. Prediction of phosphate sorption in soils from regular solid-solution theory. *Soil Sci. Soc. Am. J.* (49):578–583.

Bloom, P. R. 1983. The kinetics of gibbsite dissolution in nitric acid. *Soil Sci. Soc. Am. J.* (47):164–168.

Bloom, P. R. and M. S. Erich. 1987. Effect of solution composition on the rate and mechanism of gibbsite dissolution in acid soil solutions. *Soil Sci. Soc. Am. J.* (51):1131–1136.

Bloom, P. R. and D. F. Grigal. 1985. Modeling soil response to acidic deposition in nonsulfate adsorbing soils. *J. Environ. Qual.* (14):489–495.

Bloom, P. R. and M. B. McBride. 1979. Metal ion binding and exchange with hydrogen ions in acid-washed peat. *Soil Sci. Soc. Am. J.* (43):687–692.

Bloom, P. R., M. B. McBride, and R. M. Weaver. 1979a. Aluminum organic matter in acid soils: buffering and solution aluminum activity. *Soil Sci. Soc. Am. J.* (43):488–493.

Bloom, P. R., M. B. McBride, and R. M. Weaver. 1979b. Aluminum organic matter in acid soils. Salt-extractable aluminum. *Soil Sci. Soc. Am. J.* (43):813–815.

Bloom, P. R., R. M. Weaver, and M. B. McBride. 1978. The spectrophotometric and fluorometric determination of aluminum with 8-hydroxyquinoline and butyl acetate extraction. *Soil Sci. Soc. Am. J.* (42):713–716.

Bohn, H. L. 1968. EMF of inert electrodes in soil suspension. *Soil Sci. Soc. Am. Proc.* (32): 211–215.

Bohn, H. L. 1969. The EMF of platinum electrodes in dilute solutions and its relation to pH. *Soil Sci. Soc. Am. Proc.* (33):639–640.

Bohn, H. L. 1970. Comparisons of measured and theoretical Mn^{+2} concentrations in soil suspensions. *Soil Sci. Soc. Am. Proc.* (34):195–197.

Bohn, H. L. and R. K. Bohn. 1987. Solid activities of trace elements in soils. *Soil Sci.* (143): 398–403.

Bohn, H. L., B. L. McNeal, and G. A. O'Connor. 1979. *Soil Chemistry.* New York: Wiley.

Bohn, H. L., B. L. McNeal, and G. A. O'Conner. 1985. *Soil Chemistry* (2nd Ed). New York: Wiley.

Boll, J., T. S. Steenhuis, and J. S. Selker. 1992. Fiberglass wicks for sampling of water and solutes in the vadose zone. *Soil Sci. Soc. Am. J.* (56):701–707.

Booltink, H. W. G. and J. Bouma. 1991. Morphological characterization of bypass flow in well-structured clay soil. *Soil Sci. Soc. Am. J.* (55):1249–1254.

Bouchard, D. C. and T. L. Lavy. 1985. Hexazinone adsorption-desorption with soil and organic adsorbants. *J. Environ. Qual.* (14):181–186.

Brown, K. W., J. C. Thomas, and M. W. Holder. 1989. Development of a capillary wick unsaturated zone pore water sampler. USEPA/600/S4-88/001.

Bourgeois, W. W. and L. M. Lavkulich. 1972a. Application of acrylic plastic tension lysimeters to sloping land. *Can. J. Soil Sci.* (52): 288–290.

Bourgeois, W. W. and L. M. Lavkulich. 1972b. A study of forest soils and leachates on sloping topography using a tension lysimeter. *Can. J. Soil Sci.* (52):375–391.

Bowen, H. J. M. 1966. *Trace Elements in Biochemistry*. London: Academic Press.

Brady, N. C. 1974. *The Nature and Properties of Soils* (8th Ed.) New York: Macmillan.

Brenes, E. and R. W. Pearson. 1973. Root responses of three *Gramineae* species to soil acidity in an Oxisol and Ultisol. *Soil Sci.* (116):295–302.

Brogan, J. C. 1964. The effect of humic acid on aluminum toxicity. In *Proc. 8th International Congress Soil Sci.* (pp. 227–233). Bucharest, Romania.

Bruckert, S. 1970. Influence des composes organiques solubles sur la pedogenese en millieu acide. I. *Ann. Agron.* (21):421–472.

Brusseau, M. L. and P. S. C. Rao. 1989. The influence of sorbate-organic matter interactions on sorption nonequilibrium. *Chemosphere* (18):1691–1706.

Brusseau, M. L. and P. S. C. Rao. 1991. Influence of sorbate structure on nonequilibrium sorption of organic compounds. *Environ. Sci. Technol.* (25):1501–1506.

Budd, W. W., A. H. Johnson, J. B. Huss, and R. S. Turner. 1981. Aluminum in precipitation, streams, and shallow groundwater in the New Jersey Pine Barrens. *Water Resour. Res.* (17):1179–1183.

Burd, J. S. and J. C. Martin. 1923. Water displacement of soils and the soil solution. *J. Agric. Sci.* (13):265–295.

Burgess, J. 1988. *Ions in Solution*. Chichester, England: Ellis Horwood Limited.

Burgess, P. S. 1922. The soil solution, extracted by Lipman's direct-pressure method, compared with 1:5 water extracts. *Soil Sci.* (14):191–216.

Burnette, R. B. and A. P. Schwab. 1987. A computer program to aid in teaching diffuse layer theory. *J. Agron. Ed.* (16):30–33.

Burrows, W. D. 1977. Aquatic aluminum: Chemistry, toxicology, and environmental prevalence. *CRC Crit. Rev. Environ. Control* (7):167–216.

Cameron, F. K. 1911. *The Soil Solution: The Nutrient Medium for Plant Growth*. Easton, PA: Chemical Publishing.

Cameron, K. C., D. F. Harrison, N. P. Smith, and C. D. A. McLay. 1990. A method to prevent edge-flow in undisturbed soil cores and lysimeters. *Aust. J. Soil Res.* (28):879–886.

Cameron, R. C., G. S. P. Ritchie, and A. D. Robson. 1986. Relative toxicities of inorganic aluminum complexes to barley. *Soil Soc. Am. J.* (50):1231–1236.

Cameron, K. C., N. P. Smith, C. D. A. McLay, P. M. Fraser, R. J. McPherson, D. F. Harrison, and P. Harbottle. 1992. Lysimeters without edge flow: An improved design and sampling procedure. *Soil Sci. Soc. Am. J.* (56):1625–1628.

Carlsen, L., B. S. Jensen, and K. Nilsson. 1984. *The Migration Behavior of Technetium*. C.E.C. Rpt. EUR 9543 EN.

Carski, T. H. and D. L. Sparks. 1985. A modified miscible displacement technique for investigating adsorption-desorption kinetics in soil. *Soil Sci. Soc. Am. J.* (49):1114–1116.

Cassel, D. K., T. H. Krueger, F. W. Schroer, and E. B. Norum. 1974. Solute movement through disturbed and undisturbed soil cores. *Soil Sci. Soc. Am. Proc.* (38):36–40.

Castellan, G. W. 1971. *Physical Chemistry* (2nd Ed.). Reading, MA: Addison-Wesley.

Checkai, R. T. and W. A. Norvell. 1992. A recirculating resin-buffered hydroponic system for controlling nutrient ion activities. *J. Plant Nutr.* (15):871–892.

Chesworth, W. 1973. The parent rock effect in the genesis of soil. *Geoderma* (10):215–225.

Chien, S. H. and W. R. Clayton. 1980. Application of Elovich equation to the kinetics of phosphate release and sorption in soils. *Soil Sci. Soc. Am. J.* (44):265–268.

Childs, C. R., R. L. Profitt, and R. Lee. 1983. Movement of aluminum as an inorganic complex in some podzolized soils, New Zealand. *Geoderma* (29):139–155.

Chino, M. and A. Baba. 1981. The effects of some environmental factors on the partitioning of zinc and cadmium between roots and tops of rice plants. *J. Plant Nutr.* (3):203–214.

Christophersen, N. and H. M. Seip. 1982. A model for streamwater chemistry at Birkens, Norway. *Water Resour. Res.* (18):977–996.

Cochran, P. H., G. M. Marison, and A. L. Leaf. 1970. Variation in tension lysimeter leachate volumes. *Soil Sci. Soc. Am. J.* (34):309–311.

Cole, D. W. 1958. Alundum tension lysimeter. *Soil Sci.* (85):293–296.

Cole, D. W., S. P. Gessel, and E. E. Held. 1961. Tension lysimeter studies of ion and moisture movement in glacial till and coral atoll soils. *Soil Sci. Soc. Am. Proc.* (25):321–325.

Collins, J. F. and S. W. Buol. 1970. Effects of fluctuations in the Eh-pH environment on iron and/or manganese equilibria. *Soil Sci.* (110):111–117.

Cosby, C. J., G. M. Hornberger, J. N. Galloway, and R. F. Wright. 1985a. Modeling the effects of acid deposition: Assessment of a lumped parameter model of soil water and steamwater chemistry. *Water Resour. Res.* (21):51–63.

Cosby, B. J., G. M. Hornberger, N. J. Galloway, and R. F. Wright. 1985b. Time scales of catchment acidification. *Environ. Sci. Technol.* (19):1144–1149.

Cotton, F. A. and G. Wilkerson. 1966. *Advanced Inorganic Chemistry* (2nd Ed.). New York: Wiley.

Coulson, C. B., R. I. Davies, and D. A. Lewis. 1960. Polyphenols in plant, humus and soil. I. Polyphenols of leaves, litter and superficial humus from mull and mor sites. *J. Soil Sci.* (11):20–29.

Creasey, C. L. and S. J. Dreiss. 1988. Porous cup samplers: Cleaning procedures and potential sample bias from trace element contamination. *Soil Sci.* (145):93–101.

Cresser, M., K. Killham, and T. Edwards. 1993. *Soil Chemistry and Its Applications*. Cambridge: Cambridge Univ. Press.

Cronan, C. S. 1980. Solution chemistry of a New Hampshire subalpine ecosystem: A biogeochemical analysis. *Oikos* (34):272–281.

Cronan, C. S., J. M. Kelly, C. I. Scholfield, and R. A. Goldstein. 1987. Aluminum geochemistry and tree toxicity in forests exposed to acidic deposition. In R. Perry et al. (Eds.) *Acid Rain: Scientific and Technical Advances* (pp. 649–656). London: Selper Ltd.

Cronan, C. S. and C. L. Scholfield. 1979. Aluminum leaching response to acid precipitation effects on high-elevation watersheds in the Northeast. *Science* (204):304–305.

Cronan, C. S., W. J. Walker, and P. R. Bloom. 1986. Predicting aqueous aluminum concentrations in natural waters. *Nature* (324):'140–143.

Curtin, D. and G. W. Smillie. 1983. Soil solution composition as affected by liming and incubation. *Soil Sci. Soc. Am. J.* (47):701–707.

Dahlgren, R. A. and F. C. Ugolini. 1989. Aluminum fractionation of soil solutions from unperturbed and tephra-treated Spodosols, Cascade Range, Washington, USA. *Soil Sci. Soc. Am. J.* (53):559–566.

David, M. B. and C. T. Driscoll. 1984. Aluminum speciation and equilibria in soil solutions of a Haplorthod in the Adirondack Mountains (New York, U.S.A.). *Geoderma* (33):297–318.

David, M. B. and G. F. Vance. 1989. Generation of soil solution acid-neutralizing capacity by addition of dissolved inorganic carbon. *Environ. Sci. Technol.* (23):1021–1024.

Davies, B. E. and R. I. Davies. 1963. A simple centrifugation method for obtaining small samples of soil solution. *Nature* (198):216–217.

Davies, R. J. 1971. Relation of polyphenols to decomposition of organic matter and to pedogenic processes. *Soil Sci.* (111):80–85.

DiPascale, G. and A. Violante. 1986. Influence of phosphate ions on the extraction of aluminum by 8-hydroxyquinoline from OH-Al suspensions. *Can. J. Soil Sci.* (66):573–579.

Driscoll, C. T. 1984. A procedure for the fractionation of aqueous aluminum in dilute acidic waters. *Internat. J. Environ. Anal. Chem.* (16):267–283.

Driscoll, C. T., Jr., J. P. Baker, J. J. Bisogni, Jr., and C. L. Schofield. 1980. Effect of aluminum speciation on fish in dilute acidified waters. *Nature* (284):161–164.

Driscoll, C. T. and R. M. Newton. 1985. Chemical characteristics of Adirondack lakes. *Environ. Sci. Technol.* (19):1018–1024.

Duchaufour, P. and L. Z. Rousseau. 1960. Les phénomènes d'intoxication des plantules de résineux par le manganè dans les humus forestiers. *Rev. For. France* (11):835–847.

Dudley, L. M. and B. L. McNeal. 1987. A model for electrostatic interactions among charged sites of water-soluble, organic polyions: 1. Description and sensitivity. *Soil Sci.* (143):329–340.

Eaton, F. M., R. B. Harding, and T. J. Ganje. 1960. Soil solution extractions at tenth-bar moisture percentages. *Soil Sci.* (90):253–258.

Edmeades, D. C., D. M. Wheeler, and O. E. Clinton. 1985. The chemical composition and ionic strength of soil solutions from New Zealand topsoils. *Aust. J. Soil Res.* (23):151–165.

Edmunds, W. M. and A. H. Bath. 1976. Centrifuge extraction and chemical analysis of interstitial waters. *Environ. Sci. Technol.* (10):467–472.

Elkhatib, E. A., O. L. Bennett, V. C. Baligar, and R. J. Wright. 1986. A centrifuge method for obtaining soil solution using an immiscible liquid. *Soil Sci. Soc. Am. J.* (50):297–299.

Elkhatib, E. A., J. L. Hern, and T. E. Staley. 1987. A rapid centrifugation method for obtaining soil solution. *Soil Sci. Soc. Am. J.* (51):578–583.

Elrick, D. E. and L. K. French. 1966. Miscible displacement patterns on disturbed and undisturbed soil cores. *Soil Sci. Soc. Am. Proc.* (30):153–156.

Emmerich, W. E., L. J. Lund, A. L. Page, and A. C. Chang. 1982a. Movement of heavy metals in sewage sludge-treated soils. *J. Environ. Qual.* (11):174–178.

Emmerich, W. E., L. J. Lund, A. L. Page, and A. C. Chang. 1982b. Predicted solution phase forms of heavy metals in sewage sludge-treated soils. *J. Environ. Qual.* (11):182–186.

Epstein, E. 1972. *Mineral Nutrition of Plants: Principles and Perspectives.* New York: Wiley.

Eriksson, E. 1981. Aluminum in groundwater possible solution equilibria. *Nordic Hydrol.* (12):43–50.

Evans, A., Jr. 1986. Effects of dissolved organic carbon and sulfate on aluminum mobilization in forest soil columns. *Soil Sci. Soc. Am. J.* (50):1576–1578.

Evans, A., Jr. and L. W. Zelazny. 1986. Determination of inorganic mononuclear aluminum by selected chelation using crown ethers. *Soil Sci. Soc. Am. J.* (50):910–913.

Exner, O. 1972. The Hammett equation—The present position. In N. B. Chapman and J. Shorter (Eds.), *Advances in Linear Free Energy Relationships* (pp. 1–69). New York: Plenum Press.

Farmer, V. C., J. D. Russell, and M. L. Berrow. 1980. Imogolite and proto-imogolite allophane in spodic horizons: Evidence for a mobile aluminum silicate complex in podzol formation. *J. Soil Sci.* (31):673–684.

Fitch, A. and F. J. Stevenson. 1984. Comparisons of models for determining stability constants of metal complexes with humic substances. *Soil Sci. Soc. Am. J.* (48):1044–1049.

Firestone, M. K., K. Killham, and J. G. McColl. 1983. Fungal toxicity of mobilized soil aluminum and manganese. *Appl. Environ. Microbiol.* (48):556–560.

Fontaine, D. D., R. G. Lehmann, and J. R. Miller. 1991. Soil adsorption of neutral and anionic forms of a sulfonamide herbicide, flumetsulam. *J. Environ. Qual.* (20):759–762.

Forstner, U. 1981. Metal transfer between solid and aqueous phases. In U. Forstner and G. T. W. Wittman (Eds.), *Metal Pollution in the Aquatic Environment* (pp. 197–270). New York: Springer-Verlag.

Fox, T. R. and N. B. Comerford. 1990. Low-molecular-weight organic acids in selected forest soils of the Southeastern USA. *Soil Sci. Soc. Am. J.* (54):1139–1144.

Foy, C. D. 1974. Effects of aluminum on plant growth. In E. W. Carson (Ed.), *The Plant Root and Its Environment* (pp. 601–624). Charlottesville: Univ. Press of Virginia.

Foy, C. D. 1976. General principles involved in screening plants for aluminum and manganese tolerance. In *Proc. of Workshop on Plant Adaptation to Mineral Stress in Problem Soils* (pp. 255–267). Beltsville, MD, 22–23 November 1976. Cornell Univ. Agric. Exp. Sta., Ithaca, NY.

Foy, C. D. 1983. The physiology of plant adaptation to mineral stress. *Iowa State J. Res.* (57):355–391.

Foy, C. D. 1984. Physiological effects of hydrogen, aluminum, and manganese toxicities in acid soil. In F. Adams (Ed.), *Soil Acidity and Liming,* 2nd ed. (pp. 57–97). Agronomy Monograph 12. Madison, WI: American Society of Agronomy.

Freeman, R. A. and W. H. Everhart. 1971. Toxicity of aluminum hydroxide complexes in neutral and basic media to rainbow trout. *Trans. Am. Fish Soc.* (100):644–658.

Frink, C. R. and M. Peech. 1962. Determination of aluminum in soil extracts. *Soil Sci.* (93): 317–324.

Fu-Yong, H. L. Bohn, J. Brito, and J. Prenzel. 1992. Solid activities of aluminum phosphate and hydroxide in acid soils. *Soil Sci. Soc. Am. J.* (56):59–62.

Gambrell, R. P. and W. H. Patrick, Jr. 1982. Manganese. In A. L. Page et al. (Eds.), *Methods of Soil Analysis, Part 2, Chemical and Microbiological Properties,* 2nd ed. (pp. 313–322). Agronomy Monograph 9. Madison, WI: American Society of Agronomy.

Garcia-Miragaya, J. and A. L. Page. 1976. Influence of ionic strength and inorganic complex formation on the sorption of trace amounts of Cd by montmorillonite. *Soil Sci. Soc. Am. J.* (40):658–663.

Garrels, R. M. and C. L. Christ. 1965. *Solutions, Minerals, and Equilibria.* San Francisco: Freeman, Cooper.

Geering, H. R., J. F. Hodgson, and C. Sdano. 1969. Micronutrient cation complexes in soil solution. IV. The chemical state of manganese in soil solution. *Soil Sci. Soc. Am. Proc.* (33):81–85.

Gerrtise, R. G. and J. A. Adeney. 1992. Tracers in recharge—Effects of partitioning in soils. *J. Hydrol.* (131):255–268.

Gillman, G. P. 1976. *A Centrifuge Method for Obtaining Soil Solution.* Div. Soils Report No. 16, CSIRO, Australia.

Gillman, G. P. and L. C. Bell. 1978. Soil solution studies on weathered soils from tropical North Queensland. *Aust. J. Soil Res.* (16):67–77.

Gillman, G. P. and M. E. Sumner. 1987. Surface charge characterization and soil solution composition of four soils from the Southern Piedmont of Georgia. *Soil Sci. Soc. Am. J.* (51):589–594.

Girling, C. A. and P. J. Peterson. 1981. The significance of the cadmium species in uptake and metabolism of cadmium in crop plants. *J. Plant Nutr.* (3):707–720.

Goetz, A. J., R. H. Walker, G. Wehtje, and B. F. Hajek. 1989. Sorption and mobility of chlorimuron on Alabama soils. *Weed Sci.* (37):428–433.

Goetz, A. J., A. Wehtje, R. J. Walker, and B. Hajek. 1986. Soil solution and mobility characterization of imazaquin. *Weed Sci.* (34):788–793.

Goh, T. B. and P. M. Huang. 1984. Formation of hydroxy-Al-montmorillonite complexes as influenced by citric acid. *Can. J. Soil Sci.* (64):411–421.

Goh, T. B. and P. M. Huang. 1985. Changes in the thermal stability and acidic characteristics of hydroxy-Al-montmorillonite complexes formed in the presence of citric acid. *Can. J. Soil Sci.* (65):519–522.

Goldberg, S. P. and K. A. Smith. 1984. Soil manganese: E values, distribution of manganese-54 among soil fractions, and effects of drying. *Soil Sci. Soc. Am. J.* (48):559–564.

Grandstaff, D. E. 1986. The dissolution rate of fosteritic olivine from Hawaiian beach sand. In S. M. Coleman and D. P. Dethier (Eds.), *Rates of Chemical Weathering of Rocks and Minerals* (pp. 41–60). Orlando, FL: Academic Press.

Green, R. E. and S. R. Obien. 1969. Herbicide equilibrium in soils in relation to soil water content. *Weed Sci.* (17):514–519.

Griffin, R. A. and J. J. Jurinak. 1973. Estimation of activity coefficients from the electrical conductivity of natural aquatic systems and soil extracts. *Soil Sci.* (116):26–30.

Grover, B. L. and R. E. Lamborn. 1970. Preparation of porous ceramic cups to be used for extraction of soil water having low solute concentrations. *Soil Sci. Soc. Am. Proc.* (34): 706–708.

Guggenberger, G. and W. Zech. 1992. Sorption of dissolved organic carbon by ceramic P 80 suction cups. *Z. Planzenernähr. Bodenk.* (155):151–155.

Gustafson, D. I. and L. R. Holden. 1990. Nonlinear pesticide dissipation in soil: A new model based on spatial variability. *Environ. Sci. Technol.* (24):1032–1038.

Haines, B. L., J. B. Waide, and R. L. Todd. 1982. Soil solution nutrient concentrations sampled with tension and zero-tension lysimeters: Report of discrepancies. *Soil Sci. Soc. Am. J.* (46):658–661.

Hamaker, J. W. and C. A. I. Goring. 1976. Turnover of pesticide residues in soil. In. D. D. Kaufman et al. (Eds.), *Bound and Conjugated Pesticide Residues* (pp. 219–243). Washington, DC: American Chemical Society.

Hansen, E. and R. Harris. 1975. Validity of soil-water samples collected with porous ceramic cups. *Soil Sci. Soc. Am. Proc.* (39):528–536.

Hantschel, R., M. Kaupenjohann, R. Horn, J. Gradl, and W. Zech. 1988. Ecologically important differences between equilibrium and percolation soil extracts, Barvaria. *Geoderma* (43):213–227.

Hardin, S. D. 1988. *Effect of Rates of Applied Phosphorus and Soil Storage Temperature on P Availability of a Low-P Loess-Derived Alfisol.* MS Thesis. Knoxville: Univ. of Tennessee.

Hardin, S. D., D. D. Howard, and J. Wolt. 1989. Critical soil phosphorus of a low-P loess-derived soil as affected by storage temperature. *Comm. Soil Sci. Plant. Anal.* (20):1525–1543.

Hargrove, W. L. and G. W. Thomas. 1982. Titration properties of Al-organic matter. *Soil Sci.* (134):216–225.

Harned, H. S. and B. B. Owen. 1958. *The Physical Chemistry of Electrolytic Solutions.* New York: Reinhold Pub.

Harrold, L. L. and F. R. Dreibelbis. 1967. *Evaluation of Agricultural Hydrology by Monolith Lysimeters.* USDA Tech. Bull. No. 1376, Washington, DC.

Hay, G. W., J. H. James, and G. W. vanLoon. 1985. Solubilization effects of simulated acid rain on the organic matter of forest soil: Preliminary results. *Soil Sci.* (139):422–430.

Haynes, R. J. and P. H. Williams. 1992. Changes in soil solution composition and pH in urine-affected areas of pasture. *J. Soil Sci.* (43):323–334.

Hem, J. D. 1972. Chemical factors that influence the availability of iron and manganese in aqueous systems. *Geol. Soc. Am. Bull.* (83):443–450.

Hendershot, W. H., S. Savoie, and F. Courchesne. 1992. Simulation of stream-water chemistry with soil solution and groundwater flow contributions. *J. Hydrol.* (136):237–252.

Hendrickson, L. L. and R. B. Corey. 1983. A chelating-resin method for characterizing soluble metal complexes. *Soil Sci. Soc. Am. J.* (47):467–474.

Henrot, J. 1988. *Behavior of Technetium in Soil: Sorption-Desorption Processes.* Ph.D. Thesis. Knoxville: Univ. of Tennessee.

Hepper, C. M. 1979. Germination and growth of *Glomus caledonius* spores: The effects of inhibitors and nutrients. *Soil Biol. Biochem.* (11);269–277.

Hern, J. A., G. K. Rutherford, and G. W. vanLoon. 1985. Chemical and pedogenetic effects of simulated acid precipitation on two eastern Canadian forest soils. I. Nonmetals. *Can. J. For. Res.* (15):839–847.

Hilgard, E. W. 1912. *Soils. Their Formation, Properties, Composition, and Relation to Climate and Plant Growth.* New York: Macmillan.

Hirsch, D. and A. Banin. 1990. Cadmium speciation in soil solution. *J. Environ. Qual.* (19):366–372.

Hirsch, D., S. Nir, and A. Banin. 1989. Prediction of cadmium complexation in solution and adsorption to montmorillonite. *Soil Sci. Soc. Am. J.* (53):716–721.

Hoagland, D. R., J. C. Martin, and G. R. Stewart. 1920. Relation of the soil solution to the soil extract. *J. Agric. Res.* (20):381–395.

Hodges, S. C. 1987. Aluminum speciation: A comparison of five methods. *Soil Sci. Soc. Am. J.* (51):57–64.

Hodges, S. C. and G. C. Johnson. 1987. Kinetics of sulfate adsorption and desorption by Cecil soil using miscible displacement. *Soil Sci. Soc. Am. J.* (51):323–331.

Högfeldt, E. 1979. *Stability Constants of Metal-Ion Complexes. Part A: Inorganic Ligands.* IUPAC Chem. Data Series No. 21. Oxford: Pergamon Press.

Holder, M., K. W. Brown, J. C. Thomas, D. Zabcik, and H. E. Murray. 1991. Capillary-wick unsaturated zone pore water sampler. *Soil Sci. Soc. Am. J.* (55):1195–1202.

Holford, I. C. R., R. W. M. Wedderburn, and G. E. G. Mattingly. 1974. A Langmuir two-surface equation as a model for phosphate adsorption by soils. *J. Soil Sci.* (25):242–55.

Holmgren, G. G. S., R. L. June, and R. C. Geschwender. 1977. A mechanically controlled variable rate leaching device. *Soil Sci. Soc. Am. J.* (41):1207–1208.

Hooper, R. P. and C. A. Shoemaker. 1985. Aluminum mobilization in an acidic headwater stream: Temporal variation and mineral dissolution disequilibria. *Science* (229):463–465.

Howard, D. D. and F. Adams. 1965. Calcium requirement for penetration of subsoils by primary cotton roots. *Soil Sci. Soc. Am. Proc.* (29):558–562.

Hsu, P. H. 1979. Effect of phosphate and silicate on the crystallization of gibbsite from OH-Al solutions. *Soil Sci.* (127):219–226.

Huang, P. M. and M. Schnitzer. 1986. *Interactions of Soil Minerals with Natural Organics and Microbes.* Madison, WI: Soil Science Society of America.

Huang, P. M. and A. Violante. 1986. Influence of organic acids on crystallization and surface properties of precipitation products of aluminum. In P. M. Huang and M. Schnitzer (Eds.), *Interactions of Soil Minerals with Natural Organics and Microbes* (pp. 159–222). Madison, WI: Soil Science Society of America.

Hue, N. V. 1992. Correcting soil acidity of a highly weathered Ultisol with chicken manure and sewage sludge. *Comm. Soil Sci. Plant. Anal.* (23):241–264.

Hue, N. V., G. R. Craddock, and F. Adams. 1986. Effect of organic acids on aluminum toxicity in subsoils. *Soil Sci. Soc. Am. J.* (50):28–34.

Hutchinson, T. C., L. Bozic, and G. Munoz-Vega. 1986. Responses of five species of conifer seedlings to aluminum stress. *Water Air Soil Pollution* (31):283–294.

Hutterman, A. and B. Ulrich. 1984. Solid phase-solution-root interactions in soils subjects to acid deposition. *Phil. Trans. R. Soc. London* (305):353–368.

Ingestad, T. 1964. Growth and boron and manganese status of birch seedlings grown in nutrient solution. In C. Bould et al. (Eds.), *Plant Analysis and Fertilizer Problems IV* (pp. 169–173). East Lansing, MI: American Society of Horticultural Science.

Jackson, M. L. 1963. Aluminum bonding in soils: A unifying principle in soil science. *Soil Sci. Soc. Am. Proc.* (27):1–10.

Jackson, M. L. and G. D. Sherman. 1953. Chemical weathering of minerals in soils. *Adv. Agron.* (5):219–318.

Jacobs, L. W., D. R. Keeney, and L. M. Walsh. 1970. Arsenic residue toxicity to vegetable crops grown on Plainfield sand. *Agron. J.* (62):588–591.

Jacobs, L. W., J. H. Phillips, and M. J. Zabik. 1981. Toxic organic chemicals and elements in Michigan sewage sludges—Significance for application to cropland. In *Proc. 4th Annual Madison Conf. of Applied Res. and Practice on Municipal and Industrial Waste,* 28–30 Sept. 1981, Madison, WI. (pp. 478–486). Madison: Univ. of Wisconsin.

James, B. R., C. J. Clark, and S. J. Rhia. 1983. An 8-hydroxyquinoline method for labile and total aluminum in soil extracts. *Soil Sci. Soc. Am. J.* (47):893–897.

James, B. R. and S. J. Rhia. 1984. Soluble aluminum in acidified organic horizons of forest soils. *Can. J. Soil Sci.* (64):637–646.

Jardine, P. M., G. V. Wilson, R. J. Luxmoore, and J. F. McCarthy. 1989. Transport of inorganic and natural organic tracers through an isolated pedon in a forest watershed. *Soil Sci. Soc. Am. J.* (53):317–323.

Jardine, P. M. and L. W. Zelazny. 1986. Mononuclear and polynuclear aluminum speciation through differential kinetic reactions with ferron. *Soil Sci. Soc. Am. J.* (50):895–900.

Jardine, P. M., L. W. Zelazny, and A. Evans, Jr. 1986. Solution aluminum anomalies from various filtering materials. *Soil Sci. Soc. Am. J.* (50):891–894.

Jarrell, W. M., A. L. Page, and A. A. Elseewi. 1980. Molybdenum in the environment. *Residue Rev.* (74):1–43.

Jarvis, S. C. 1984. The forms of occurrence of manganese in some acidic soils. *J. Soil Sci.* (35):421–429.

Jemison, J. M., Jr., and R. H. Fox. 1992. Estimation of zero-tension pan lysimeter collection efficiency. *Soil Sci.* (154):85–94.

Jenne, E. D. 1979. *Chemical Modeling in Aqueous Systems*. ACS Symposium Series No. 93. Washington, DC: American Chemical Society.

Joffe, J. S. 1929. A new type of lysimeter at the New Jersey Agricultural Experiment Station. *Science* (70):147–148.

Joffe, J. S. 1932. Lysimeter studies. I. Moisture percolation through the soil profile. *Soil Sci.* (34):123–143.

Joffe, J. S. 1949. *Pedology*. New Brunswick, NJ: Pedology Publications.

Johnson, D. W. and D. E. Todd. 1984. Effects of acid irrigation on carbon dioxide evolution, extractable nitrogen, phosphorus, and aluminum in a deciduous forest soil. *Soil Sci. Soc. Am. J.* (48):664–666.

Johnson, K. S. and R. M. Pytkowicz. 1978. Ion association of H^+, Na^+, K^+, Mg^{2+}, and Ca^{2+} with Cl^- at 25°C. *Am. J. Sci.* (278):1428.

Johnson, L. R. and A. E. Hiltbold. 1969. Arsenic content of soils and crops following use of methanarsonate herbicides. *Soil Sci. Soc. Am. J.* (33):279–282.

Johnson, N. L. and S. Kotz. 1970. *Continuous Univariate Distributions—1*. New York: Wiley.

Johnson, N. M., C. T. Driscoll, J. S. Eaton, G. E. Likens, and W. H. McDowell. 1981. "Acid rain," dissolved aluminum, and chemical weathering at the Hubbard Brook Experimental Forest, New Hampshire. *Geochim Cosmochim Acta* (45):1421–1437.

Jordan, C. F. 1968. A simple, tension-free lysimeter. *Soil Sci.* (105):81–86.

Joslin, J. D. and M. H. Wolfe. 1992. Red spruce soil solution chemistry and root distribution across a cloud water deposition gradient. *Can. J. For. Res.* (22):893–904.

Jury, W. A., W. R. Gardner, and W. H. Gardner. 1991. *Soil Physics* (5th Ed.). New York: Wiley.

Kamprath, E. J. 1984. Crop response to lime on soils in the tropics. In F. Adams (Ed.), *Soil Acidity and Liming,* 2nd ed. (pp. 349–368). Agronomy Monograph 12. Madison, WI: American Society of Agronomy.

Karanthanasis, A. D. 1989. Soil solution: A sensitive indicator of mineral stability in pedogenic environments. In S. S. Augustithis (Ed.), *Weathering: Its Products and Deposits* (pp. 157–195). Athens, Greece: Theophrastus.

Karanthanasis, A. D., V. P. Evangelou, and Y. L. Thompson. 1988. Aluminum and iron equilibria in soil solutions and surface waters of acid mine watersheds. *Soil Sci. Soc. Am. J.* (17):534–543.

Kennedy, V. C., G. W. Zellwerger, and B. F. Jones. 1974. Filter pore-size effects on the analysis of Al, Fe, Mn, and Ti in water. *Water Resour. Res.* (10):785–790.

Keyser, H. H. and D. N. Munns. 1979a. Effects of calcium, manganese, and aluminum on growth of rhizobia in acid media. *Soil Sci. Soc. Am. J.* (43):500–503.

Keyser, H. H. and D. N. Munns. 1979b. Tolerance of rhizobia to acidity, aluminum, and phosphate. *Soil Sci. Soc. Am. J.* (43):519–523.

Khanna, P. K. and F. Beese. 1978. The behavior of sulfate on salt input in podzolic brown earth. *Soil Sci.* (125):16–22.

Khanna, P. K., J. Prenzel, K. J. Meiwes, B. Ulrich, and E. Matzner. 1987. Dynamics of sulfate retention by acid forest soils in an acidic deposition environment. *Soil Sci. Soc. Am. J.* (51):446–452.

Khasawneh, F. E. 1971. Solution ion activity and plant growth. *Soil Sci. Soc. Am. Proc.* (35):426–436.

Khasawneh, F. E. and F. Adams. 1967. Effect of dilution on calcium and potassium contents of soil solutions. *Soil Sci. Soc. Am. Proc.* (31):172–176.

Khattak, R. A., G. H. Haghnia, R. L. Mikkelsen, A. L. Page, and G. R. Bradford. 1989. Influence of binary interactions of arsenate, molybdate, and selenate on yield and composition of alfalfa. *J. Environ. Qual.* (18):355–360.

Kielland, J. 1937. Individual activity coefficients of ions in aqueous solutions. *J. Am. Chem. Soc.* (59):1675–1678.

Kinniburgh, D. G. and D. L. Miles. 1983. Extraction and chemical analysis of interstitial water from soils and rocks. *Environ. Sci. Technol.* (17):362–368.

Kittrick, J. A. 1966. The free energy of formation of gibbsite and $Al(OH)_4^-$ from solubility measurements. *Soil Sci. Soc. Am. Proc.* (30):595–597.

Kittrick, J. A. 1969. Soil minerals in the Al_2O_3-SiO_2-H_2O system and a theory of their formation. *Clays and Clay Minerals* (17):157–166.

Kittrick, J. A. 1980. Gibbsite and kaolinite solubilities by immiscible displacement of equilibratium solutions. *Soil Sci. Soc. Am. J.* (44):139–142.

Kittrick, J. A. 1983. Accuracy of several immiscible displacement liquids. *Soil Sci. Soc. Am. J.* (47):1045–1047.

Ko, W.H. and F. K. Hora. 1972. Identification of the Al ion as a soil fungitoxin. *Soil Sci.* (113):42–45.

Kohnke, H. 1968. *Soil Physics*. New York: McGraw-Hill.

Kohnke, H., F. R. Dreibelbis, and J. M. Davidson. 1940. *A Survey and Discussion of Lysimeters and a Bibliography on Their Construction and Performance*. USDA Misc. Publ. No. 374, Washington, DC.

Krug, E. C. and C. R. Frink. 1983. Acid rain on acid soil: A new perspective. *Science* (221): 520–525.

Krug, E. C. and P. J. Isaacson. 1984. Comparison of water and dilute acid treatment on organic and inorganic chemistry of leachate from organic-rich horizons of an acid forest soil. *Soil Sci.* (137):370–378.

Ladd, J. N. and J. H. A. Butler. 1975. Humus-enzyme systems and synthetic organic polymers-enzyme analogs. In E. A. Paul and A. D. McLaren (Eds.), *Soil Biochemistry* (pp. 143–194). New York: Marcel Dekker.

Laskowski, D. A., P. M. Tillotson, D. D. Fontaine, and E. J. Martin. 1990. Probability modeling. *Phil. Trans. R. Soc. Lond. B.* (329):383–389.

Lee, E. H., H. E. Heggestad, and J. E. Bennett. 1982. Effects of sulfur dioxide fumigation in open-top field chambers on soil acidification and exchangeable aluminum. *J. Environ. Qual.* (11):99–102.

Leffelaar, P. A., A. Kamphorst, and R. Pal. 1983. Nomographic estimation of activity co-efficients from electrical conductivity data of soil extracts. *J. Indian Soc. Soil Sci.* (31): 20–27.

Lehninger, A. L. 1970. *Biochemistry.* New York: Worth.

Le Roux, J. and M. E. Sumner. 1967. Studies on the soil solution of various Natal soils. *Geoderma* (1):125–130.

Lewis, G. N. and M. Randall. 1961. *Thermodynamics* (2nd Ed.). New York: McGraw-Hill.

Lind, C. J. and J. D. Hem. 1975. *Effects of Organic Solutes on Chemical Reactions of Aluminum.* U.S. Geol. Survey Water Supply Paper 1827–G.

Lindberg, R. D. and D. D. Runnells. 1984. Ground water redox reactions: An analysis of equilibrium state applied to Eh measurements and geochemical modeling. *Science* (225): 925–927.

Lindberg, S. E., R. C. Harris, and R. R. Turner. 1982. Atmospheric deposition of metals to forest vegetation. *Science* (215):1609–1611.

Lindsay, W. L. 1979. *Chemical Equilibria in Soils.* New York: Wiley.

Litaor, M. I. 1988. Review of soil solution samplers. *Water Resour. Res.* (24):727–733.

Luce, R. W., R. W. Bartlett, and G. A. Parks. 1972. Dissolution kinetics of magnesium silicates. *Geochim Cosmochim Acta* 36:36–50.

Lund, Z. F. 1970. The effect of calcium and its relation to several cations in soybean root growth. *Soil Sci. Soc. Am. Proc.* (34):456–459.

Luxmoore, R. J. 1981. Micro-, meso-, and macroporosity of soil. *Soil Sci. Soc. Am. J.* (45): 671.

Madhun, Y. A., J. L. Young, and V. H. Freed. 1986. Binding of herbicides by water-soluble organic materials in soil. *J. Environ. Qual.* (15):64–68.

Magistad, O. C. 1925. The aluminum content of the soil solution and its relation to soil reaction and plant growth. *Soil Sci.* (20):181–225.

Mahler, R. J., F. T. Bingham, G. Sposito, and A. L. Page. 1980. Cadmium-enriched sewage sludge application to acid and calcareous soils: Relation between treatment, cadmium in saturation extracts, and cadmium uptake. *J. Environ. Qual.* (9):359–364.

Majka, J. T. and T. L. Lavy. 1977. Adsorption, mobility, and degradation of cyanazine and diuron in soils. *Weed Sci.* (25):401–406.

Marbut, C. F. 1935. Soils of the United States. Part III. In *Atlas of American Agriculture* (pp. 1–98). Washington, DC: United States Department of Agriculture.

Marion, G. M. and G. L. Babcock. 1976. Predicting specific conductance and salt concentration in dilute aqueous solutions. *Soil Sci.* (122):181–187.

Marshall, C. E., M. Y. Chowdhury, and W. J. Upchurch. 1973. Lysimetric and chemical investigations of pedological changes. Part 2: Equilibration of profile samples with aqueous solutions. *Soil Sci.* (116):336–358.

Martin, J. P. and K. Haider. 1986. Influence of mineral colloids on turnover rates of soil organic carbon. In P. M. Huang and M. Schnitzer (Eds.), *Interactions of Soil Minerals with Natural Organics and Microbes* (pp. 283–304). Madison, WI: Soil Science Society of America.

Mattigod, S. V. and G. Sposito. 1979. Chemical modeling of trace metal equilibria in contaminated soil solutions using the computer program GEOCHEM. In E. A. Jenne (Ed.), *Chemical Modeling in Aqueous Systems* (pp. 837–856). ACS Symposium Series No. 93. Washington, DC: American Chemical Society.

Matzner, E. and B. Ulrich. 1985. "Waldsterben": Our dying forests. II. Implications of the chemical soil conditions for forest decline. *Experientia* (41):578–584.

May, H. M., P. A. Helmke, and M. L. Jackson. 1979. Gibbsite solubility and thermodynamic properties of hydroxy-aluminum ions in aqueous solution at 25 C. *Geochim Cosmochim Acta* (43):861–868.

Mayer, R. and B. Ulrich. 1977. Acidity of precipitation as influenced by the filtering of atmospheric sulphur and nitrogen compounds—Its role in the element balance and effect on soil. *Water Air Soil Pollution* (7):409–416.

McCall, P. J. and G. L. Agin. 1985. Desorption kinetics of picloram as affected by residence time in the soil. *Environ. Toxicol. Chem.* (4):37–44.

McCarthy, J. F. and J. M. Zachara. 1989. Subsurface transport of contaminants. *Environ. Sci. Technol.* (23):496–502.

McCormick, L. H. and K. C. Steiner. 1978. Variation in aluminum tolerance among six genera of trees. *Forest Sci.* (24):565–568.

McFee, W. W. and C. S. Cronan. 1982. The action of wet and dry deposition components of acid precipitation on litter and soil. In F. M. D'Itri (Ed.), *Acid Precipitation Effects on Ecological Systems* (pp. 435–451). Ann Arbor, MI: Ann Arbor Science.

McGuire, P. E., B. Lowery, and P. A. Helmke. 1992. Potential sampling error: Trace metal adsorption on vacuum porous cup samplers. *Soil Sci. Soc. Am. J.* (56):74–82.

McKeague, J. A., M. V. Cheshire, F. Andreux, and J. Berthelin. 1986. Organo-mineral complexes in relation to pedogenesis. In P. M. Huang and M. Schnitzer (Eds.), *Interactions of Soil Minerals with Natural Organics and Microbes* (pp. 549–592). Madison, WI: Soil Science Society of America.

McLean, E. O. 1965. Aluminum. In C. A. Black et al. (Eds.), *Methods of Soil Analysis, Part 2, Chemical and Microbiological Properties* (pp. 978–998). Agronomy Monograph 9. Madison, WI: American Society of Agronomy.

McMahon, M. A. and G. W. Thomas. 1974. Chloride and tritiated water flow in disturbed and undisturbed soil cores. *Soil Sci. Soc. Am. Proc.* (38):727–732.

Melchior, D. C. and R. L. Bassett. 1990. *Chemical Modeling of Aqueous Systems II.* ACS Symposium Series No. 416. Washington, DC: American Chemical Society.

Mengel, K. and E. A. Kirkby. 1979. *Principles of Plant Nutrition.* Bern, Switzerland: International Potash Institute.

Menzies, N. W. and L. C. Bell. 1988. Evaluation of the influence of sample preparation and extraction technique on soil solution composition. *Aust. J. Soil Res.* (26):451–464.

Michael, J. L., D. G. Neary, and M. J. M. Wells. 1989. Picloram movement in soil solution and streamflow from a Coastal Plain forest. *J. Environ. Qual.* (18):89–95.

Michalas, F., V. Glavac, and H. Parlar. 1992. The detection of aluminum complexes in forest soil solutions and beech xylem saps. *Fresenius J. Anal. Chem.* (343):308–312.

Mikkelsen, R. L., G. H. Haghnia, and A. L. Page. 1987. Effects of pH and selenium oxidation state on the selenium accumulation and yield of alfalfa. *J. Plant Nutr.* (10):937–950.

Misra, U. K., R. W. Blanchar, and W. J. Upchurch. 1974. Aluminum content of soil extracts as a function of pH and ionic strength. *Soil Sci. Soc. Am. Proc.* (38):897–902.

Monod, J. 1949. The growth of bacterial cultures. *Ann. Rev. Microbiol.* (3):371–394.

Moore, J. W. and R. G. Pearson. 1981. *Kinetics and Mechanism. A Study of Homogeneous Chemical Reactions.* New York: Wiley.

Morel, F. and J. Morgan. 1972. A numerical method for computing equilibria in aqueous chemical systems. *Environ. Sci. Technol.* (6):58–67.

Morgan, J. F. 1916. The soil solution obtained by the oil pressure method. *Mich. Agric. Exp. Sta. Tech. Bull.* (28):1–38.

Morris, H. D. 1949. The soluble manganese content of acid soils and its relation to the growth and manganese content of sweet clover and lespedeza. *Soil Sci. Soc. Am. Proc.* (13):362–371.

Morrison, R. D. and B. Lowery. 1990. Effect of cup properties, sampler geometry and vacuum on the sampling rate of porous cup samplers. *Soil Sci.* (149):308–316.

Moss, P. 1963. Some aspects of the cation status of soil moisture. *Plant Soil* (18):99–113.

Motojima, K. and N. Ishiwatari. 1965. Determination of microamount of aluminum, chromium, copper, iron, manganese, molybdenum, and nickel in pure water by extraction photometry. *J. Nuclear Sci. Technol.* (2):13–17.

Mubarak, A. and R. A. Olsen. 1976a. Immiscible displacement of the soil solution by centrifugation. *Soil Sci. Soc. Am. J.* (40):329–341.

Mubarak, A. and R. A. Olsen. 1976b. An improved technique for measuring soil pH. *Soil Sci. Soc. Am. J.* (40):880–882.

Mulder, J., J. J. M. van Grinsven, and N. van Breemen. 1987. Impacts of acid atmospheric deposition on wood land soils in the Netherlands: III. Aluminum chemistry. *Soil Sci. Soc. Am. J.* (51):1640–1646.

Mullins, G. L. and L. E. Sommers. 1983. *Chemical Speciation of Leachates from Waste Disposal Sites.* Purdue Univ. Water Resour. Res. Center Tech. Rpt. 158, West Lafayette, IN.

Nair, V. D. and J. Prenzel. 1978. Calculations of equilibrium concentration of mono- and polynuclear hydroxyaluminum species at different pH and total aluminum concentrations. *Z. Pflanzenernähr Bodenk.* (141):741–751.

National Research Council. 1980. *Mineral Tolerance of Domestic Animals.* Washington, DC: National Academy of Sciences.

Neal, C. 1988. Aluminum solubility relationships in acid waters—A practical example of the need for a radical reappraisal. *J. Hydrol.* (104):141–159.

Neary, A. J. and F. Tomassini. 1985. Preparation of alundum/ceramic plate tension lysimeters for soil water collection. *Can. J. Soil Sci.* (65):169–177.

Neilson, D. R. and J. Bouma (Eds.). 1985. *Soil Spatial Variability*. Proc. of a Workshop of the International Soil Science Society and the Soil Science Society of America, 30 Nov.–1 Dec. 1984, Las Vegas, Nevada. Netherlands: Pudoc Wageningen.

Ng Kee Kwong, N. F. and P. M. Huang. 1979a. The relative influence of low-molecular weight, complexing organic acids on the hydrolysis of aluminum. *Soil Sci.* (128):337–342.

Ng Kee Kwong, N. F. and P. M. Huang. 1979b. Surface reactivity of aluminum hydroxides precipitated in the presence of low molecular weight organic acids. *Soil Sci. Soc. Am. J.* (43):1107–1113.

Ng Kee Kwong, N. F. and P. M. Huang. 1981. Comparison of the influence of tannic acid and selected low-molecular-weight organic acids on precipitation products of aluminum. *Geoderma* (26):179–193.

Nilsson, S. I. and B. Bergkvist. 1983. Aluminum chemistry and acidification processes in a shallow podzol on the Swedish west coast. *Water Air Soil Pollution* (20):311–329.

Nordstrom, D. K. 1982. The effect of sulfate on aluminum concentrations in natural waters: Some stability relations in the system Al_2O_3–SO_3–H_2O at 298 K. *Geochim Cosmochim Acta* (46):681–692.

Nordstrom, D. K., L. N. Plummer, T. M. L. Wigley, T. J. Wolery, J. W. Ball, E. A. Jenne, R. L. Bassett, D. A. Crerar, T. M. Florence, B. Fritz, M. Hoffman, G. R. Holdren, Jr., G. M. Lafon, S. V. Mattigod, R. E. McDuff, F. Morel, M. M. Reddy, G. Sposito, and J. Thrailkill. 1979. A comparison of computerized chemical models for equilibrium calculations in aqueous systems. In E. A. Jenne (Ed.), *Chemical Modeling in Aqueous Systems* (pp. 857–892). ACS Symposium Series No. 93. Washington, DC: American Chemical Society.

Norvell, W. A. and W. L. Lindsay. 1982. Estimation of the concentration of Fe^{3+} and the $(Fe^{3+})(OH^-)_3$ ion product from equilibria of EDTA in soil. *Soil Sci. Soc. Am. J.* (46):710–715.

Nye, P. H. and P. B. Tinker. 1977. *Solute Movement in the Soil-Root System*. Berkeley: Univ. of California Press.

O'Dell, J. D., J. D. Wolt, and P. M. Jardine. 1992. Transport of imazethapyr in undisturbed soil columns. *Soil Sci. Soc. Am. J.* (56):1711–1715.

Ohno, T., E. I. Sucoff, M. S. Ehrich, P. Bloom, C. A. Buschena, and R. K. Dixon. 1988. Growth and nutrient content of red spruce seedlings in soil amended with aluminum. *J. Environ. Qual.* (17):666–672.

Olomu, M. D., G. J. Racz, and C. M. Cho. 1973. Effect of flooding on the Eh, pH, and concentrations of Fe and Mn in several Manitoba soils. *Soil Sci. Soc. Am. Proc.* (37):220–224.

Ononye, A. I., J. G. Graveel, and J. D. Wolt. 1987. Direct determination of benzidine in unaltered soil solution by liquid chromatography. *Bull. Environ. Contamin. Toxicol.* (39):524–532.

Oremland, R. S., J. T. Hollibaugh, A. S. Maest, T. S. Presser, L. G. Miller, and C. W. Culbertson. 1989. Selenate reduction to elemental selenium by anaerobic bacteria in sediments and culture: Biogeochemical significance of a novel, sulfate-independent respiration. *Appl. Environ. Microbiol.* (55):2333–2343.

Owen, B. B. and S. R. Brinkley, Jr. 1941. Calculation of the effect of pressure upon ionic equilibria in pure water and salt solutions. *Chem. Rev.* (29):461–474.

Paces, T. 1978. Reversible control of aqueous aluminum and silica during irreversible evolution of natural waters. *Geochim Cosmochim Acta* (42):1487–1493.

Page, A. L., F. T. Bingham, and C. Nelson. 1992. Cadmium adsorption and growth of various plant species as influenced by solution cadmium concentration. *J. Environ. Qual.* (1): 288–291.

Parfitt, R. L. and M. Saigusa. 1985. Allophane and humus-aluminum in Spodosols and Andepts formed from the same volcanic ash beds in New Zealand. *Soil Sci.* (139):149–155.

Parker, F. W. 1921. Methods of studying the concentration and composition of the soil solution. *Soil Sci.* (12):209–232.

Parker, J. C. and M. Th. van Genuchten. 1984. *Determining Transport Parameters from Laboratory and Field Tracer Experiments.* Virginia Agric. Exper. Sta. Bull. No. 84-3.

Pasricha, N. S. 1987. Predicting ionic strength from specific conductance in aqueous soil solutions. *Soil Sci.* (143):92–96.

Patterson, M. G., G. A. Buchanan, R. H. Walker, and R. M. Patterson. 1982. Fluometuron in soil solutions as an indicator of its efficacy in three soils. *Weed Sci.* (30):688–691.

Pavan, M. A., F. T. Bingham, and P. F. Pratt. 1982. Toxicity of aluminum to coffee in Ultisols and Oxisols amended with $CaCO_3$, $MgCO_3$, and $CaSO_4 \cdot 2H_2O$. *Soil Sci. Soc. Am. J.* (46):1201–1207.

Pearson, R. G. 1973. *Hard and Soft Acids and Bases.* Stroudsburg, PA: Dowden, Hutchinson, and Ross.

Peech, M. 1965. Hydrogen-ion activity. In C. A. Black (Ed.), *Methods of Soil Analysis. Part 2: Chemical and Microbiological Properties* (pp. 914–926). Madison, WI: American Society of Agronomy.

Pennington, K. L., S. S. Harper, and W. C. Koskinen. 1991. Interaction of herbicides with water-soluble organic matter. *Weed Sci.* (39):667–672.

Pepper, I. L., D. F. Bezdicek, A. S. Baker, and J. M. Sims. 1983. Silage corn uptake of sludge-applied zinc and cadmium as affected by soil pH. *J. Environ. Qual.* (12):270–275.

Perrin, D. D. 1979. *Stability Constants of Metal-Ion Complexes. Part B: Organic Ligands.* IUPAC Chem. Data Series No. 22. Oxford: Pergamon Press.

Perrin-Ganier, M. Schiavon, J. M. Portal, C. Breuzin, and M. Babut. 1993. Porous cups for pesticides monitoring in soil solution—Laboratory tests. *Chemosphere* (26):2231–2239.

Phillips, I. R. and W. J. Bond. 1989. Extraction procedure for determining soil solution and exchangeable ions on the same soil sample. *Soil Sci. Soc. Am. J.* (53):1294–1297.

Plankey, B. J. and H. H. Patterson. 1987. Kinetics of aluminum-fulvic acid complexation in acidic waters. *Environ. Sci. Technol.* (21):595–601.

Plankey, B. J. and H. H. Patterson. 1988. Effect of fulvic acid on the kinetics of aluminum floride complexation in acidic waters. *Environ. Sci. Technol.* (22):1454–1459.

Plankey, B. J., H. H. Patterson, and C. S. Cronan. 1986. Kinetics of aluminum fluoride complexation in acidic waters. *Environ. Sci. Technol.* (20):160–165.

Pohlman, A. A. and J. A. McCall. 1986. Kinetics of metal dissolution from forest soils by soluble organic acids. *J. Environ. Qual.* (15):86–92.

Pohlman, A. A. and J. G. McColl. 1989. Organic oxidation and manganese and aluminum mobilization in forest soils. *Soil Sci. Soc. Am. J.* (53):686–690.

Poletika, N. N., K. Roth, and W. A. Jury. 1992. Interpretation of solute transport data obtained with fiberglass wick soil solution samplers. *Soil Sci. Soc. Am. J.* (56):1751–1753.

Ponnamperuma, F. N., E. M. Tianco, and T. A. Loy. 1966. Ionic strengths of the solutions of flooded soils and other natural aqueous solutions from specific conductance. *Soil Sci.* (102):408–413.

Powell, P. E., G. R. Cline, C. P. P. Reid, and P. J. Szaniszlo. 1980. Occurrence of hydroxyamate siderophore iron chelators in soils. *Nature* (287):833–834.

Powell, P. E., P. J. Szaniszlo, G. R. Cline, and C. P. P. Reid. 1982. Hydroxyamate siderophores in the iron nutrients of plants. *J. Plant Nutr.* (5):653–673.

Presser, T. S. and J. Barnes. 1985. *Dissolved Constituents Including Selenium in Waters in the Vicinity of Kesterson National Wildlife Refuge and the West Grasslands, Fresno and Merced Counties, California.* U.S. Geological Survey Water Resources Invest. Rep. 85–4220. Washington, DC: U.S. Gov. Print. Office.

Pytkowicz, R. M. 1979. *Activity Coefficients in Electrolyte Solutions.* Boca Raton, FL: CRC Press.

Qian, P. and J. D. Wolt. 1990. Effects of drying and time of incubation on the composition of displaced soil solution. *Soil Sci.* (149):367–373.

Qian, P., J. D. Wolt, and D. D. Tyler. 1994. Tillage and time of nitrogen fertilization effects on soil solution composition. *Soil Sci.* (in press).

Qualls, R. G. and B. L. Haines. 1992. Biodegradability of dissolved organic matter in forest throughfall, soil solution, and stream water. *Soil Sci. Soc. Am. J.* (56):578–586.

Rao, D. N. and D. S. Mikkelsen. 1977. Effect of rice straw additions on production of organic acids in a flooded soil. *Plant Soil* (47):303–311.

Rasmussen, R. K. 1989. Aluminum contamination and other changes of acid soil solution isolated by means of porcelain suction-cups. *J. Soil Sci.* (40):95–101.

Reeve, R. C. and E. J. Doering. 1965. Sampling the soil solution for salinity appraisal. *Soil Sci.* (99):339–344.

Reid, D. A. 1976. Genetic potentials for solving problems of soil mineral stress: Aluminum and manganese toxicities in the cereal grains. In *Proc. of Workshop on Plant Adaptation to Mineral Stress in Problem Soils* (pp. 55–64). Beltsville, MD, 22–23 November 1976. Cornell Univ. Agric. Exp. Sta., Ithaca, NY.

Reilly, P. M. and G. E. Blau. 1974. The use of statistical methods to build mathematical models of chemical reacting systems. *Can. J. Chem. Eng.* (52):289–299.

Reitemeier, R. F. and L. A. Richards. 1944. Reliability of the pressure-membrane method for extraction of soil solution. *Soil Sci.* (57):119–135.

Reynolds, B. 1984. A simple method for the extraction of soil solution by high speed centrifugation. *Plant Soil* (78):437–440.

Rhodes, E. R. and W. L. Lindsay. 1978. Solubility of aluminum in soils of the humid tropics. *J. Soil Sci.* (29):324–330.

Rich, C. I. 1968. Hydroxy interlayers in expansible layer silicates. *Clays and Clay Minerals* (16):15–30.

Richard, T. L. and T. S. Steenhuis. 1988. Tile drain sampling of preferential flow on a field scale. *J. Contam. Hydrol.* (3):307–325.

Richards, L. A. 1941. A pressure-membrane extraction apparatus for soil solution. *Soil Sci.* (51):377–386.

Richburg, J. S. and F. Adams. 1970. Solubility and hydrolysis of aluminum in soil solutions and saturated-paste extracts. *Soil Sci. Soc. Am. Proc.* (34):728–734.

Richter, J. 1987. *The Soil as a Reactor.* Cremlingen, Germany: Catena Verlag.

Ritchie, G. S. P., M. P. Nelson, and M. G. Whitten. 1988. The estimation of free aluminum and the complexation between fluoride and humate anions for aluminum. *Comm. Soil Sci. Plant. Anal.* (19):857–871.

Robbins, C. W. 1985. The $CaCO_3$-CO_2-H_2O system in soils. *J. Agron. Ed.* (14):3–7.

Roberson, C. E. and J. D. Hem. 1969. *Solubility of Aluminum in the Presence of Hydroxide, Fluoride, and Sulfate.* U.S. Geol. Survey Water Supply Paper 1827-C.

Ross, D. S. and R. J. Bartlett. 1990. Effects of extraction methods and sample storage on properties of solutions obtained from forested Spodosols. *J. Environ. Qual.* (19):108–113.

Rousseau, L. Z. 1960. De l'influence du type d'humus sur le developement des plantules sapins dans les vosges. *Annales de L'Ecole Nationale Dans Eaux et Forêts et de la Station de Recherches et Expériences.* Nancy, France (17):13–118.

Ruess, J. O. 1983. Implications of the calcium-aluminum exchange system for the effect of acid precipitation on soils. *J. Environ. Qual.* (12):591–595.

Russell, A. E. and J. J. Ewel. 1985. Leaching from a tropical Andept during big storms: A comparison of three methods. *Soil Sci.* (139):181–189.

Russell, E. W. 1973. *Soil Conditions and Plant Growth* (10th Ed.). New York: Longman.

Rutherford, G. K., G. W. vanLoon, and J. A. Hern. 1985. Chemical and pedogenetic effects of simulated acid precipitation on two eastern Canadian forest soils. II. Metals. *Can. J. For. Res.* (15):848–854.

Sadana, U. S. and P. N. Takkar. 1988. Effect of sodicity and zinc on soil solution chemistry of manganese under submerged conditions. *J. Agric. Sci.* (111):51–55.

Sadiq, M. and W. L. Lindsay. 1979. *Selection of Standard Free Energies of Formation for Use in Soil Chemistry.* Colorado State Univ. Tech. Bull. 134.

Saffigna, P. G., D. R. Keeney and C. B. Tanner. 1977. Lysimeter and field measurements of chloride and bromide leaching in an uncultivated loamy sand. *Soil Sci. Soc. Am. J.* (41):478–482.

Sanders, J. R. 1983. The effect of pH on the total and free ionic concentration of manganese, zinc, and cobalt in soil solutions. *J. Soil Sci.* (34):315–323.

Sannigrahi, A. K., P. C. Bishagee, and S. K. Gupta. 1983. Distribution of different forms of iron and manganese in some lateritic soils of West Bengal under different forest vegetations. *Indian Agric.* (27):85–91.

Sato, K. and I. Yamane. 1973. Studies on soil solution: 2. Composition of soil solution. *J. Sci. Soil Manure* (44):246–250.

Schacklette, H. T. and J. G. Boerngen. 1984. *Element Concentrations in Soils and Other Surficial Materials of the Coterminous United States.* U.S. Geol. Survey Prof. Paper 1270.

Schecher, W. D. and D. C. McAvoy. 1992. MINEQL+: A software environment for chemical equilibrium modeling. *Comput., Environ. and Urban Systems* (16):65–76.

Scheir, G. A. 1985. Response of red spruce and balsam fir seedlings to aluminum toxicity in nutrient solutions. *Can. J. For. Res.* (15):29–33.

Schnitzer, M. 1978. Humic substances: Chemistry and reactions. In M. Schnitzer and S. U. Khan (Eds.), *Soil Organic Matter* (pp. 1–64). New York: Elsevier.

Schnitzer, M. 1986. Binding of humic substances by soil mineral colloids. In P. M. Huang and M. Schnitzer (Eds.), *Interactions of Soil Minerals with Natural Organics and Microbes* (pp. 77–102). Madison, WI: Soil Science Society of America.

Schofield, R. K. 1947. A ratio law governing the equilibrium of cations in the soil solution. *Proc. 11th International Congr. Pure Appl. Chem.* (3):257–261.

Schofield, R. K. and A. W. Taylor. 1955. Measurements of the activities of bases in soils. *J. Soil Sci.* (6):137–146.

Scott, D. C. and J. B. Weber. 1967. Herbicide phytotoxicity as influenced by adsorption. *Soil Sci.* (104):151–158.

Scow, K. M., S. Simkins, and M. Alexander. 1986. Kinetics of mineralization of organic compounds at low concentrations in soil. *Appl. Environ. Microbiol.* (51):1028–1035.

Scribner, S. L., T. R. Benzing, S. Sun, and S. A. Boyd. 1992. Desorption and bioavailability of aged simizine residues in soil from a continuous corn field. *J. Environ. Qual.* (21): 115–120.

Shorey, E. C. 1913. Some organic soil constituents. *USDA Bur. Soils Bull.* (88):5–41.

Siegrist, R. L. and P. D. Jenssen. 1990. Evaluation of sampling method effects on volatile organic compound measurements in contaminated soils. *Environ. Sci. Technol.* (24): 1387–1392.

Sikora, F. J., J. P. Copeland, G. L. Mullins, and J. M. Bartos. 1991. Phosphorus dissolution kinetics and bioavailability of water-insoluble fractions from monoammonium phosphate fertilizers. *Soil Sci. Soc. Am. J.* (55):362–368.

Sikora, F. J. and M. B. McBride. 1989. Aluminum complexation by catechol as determined by ultraviolet spectrophotometry. *Environ. Sci. Technol.* (23):349–356.

Sikora, F. J. and J. D. Wolt. 1986. Effect of cadmium- and zinc-treated sludge on yield and cadmium-zinc uptake of corn. *J. Environ. Qual.* (15):340–345.

Silkworth, D. R. and D. F. Grigal. 1981. Field comparison of soil solution samplers. *Soil Sci. Soc. Am. J.* (45):440–442.

Simkins, S. and M. Alexander. 1984. Models for mineralization kinetics with variables of substrate concentration and population density. *Appl. Environ. Microbiol.* (47):1299–1306.

Sims, G. K., M. Radosevich, X. T. He, and S. J. Traina. 1991. The Effects of Sorption on the Bioavailability of Pesticides. In W. B. Betts (ed.) *Biodegradation: Natural and Synthetic Materials* (pp. 119–137). London: Springer-Verlag.

Sims, G. K., J. D. Wolt, and R. G. Lehmann. 1992. *Bioavailability of Sorbed Pesticides and Other Xenobiotic Molecules.* Proc. International Symp. on Environ. Aspects of Pesticide Microbiology. August 17–21, Sigtuna, Sweden.

Singh, S. S. and J. E. Brydon. 1967. Precipitation of aluminum by calcium hydroxide in the presence of Wyoming bentonite and sulfate ions. *Soil Sci.* (103):162–168.

Singh, S. S. and J. E. Brydon. 1970. Activity of aluminum hydroxy sulfate and the stability of hydroxy aluminum interlayers in montmorrillonite. *Clays and Clay Minerals* (7):114–124.

Singh, S. S. and N. M. Miles. 1978. Effect of sulfate ions on the stability of an aluminum-interlayered Wyoming bentonite. *Soil Sci.* (126):323–329.

Singh, U. and G. Uehara. 1986. Electrochemistry of the double-layer: Principles and applications to soils. In D. L. Sparks (Ed.), *Soil Physical Chemistry* (pp. 1–38). Boca Raton, FL: CRC Press.

Sivasubramaniam, S. and O. Talibudeen. 1972. Potassium-aluminum exchange in acid soils. I. Kinetics. *J. Soil Sci.* (23):163–176.

Skopp, J. 1986. Analysis of time-dependent chemical processes in soil. *J. Environ. Qual.* (15):205–213.

Smith, R. W. and J. D. Hem. 1972. *Effect of Aging on Aluminum Hydroxide Complexes in Dilute Aqueous Solutions.* U.S. Geol. Survey Water Supply Paper 827-D.

Soil Science Society of America (SSSA). 1987. *Glossary of Soil Science Terms.* Madison, WI: Soil Science Society of America.

Soileau, J. M. and R. D. Hauck. 1987. *A Historical View of U.S. Lysimetry Research with Emphasis on Fertilizer Losses.* Proc. International Conf. on Infiltration Dev. and Applic. Jan. 6–9, Honolulu, HI.

Sommers, L. E. 1977. Chemical composition of sewage sludges and analysis of their potential use as fertilizers. *J. Environ. Qual.* (6):225–239.

Sommers, L. E. 1980. Toxic metals in agricultural crops. In G. Bitton, B. L. Damron, G. T. Edds, and J. M. Davidson (Eds.), *Sludge—Health Risks of Land Application* (pp. 105–140). Ann Arbor, MI: Ann Arbor Sci. Pub.

Sommers, L. E. and W. L. Lindsay. 1979. Effect of pH and redox on predicted heavy metal-chelate equilibria in soils. *Soil Sci. Soc. Am. J.* (43):39–47.

Sparks, D. L. 1984. Ion activities: An historical and theoretical overview. *Soil Sci. Soc. Am. J.* (48):514–518.

Sparks, D. L. 1986. Kinetics of reactions in pure and mixed systems. In D. L. Sparks (Ed.), *Soil Physical Chemistry* (pp. 83–145). Boca Raton, FL: CRC Press.

Sparks, D. L. and P. M. Jardine. 1981. Thermodynamics of potassium exchange in soil using a kinetics approach. *Soil Sci. Soc. Am. J.* (45):1094–1099.

Sposito, G. 1981. *The Thermodynamics of Soil Solutions.* New York: Oxford/Clarendon Press.

Sposito, G. 1984. The future of an illusion: Ion activities in soil solutions. *Soil Sci. Soc. Am. J.* (48):531–536.

Sposito, G., F. T. Bingham, S. S. Yadar, and C. A. Inouye. 1982. Trace metal complexation by fulvic acid extracted from sewage sludges: II. Development of chemical models. *Soil Sci. Soc. Am. J.* (46):51–56.

Sposito, G., K. M. Hortzclaw, and C. S. LeVesque-Madore. 1981. Trace metal complexation by fulvic acid from sewage sludge: I. Determination of stability constants and linear correlation analysis. *Soil Sci. Soc. Am. J.* (45):465–468.

Sposito, G. and S. V. Mattigod. 1980. *GEOCHEM: A Computer Program for the Calculation of Chemical Equilibria in Soil Solutions and Other Natural Water Systems.* The Kearney Foundation of Soil Science, Univ. California, Riverside.

Stalmans, M., A. Maes, and A. Cremers. 1986. Role of organic matter as a geochemical sink for technetium in soils and sediments. In G. Desmet and C. Myttenaere (Eds.), *Technetium in the Environment* (pp. 91–113). London: Elsevier.

Steiner, K. C., J. R. Barbour, and L. H. McCormick. 1984. Response of *Populus* hybrids to aluminum toxicity. *Forest Sci.* (30):404–410.

Steiner, K. C., L. H. McCormick, and D. S. Canavera. 1980. Differential response of paper birch provenances to aluminum in solution culture. *Can. J. For. Res.* (10):25–29.

Stevenson, F. J. 1967. Organic acid in soil. In A. D. McLaren and G. H. Peterson (Eds.), *Soil Biochemistry* (pp. 119–146). New York: Wiley.

Stevenson, F. J. 1972. Organic matter reactions involving herbicides in soil. *J. Environ. Qual.* (1):333–343.

Stevenson, F. J. 1982. *Humus Chemistry.* New York: Wiley.

Stevenson, F. J. and M. S. Ardakani. 1972. Organic matter reactions involving micronutrients in soils. In J. J. Mortvedt (Ed.), *Micronutrients in Agriculture* (pp. 79–114). Madison, WI: Soil Science Society of America.

Stevenson, F. J. and A. Fitch. 1986. Chemistry of complexation of metal ions with soil solution organics. In P. M. Huang and M. Schnitzer (Eds.), *Interactions of Soil Minerals with Natural Organics and Microbes* (pp. 29–58). Madison, WI: Soil Science Society of America.

Stuanes, A. O. 1983. Possible indirect long-term effects of acid precipitation on forest growth. *Aquilo Ser. Bot.* (19):50–63.

Stone, A. T. and J. J. Morgan. 1984. Reduction and dissolution of manganese (III) and manganese (IV) oxides by organics: 2. Survey of the reactivity of organics. *Environ. Sci. Technol.* (18):617–624.

Stumm, W. and J. J. Morgan. 1981. *Aquatic Chemistry* (2nd Ed.). New York: Wiley.

Suarez, D. L. 1986. A soil water extractor that minimizes CO_2 degassing and pH errors. *Water Resour. Res.* (22):876–880.

Suarez, D. L. 1987. Prediction of pH errors in soil-water extractors due to degassing. *Soil Sci. Soc. Am. J.* (51):64–67.

Sun, M. S., D. K. Harriss, and V. R. Magnuson. 1980. Activity corrections for ionic equilibria in aqueous solutions. *Can. J. Chem.* (58):1253–1257.

Swaine, D. J. 1969. *The Trace Element Content of Soils.* Commonwealth Bureau of Soil Sci. Tech. Commun. No. 148, Farnham Royal, Bucks, England.

Swallow, C. W., D. E. Kissel, and C. E. Owensby. 1987. Soil coring machine for microplots and large soil cores. *Agron. J.* (79):756–758.

Tackett, J. L., E. Burnett, and D. W. Fryrear. 1965. A rapid procedure for securing large, undisturbed soil cores. *Soil Sci. Soc. Amer. Proc.* (29):218–220.

Thomas, G. W. 1974. Chemical reactions controlling soil solution electrolyte concentration. In E. W. Carson (Ed.), *The Plant Root and Its Environment* (pp. 483–506). Charlottesville: Univ. of Virginia Press.

Thomas, G. W. and W. L. Hargrove. 1984. The chemistry of soil acidity. In F. Adams (Ed.), *Soil Acidity and Liming,* 2nd ed. (pp. 3–56). Agronomy Monograph 12. Madison, WI: American Society of Agronomy.

Thompson, G. W. and R. J. Medve. 1984. Effects of aluminum and manganese on the growth of ectomycorrhizal fungi. *Appl. Environ. Microbiol.* (48):556–560.

Thornton, F. C., M. Schaedle, and D. J. Raynal. 1987. Effects of aluminum on red spruce seedlings in solution culture. *Environ. Exp. Bot.* (27):489–498.

Thornton, F. C., M. Schaedle, and D. J. Raynal. 1989. Tolerance of red oak and American and European beech seedlings to aluminum. *J. Environ. Qual.* (18):541–545.

Thornton, F. C., M. Schaedle, and J. Raynal, and C. Zipperer. 1986. Effect of aluminum on honeylocust (*Gleditsia triacanthos* L.) seedlings in solution culture. *J. Exp. Bot.* (37): 775–785.

Thurman, E. M. 1985. Humic substances in groundwater. In G. R. Aiken, D. M. McKnight, R. L. Wershaw, and P. MacCarthy (Eds.), *Humic Substances in Soil, Sediment, and Water* (pp. 87–103). New York: Wiley.

Till, A. R. and T. P. McCabe. 1976. Sulfur leaching and lysimeter characterization. *Soil Sci.* (121):44–47.

Tokashiki, Y., J. B. Dixon, and D. C. Golden. 1986. Manganese oxide analysis in soils by combined x-ray diffraction and selective dissolution methods. *Soil Sci. Soc. Am. J.* (50): 1079–1084.

Tokunaga, T. K. and S. M. Benson. 1992. Selenium in Kesterson Reservoir: Ephemeral pools formed by groundwater rise: I. A field study. *J. Environ. Qual.* (21):246–251.

Tsai, P. P. and P. H. Hsu. 1984. Studies of aged OH-Al solutions using kinetics of Al-ferron reactions and sulfate precipitation. *Soil Sci. Soc. Am. J.* (48):59–65.

Turner, R. C. 1969. Three forms of aluminum in aqueous Systems determined by 8-quinolinolate extraction methods. *Can. J. Chem.* (47):2521–2527.

Turner, R. C. 1971. Kinetics of reactions of 8-quinolinol and acetate with hydroxyaluminum species in aqueous solutions. 2. Initial solid phases. *Can. J. Chem.* (49):1688–1690.

Turner, R. C. and W. Sulaiman. 1971. Kinetics of reactions of 8-quinolinol and acetate with hydroxyaluminum species in aqueous solutions. 1. Polynuclear hydroxy-aluminum cations. *Can. J. Chem.* (49):1683–1687.

Ulrich, B. 1981a. Destasbilisierung von Waldökosystem durch Akkumulation von Luftverunreinigungen. *Forst-u. Hotzwirt.* (21):525–532.

Ulrich, B. 1981b. Okologische Gruppierung von Böden nachihren chemischen Bodenzusland. *Z. Pflazenernähr Bodenk.* (144):289–305.

Ulrich, B. 1983. Soil acidity and its relation to acid deposition. In B. Ulrich and J. Pankrath (Eds.), *Effects of Accumulation of Air Pollutants in Forest Ecosystems* (pp. 127–146). Dordrecht, Netherlands: D. Reidel.

Ulrich, B., R. Mayer, and P. K. Khanna. 1980. Chemical changes due to acid precipitation in a loess-derived soil in Central Europe. *Soil Sci.* (130):193–199.

Ulrich, B., K. J. Meiwes, N. Konig, and P. K. Khanna. 1984. Untersuchungsuerfahren and Kriterien zur Bewertung der Versauerung und iherer Folgen in Woldboden. *Forst-u. Holzwirt.* (39).

U.S. Department of Agriculture (USDA). 1979. *Animal Waste Utilization on Cropland and Pastureland.* USDA Utilization Res. Rep. No. 6. Washington, DC: U.S. Government Printing Office.

United States Salinity Laboratory Staff. 1954. *Diagnosis and Improvement of Saline and Alkali Soils.* USDA Handbook No. 60.

Upchurch, W. J., M. Y. Chowdhury, and C. E. Marshall. 1973. Lysimetric and chemical investigations of pedological changes: Part 1. Lysimeters and their drainage waters. *Soil Sci.* (116):266–281.

Vaidyanathan, L. V. and P. H. Nye. 1966. The measurement and mechanism of ion diffusion in soils: I. An exchange resin paper method for measurement of the diffusive flux and diffusion coefficient of nutrient ions in soils. *J. Soil Sci.* (17):175–183.

van Breeman, N. 1973. Dissolved aluminum in acid sulfate soils and acid mine waters. *Soil Sci. Soc. Am. J.* (37):694–697.

van Breeman, N. and R. Brinkman. 1978. Chemical equilibria and soil formation. In G. H. Bolt and M. G. M. Bruggenwert (Eds.), *Soil Chemistry: A. Basic Elements* (pp. 141–170). Amsterdam: Elsevier Sci.

van Breeman, N., C. T. Driscoll, and J. Mulder. 1984. Acidic deposition and internal proton sources in acidification of soils and water. *Nature* (307):599–604.

Van der Ploeg, R. R. and F. Beese. 1977. Model calculations for the extraction of soil water by ceramic cups and plates. *Soil Sci. Soc. Am. J.* (41):466–470.

van Pragg, H. J. and F. Weissen. 1984. The intensity factor in acid forest soils: Extraction and composition of the soil solution. *Pedologie* (34):203–214.

van Praag, H. J. and F. Weissen. 1985. Aluminum effects on spruce and beech seedlings. I. Preliminary observations on plant and soil. *Plant Soil* (83):331–338.

van Praag, H. J., F. Weissen, S. Sougnez-Remy, and G. Carletti. 1985. Aluminum effects on spruce and beech seedlings. II. Statistical analysis of sand culture experiments. *Plant Soil* (83):339–356.

Van Shipout, T. P. J., H. A. J. Van Laven, O. H. Boersma, and J. Bouma. 1987. The effect of bypass flow and internal catchment of rain on the water regime in a clay loam grassland. *J. Hydrol.* (95):1–11.

Vass, A. A., W. M. Bass, J. D. Wolt, J. E. Foss, and J. T. Ammons. 1992. Time since death determinations of human cadavers using soil solution. *J. Forensic Sci.* (37):1236–1253.

Violante, A. and M. L. Jackson. 1981. Clay influence on the crystallization of aluminum hydroxide polymorphs in the presence of citrate, sulfate, and chloride. *Geoderma* (25): 199–214.

Violante, A. and P. Violante. 1980. Influence of pH, concentration, and chelating power of organic anions on the synthesis of aluminum hydroxides and orthohydroxides in nitric acid. *Soil Sci. Soc. Am. J.* (47):164–168.

Vlamis, J. 1953. Acid soil infertility as related to soil-solution and solid-phase effects. *Soil Sci.* (75):383–394.

Wada, K. 1977. Allophane and imogolite. In J. B. Dixon and S. B. Weed (Eds.), *Minerals in Soil Environments* (pp. 603–638). Madison, WI: Soil Science Society of America.

Wada, S. and K. Wada. 1980. Formation, composition and structure of hydroxy-aluminosilicate ions. *J. Soil Sci.* (31):457–467.

Wagatsuma, T. and M. Kaneko. 1987. High toxicity of hydroxy-aluminum polymer ion to plant roots. *Soil. Sci. Plant Nutr.* (33):57–67.

Wagman, D. D., W. H. Evans, V. B. Parker, R. H. Schumm, I. Halow, S. M. Bailey, K. L. Churney, and R. L. Nuttall. 1982. The NBS tables of chemical thermodynamic properties. *J. Phys. Chem. Ref. Data* (11):Suppl 2.

Walker, D. S., J. D. O'Dell, J. D. Wolt, and G. N. Rhodes, Jr. 1990. A tractor mounted sampler for obtaining large soil columns. *Weed Technol.* (4):913–917.

Walworth, J. L. 1992. Soil drying and rewetting, or freezing and thawing, affects soil solution composition. *Soil Sci. Soc. Am. J.* (56):433–437.

Wang, M. K., M. L. White, and S. L. Hem. 1983. Influence of acetate, oxalate, and citrate anions on precipitation of aluminum hydroxide. *Clays and Clay Minerals* (31):65–68.

Wang, T. S. C., T. K. Yang, and T. T. Chuang. 1967. Soil phenolic acids as plant growth inhibitors. *Soil Sci.* (103):239–246.

Warrick, A. W. and A. Amoozegar-Fard. 1977. Soil water regimes near porous cup water samplers. *Water Resour. Res.* (13):203–207.

Weaver, G. T., P. K. Khanna, and F. Beese. 1985. Retention and transport of sulfate in a slightly acid forest soil. *Soil Sci. Soc. Am. J.* (49):746–750.

Wehjte, G., R. Dickens, J. W. Wilcut, and B. F. Hajek. 1987. Sorption and mobility of sulfometuron and imazapyr in five Alabama soils. *Weed Sci.* (35):858–864.

Wendt, J. W. 1992. A device for extracting clarified soil solutions without filter paper. *Comm. Soil Sci. Plant Anal.* (23):769–774.

Whelan, B. R. and N. J. Barrow. 1980. A study of a method for displacing soil solution by centrifuging with an immiscible liquid. *J. Environ. Qual.* (9):315–319.

White, G. N., S. B. Feldman, and L. W. Zelazny. 1990. Rates of nutrient release by mineral weathering. In A. Lucier and S. Gaines (Eds.), *Mechanisms of Forest Response to Acidic Deposition* (pp. 108–162). New York: Springer-Verlag.

White, G. N., S. B. Feldman, and L. W. Zelazny. 1993. Rates of nutrient release by mineral weathering. In A. A. Lucier and S. G. Gaines (Eds.), *Mechanisms of Forest Response to Acidic Deposition* (pp. 108–162). New York: Springer-Verlag.

White, G. N. and L. W. Zelazny. 1986. Charge properties of soil colloids. In D. L. Sparks (Ed.), *Soil Physical Chemistry* (pp. 39–81). Boca Raton, FL: CRC Press.

White, R. E. 1987. *Introduction to the Principles and Practice of Soil Science.* 2nd Ed. Oxford: Blackwell Scientific.

Whitehead, D. C. 1964. Identification of p-hydrobenzoic, vanillic, p-coumaric, and ferulic acids in soils. *Nature* (202):417–418.

Whitten, M. G., G. S. P. Ritchie, and I. R. Willett. 1992. Forms of soluble aluminum in acidic topsoils estimated by ion chromatography and 8-hydroxyquinoline and their correlation with growth of subterranian clover. *J. Soil Sci.* (43):283–293.

Wolt, J. 1987. *Soil Solution: Documentation, Source Code, and Program Key.* Tenn. Agric. Exp. Sta. Res. Rep. No. 87-19.

Wolt, J. D. 1981. Sulfate retention by acid sulfate-polluted soils in the Copper Basin area of Tennessee. *Soil Sci. Soc. Am. J.* (45):283–287.

Wolt, J. D. 1989. SOILSOLN: A program for teaching equilibria modeling of soil solution composition. *J. Agron. Ed.* (18):40–42.

Wolt, J. D. 1990. Effects of acidic deposition on the chemical form and bioavailability of soil aluminum and manganese. In A. Lucier and S. Gaines (Eds.), *Mechanisms of Forest Response to Acidic Deposition* (pp. 62–107). New York: Springer-Verlag.

Wolt, J. D. 1993. Soil solution assessment of the soil availability of xenobiotics. In K. B. Hoddinott and T. A. O'Shay (Eds.), *Applications of Agricultural Analysis in Environmental Studies, ASTM STP 1162* (pp. 71–85). Philadelphia: American Society of Testing and Materials.

Wolt, J. D. and F. Adams. 1979a. Critical levels of soil- and nutrient-solution calcium for vegetative growth and fruit development of Florunner peanuts. *Soil Sci. Soc. Am. J.* (43): 1159–1164.

Wolt, J. D. and F. Adams. 1979b. The release of sulfate from soil-applied basaluminite and alunite. *Soil Sci. Soc. Am. J.* (43):118–121.

Wolt, J. D. and J. G. Graveel. 1986. A rapid routine method for obtaining soil solution using vacuum displacement. *Soil Sci. Soc. Am. J.* (50):602–605.

Wolt, J. D., N. V. Hue, and R. L. Fox. 1992. Solution sulfate chemistry in three sulfur-retentive Hydrandepts. *Soil Sci. Soc. Am. J.* (56):89–95.

Wolt, J. D. and D. A. Lietzke. 1982. The influence of anthropogenic sulfur inputs upon soil properties in the Copper Basin region of Tennessee. *Soil Sci. Soc. Am. J.* (46):651–656.

Wolt, J. D., G. N. Rhodes, Jr., J. G. Graveel, E. M. Glosauer, M. K. Amin, and P. L. Church. 1989. Activity of imazaquin in soil solution as affected by incorporated wheat (*Triicum aestivum*) straw. *Weed Sci.* (37):254–258.

Wright, R. J. and S. F. Wright. 1987. Effects of aluminum and calcium on the growth of subterranean clover in Appalachian soils. *Soil Sci.* (143):341–348.

Yin, C. and J. P. Hasset. 1986. Gas-partitioning approach for laboratory and field studies of mirex fugacity in water. *Environ. Sci. Technol.* (20):1213–1217.

Zabowski, D. 1989. Limited release of soluble organics from roots during centrifugal extraction of soil solutions. *Soil Sci. Soc. Am. J.* (53):977–979.

Zabowski, D. and R. S. Sletten. 1991. Carbon dioxide degassing effects on the pH of Spodosol soil solutions. *Soil Sci. Soc. Am. J.* (55):1456–1461.

Zabowski, D. and F. C. Ugolini. 1990. Lysimeter and centrifuge soil solutions: Seasonal differences between methods. *Soil Sci. Soc. Am. J.* (54):1130–1135.

Zhang, P., T. J. Ganje, A. L. Page, and A. C. Chang. 1988. Growth and uptake of selenium by Swiss chard in acid and neutral soils. *J. Environ. Qual.* (17):314–316.

Index

Acid deposition, 195, 197, 198, 199, 240–244
Activity, *see also* Ion activity
  coefficient, 38
    McInnes, 40
    mean ionic, 40–41, 43
    nonelectrolytes, 302–303
    and ratio law, 172
    single ion, 40–41, 43–47, 78
  definition of, 38, 245
  ratio, 212. *See also* Ion activity ratio
  solid phase, 182, 183–184
  xenobiotic:
    ionized forms, 302, 304, 305
    un-ionized forms, 302–303
Alunite, 197, 198, 199, 238
Aluminum:
  and acid buffering, 239
  and acid deposition, 240–244
  bioavailability, 227–235
  colloidal, 93
  complexation, 94, 222–223, 227, 228, 241
  critical threshold for phytotoxicity, 211,
    213, 230–231, 233–234, 235
  in lysimeter solutions, 124, 125, 126, 135
  in soil solid phase, 236–239
  in soil solution, 146, 220–245
    distribution, 221–224
    equilibria with aluminum oxyhydroxide
      minerals, 188
    equilibria with aluminum hydroxy
      sulfate minerals, 195–198, 199, 237–238

    measurement, 224–227
    and soil solution calcium, 146, 230–231,
      243–244
    transient effects, 141
  speciation, 84
    hydrolysis and deprotonation, 156
    monomer and polymer interactions, 157,
      225–226
    pH effects, 186
    polymers, 156, 221–222, 231
    temperature and pressure effects,
      185–186
Aluminum hydroxide potential, 189, 190, 197,
  198
Ammonia:
  ammonium ion activity, 214
  soil solution, 216
Availability, 13, 209. *See also* Bioavailability
  absolute, 209
  definition, 209

Basaluminite, 197, 199, 238
Bioavailability, 6, 15, 97, 118, 209–217. *See
  also* Availability
  critical thresholds, 209
  ion sink bioavailability index, 118
  xenobiotic, 303–306
Biogeochemical availability, 218–219. *See
  also* Chemical availability

Cadmium, 255–261

Cadmium (*Continued*)
  bioavailability, 259–260
  competitive ion effects, 260–261
  ion pairs and complexes, 257–259
  reduction, 274
Calcium:
  bioavailability, 213–214, 215–216
  complexation, 94
  and ion activity ratio, 171
  in lysimeter solutions, 124, 126
  soil solution, 146, 213, 215–217
    modeled in equilibration solutions, 150,
      193–194
    and soil solution aluminum, 146,
      230–231, 243–244
    and soil solution magnesium, 216–217
  weathering, 204
Carbonate equilibria, 80
  CO_2-degassing, 80, 83, 84, 112
  mineral stability relationships, 192–195,
    196
  and soil solution pH, 80–84, 194–195
  temperature dependence, 83
Capillary-wick lysimeters, *see* Ebermayer
  lysimeters
Centrifugal displacement, 19, 104–109, 120
  comparison to lysimeter solutions,
    123–124
  and effect of dilute double layer, 15, 16, 20
  methodology:
    high pressure, 20–21, 107–108
    immiscible displacement, 96, 97, 103,
      108–109
    low pressure, 96, 104–107
  pressure effects:
    on carbonate equilibria, 82
    on solution pH, 79–80
  temperature effects
    on solution pH, 79–80
    on carbonate equilibria, 82
Chelation, 118–120
Chemical availability, 209–219, *see also*
  Availability; Bioavailability
Chemical intensity, 4, 8, 144, 305–306
  as affected by soil fertility management,
    153
  and biogeochemical availability, 218–219
  intensity response curve, 210–213
  and quantity-intensity relationships, 169
  speciation and complexation effects,
    213–214
  xenobiotics, 293
Chemical potential, 13, 33–47
  of condensed phase component, 37–38
    in ideal dilute solutions, 38
      obeying Henry's law, 37–38
      obeying Raoult's law, 37

of electrolyte solutions, 38–47
of ideal gas, 34–36
of real gas, 36
and solid phase activity, 184
standard state, 36–37, 69
  condensed phase components, 37–39
  definition, 37
  electrolytes, 39
  ideal gas, 36
  real gas, 36
  summary, 35
reference state, 36–37
  condensed phase components, 37–39
  definition, 37
  electrolytes, 39
  ideal gas, 36
  real gas, 36
of water, 16. *See also* Water potential
Chemical thermodynamics, 22. *See also*
  statics
  activation energy, 63–65
    activation barrier, 63
    Arrhenius equation, 63–65
    frequency factor, 63, 65
  complexation equilibria, 160–161, 228, 254,
    258, 264, 269
  Dalton's law of partial pressure, 81
  hydrolysis and deprotonation, 154–155
    enthalpy of hydration, 155
    equilibria, 158–159, 258, 264
  entropy-driven and enthalpy-driven
    reactions, 253, 297
  fugacity, 36
  fundamental equations of state, 27, 28,
    30–31, 77
  Henry's law, 37–38
  Kirchoff relation, 30
  ligand field stabilization energy,
    254–255
  Maxwell relations, 28–30, 68–69
  partial molar quantities, 32
  phase rule, 32–33, 188, 191–192, 192–193
  practical system, 37, 38, 69
  rational system, 69
  Raoult's law, 37
  stationary state model, 3, 24–25, 181
  steady state model, 3, 25–27
  thermodynamic data files, 165, 168
  variables of state, 28
  van't Hoff equation, 64, 206
Chemisorption, 174–180, 294
  and solid phase phosphorus activity, 184
  definition, 174
  distribution coefficient, 174, 175, 176, 178,
    179, 219, 296–300, 305
    apparent, 298, 307–308
    effect of soil moisture, 219

protonation/deportonation effects, 294–295, 310
dynamic, 177–179
Fruendlich equation, 175
isotherms:
L-curve, 176
model fitting, 176–177
Langmuir equation, 175–176
linear free energy relationship, 179–180
partition coefficient, 178, 180
Chromium:
hydroxyaquo species, 157
in lysimeter solutions, 126
Column displacement, 96, 101, 104, 109–114, 120
syringe-pressure displacement, 113–114
traditional, 7, 109–112
vacuum displacement, 111, 112–113
Complexation and exchange techniques, 118–120, 267–269
as a bioavailability index, 118
limitations, 119–120
and metal ion activity, 118
methodology, 119
Contact-exchange theory, 12
Davies equation, 44–47, 78
Debye-Hückel:
equation, 43–44, 45–46, 85
limiting law, 10, 43, 288
Dilute double layer, 14–15, 20. *See also* Electrical double layer
Disequilibrium index, 206
Disequilibrium ratio, 183, 206
Distribution coefficient, *see* Chemisorption
Dissolved organic carbon (DOC), 92–94, 146, 275–310
characterization of, 93–94
as a chemically reacting soil compartment, 92
forms, 275–280
amino acids, 278–279
fulvic acid, 281
hydroxyamate siderophores, 280
low molecular weight organic acids, 223, 276–278, 279
polyphenols, 264–266, 281
sugar acids, 279–280
in lysimeter solutions, 125, 239–240
ion complexation, 94, 221–224, 281–285
measurement, 282–284
modeling, 284–285, 285–289
in soil solution, 276, 279
xenobiotic complexation, 295
Donnan free space, 14. *See also* Electrical double layer
Dynamics, 2, 22, 47–63. *See also* Kinetics

chemisorption, 177–180
compartment models, 58–63
definition, 60
nonlinear optimization, 60–63
multicompartment models, 58–60
two-compartment models, 58, 60–62, 63, 73, 177–179
empirical models, 48, 56–60
biocontinuum, 177–179
Elovich equation, 56–57
fractional order, 56
Michaelis-Menten equation, 57–58, 69–71
Monod kinetics, 58, 71–73
Monod-with-growth-kinetics, 58, 72–73
summarized, 61
ion exchange, 173–174
mineral weathering, 202–206
xenobiotic mobility, 308–309
Ebermayer lysimeters, 121, 135–139
capillary-wick lysimeters, 136, 139, 140, 141
plate lysimeters, 136–139
trough lysimeters, 136
zero tension, 136, 137, 139
Electrical conductivity:
and ionic strength, 85–86, 87
definition, 85
soil solution and soil water extracts, 88
specific conductance, 85
Electrical double layer, 13–15
dilute double layer, 14–15
and Donnan equilibrium, 14
anion exclusion, 219
Gouy-Chapman and Stern models, 14
Electrical potential (EH), 86–92. *See also* Electrochemical potential
measurement, 88–89
interpretation of, 89
sensitivity of, 88
redox, 88, 89, 90
cadmium availability, 274
manganese availability, 201, 202
selenium availability , 271
sign convention, 86
technicium availability, 272–273
relationship to electron activity, 91
theoretical basis, 86–88
Electrochemical potential, 12–13, 75–76
Electrolytes:
and soil solution, 7, 8
theory of electrolytic solutions, 7, 10–13
Electron activity, 89–92. *See also* Electrical potential
and pH, 91–92
Equilibrium redox potential, *see* Electrical potential

Exchange current, 88

Fruendlich equation, *see* Chemisorption

Geochemical availability, 217-219. *See also* Chemical availability
Gibbsite:
  solubility:
    comparison to amorphous Al(OH)$_{3s}$, 185, 186
    temperature dependence, 185, 186
  stability relationships of aluminum oxyhydroxides, 187-188, 189, 190, 191, 198, 199, 207, 208, 236-239
Gypsum:
  free energy of formation, 31
  solubility, 182-183, 184
Guttelburg equation, 44

Hard and soft acids and bases, 250-255, 261-262
Hydration, *see* Hydrolysis
Hydrolysis, 154-157
  and deprotonation, 154, 157
  equilibria, 158-159
  and hydration:
    amphoteric species, 156
    enthalpy, 155
    thermodynamic description, 154-155
  hydroxyaquo species, 155-157
Hydroxyapatite, 198-201

Immiscible displacement, *see* Centrifugal displacement
Intensity factor, *see also* Chemical intensity
  definition, 4
Ion activity, 10, 12-13, 39-40, 154. *See also* Activity
  and bioavialability, 213, 214-217
  and exchange coefficients, 170
  ratio, 98, 99-100, 101, 103, 146, 171, 212, 214-217, 262
  single-ion, 13, 44, 75, 78
  trace metal, 251
    measurement by competitive equilibria, 118-120
Ion activity product, 39, 182-183
  calcite equilibria, 194
  effects of temperature and pressure, 185-186
Ion association, 162-164. *See also* Ion complexes and ion pairs
  Bjerrum theory, 162
  complexation equilibria, 160-161
  Eigen-Wilkens mechanism, 162
  measurement of, 164

Ion complexes, 157-164. *See also* Ion pairs
  aluminum, 222-224, 227, 228
  cadmium, 257-260
  definition, 161
  interchange rate constant, 162, 163
  monodentate ligands, 157-159, 162
Ion difference, 150
Ion exchange, 172-174. *See also* Quantity-intensity relationships
  dynamic expression, 173-174
Ion pairs, 157-164. *See also* Ion complexes
  aluminum, 222-224, 227, 228
  Bjerrum, 162
  cadmium, 257-260
  definition, 162
  effect on ion difference, 150
  effect on ionic strength, 150
  outer sphere association constant, 162, 163
Ion speciation, 153-164, 213-214
  aluminum, 225-227
  trace metals, 247-250
  xenobiotics, 294-295
Ion speciation models, 165-168, 227
  comparison of outputs, 166
  computational algorithms, 167-168
  degree of computational intensity, 165
  equilibrium constant approach, 165-166
  Gibbs free energy minimization, 165-166
  materials balance, 167
  organic carbon reactivity, 285-289, 291
  thermodynamic data files, 165, 168
Ionic strength, 15, 78-79, 85-86, 146
  definition, 85
  effect of ion pairing, 86, 146-150
  estimation by electrical conductivity, 86, 87, 150. *See also* Electrical conductivity
  soil solution and soil water extracts, 86, 88, 150
Inner sphere complexes, *see* Ion complexes
Iron:
  activity determination by complexation, 119
  complexation:
    competetive chelation equilibria, 267-269
    with glycinates, 159-161
  in lysimeter solutions, 124, 126
  solvent exchange, 164

Jurbanite, 197, 199, 238, 244

Kaolinite, 189, 190, 191, 198, 199, 207, 208
Kinetics, 22. *See also* Dynamics
  applicability to soils, 23, 48
  complex mechanisms, 52-55

consecutive reactions, 55
competing reactions, 54–55
mixed reactions, 55
opposing reactions, 52–54
definition, 47
exchange rate of reaction, 26, 68
heterogeneity, 48, 55
homogeneity, 48, 50
half-life or half-time, 51
molecularity, 49
primary mineral dissolution, 205–206
rate law, 48–49, 50, 51, 61
definition, 48
reaction order, 49–52
first order, 50–51
fractional order, 56, 174, 206
pseudo-first order, 51–52, 205–206. *See also* Second order
second order, 51–52, 265
zero order, 50

L-curve isotherms, 176
Langmuir equation, *see* Chemisorption
Lysimetry, 18, 19, *see also specific types*
definition, 121
Ebermayer, 121, 135–139
effects of construction materials, 123, 124–126
on dissolved organic carbon, 125
on trace metals, 124–125
tension lysimeters, 135
filled-in, 121, 129–130
monolith, 121, 126–129
block, 127
column, 124, 127–129
sampling rate, 133, 134
solutions:
compared to soil water extracts, 122, 123
compared to displaced soil solutions, 123–124
zero tension compared to tension plate, 128–129, 136
tension, 121, 130–135
xenobiotics, 300–301
zone of influence, 122, 130–131
Magnesium:
complexation, 94
in lysimeter solutions, 124, 126
inner sphere complexes, 163
soil solution, 216–217
Manganese, 263–267
complexation, 94
competetive chelate equilibria, 267–268
dynamics, 255–256
in lysimeter solutions, 126

soil solid phase, 263–264
soil solution, 201–202
solubility relationships, 201–202, 203, 274
Matric potential, *see* Soil water matric potential
Master variables, 74–94, 107. *See also specific variables*
dissolved organic carbon, 74, 92–94
electrical potential, 74, 86–92
ionic strength, 74, 85–86
pH, 74–84
Microcline, 191, 208
Mineral stability diagrams, 186–202
amorphous minerals, 195–198
carbonates, 192–195
layer silicates, 188–191
oxyhydroxides, 188–191
potassium aluminum silicates, 191–192
soluble salts, 192–195, 196
sparingly soluble salts, 195–198
Mineral weathering:
ion mobility, 202, 203
chemical, 181–182
physical, 181
rate, 202–206
Montmorillonite, 189, 190, 191, 207–208
Monolith lysimeters, 126–129. *See also* Lysimetry
block, 127
column, 124, 127–129, 300–301, 309
definition, 126
sidewall effects, 127–128
and solute breakthrough, 128, 129
Multisite Langmuir equation, *see* Chemisorption
Muscovite, 191, 208

Organic acids, 84
Outer sphere complexes, *see* Ion pairs

Partition coefficient, *see* Chemisorption
pe, *see* Electron activity
pe + pH, 91–92, 201, 203
Pedogenesis:
definition, 181
dynamic approaches, 181–182. *See also* Mineral weathering
static approaches, 181. *See also* Mineral stability
pH, *see also* Soil pH
definition, 74
measurement of, 75–79
electromotive force, 75, 76, 78
ionic strength effect on, 78–79
junction potential, 75

pH (*Continued*)
pressure effect on, 77–78
temperature effect on, 75, 77
Phosphate:
availability, 198, 200
intensity response curve, 212
minerals, 198–201
soil solution, 200
Plate lysimeters, *see* Ebermayer lysimeters
Podzolization, 239–240
Power function, 56, 175
Pressure membrane extraction, 103, 114–116
Preferential flow, 122, 123. *See also* Soil water
flow
lysimeter sidewall effects, 127–128, 129
measurement by block lysimetry, 127

Quantity-intensity relationships, 169–180
selectivity coefficients, 170, 173
Gapon, 170, 172
Kerr, 170, 172
Krishnamoorthy and Overstreet, 173
Vaneslow, 173
ratio law, 170–172
defined, 170
effects of dilution, 171–172

Ratio law, *see* Quantity-intensity relationships

Saturation extracts, *see* Soil water extracts
Scaling of soil processes, 2, 22–23, 139, 141,
203–204
Selenium, 270–271
Sodium:
carbonate equilibria, 194–195, 196
hydrolysis, 194
soil solution, 146
Soil:
as a flow reactor, 3
chemical reactivity, 1, 2, 3, 6, 121, 154
and dissolved organic carbon, 92
and quantity-intensity relationships, 169
kinetic description, 23
definition, 1
heterogeneity, 2, 5, 131, 142
spatial variability in soil solution
composition, 139–141
Soil fertility management, 153
effect on soil solution calcium, 150
Soil pH, *see also* pH
active acidity, 79
carbonate buffering, 80–84, 89
CO_2-degassing, 84, 112
measurement, 79–84
by immiscible displacement, 97–98, 107
moisture effect, 97, 98

suspension effect, 10–11
reserve acidity, 79
Soil solution, 2
chemistry, 4, 6
emergence as a subdiscipline, 7
composition, 18, 144–168
comprehensive analysis, 144–153
duration of displacement effects,
103–104
modeling, *see* Ion speciation models
major ions, 145–146, 146–152
moisture effect, 96–98, 102
spatial variability, 139–140, 141–142
storage effect, 102
temperature effect, 101–102
temporal variation, 19, 142
veracity of, 9–10
definition, 4, 13, 15–18, 19
conceptual, 5, 6, 7, 9, 13
operational, 5, 9, 19
electrical conductivity of, 88
and mineral stability, 189–191, 192, 195
obtaining, 9, 18, 95–120. *See also* Specific
techniques
duration of displacement effects,
103–104
effect of equilibration time, 98–101
effects on pH, 79–84, 97–98, 109
sample pretreatment effects, 98–102
soil moisture effects, 95–101, 102, 116
storage temeprature effects, 101–102
pH, *see* Soil pH
as plant growth medium, 7, 12, 146, 195,
211–212
and soil water, 15–19
"true" or "unaltered," 7, 8–10, 74, 95,
110–112, 122
watershed and field scale sampling,
139–141, 204
Soil water, 15–18, 142
characteristic function, *see* Water
characteristic function
definition, 4
flow:
bypass, 17–18, 136, 143, 300, 308
diffuse, 17–18, 122, 123, 130, 137, 143, 308
Richards equation, 142–143
internal catchment, 18, 217
lysimeter column holdup, 128–129
matric potential, 16–17
mobile water, 6, 17–18, 122–123, 131, 137,
139, 143, 217–219
pore size range, 17–18
and air entry potential, 143
capillary, 17, 18
macropore, 17–18

mesopore, 17–18
micropore, 17–18
Soil water extracts, 8, 19, 97, 120
   compared to lysimeter solutions, 122–123
   electrical conductivity of, 88
   saturation extracts, 116–118
   saturated paste extracts, 79
     methodology, 116–118
     moisture effects, 117
   soil-water slurries, 79
     methodology, 118
Solubility product, 183
   relationship to ion activity product,
     183–186
   trace metal sulfides, 254
Solute transport, 15, 18–19
Stability diagrams, *see* Mineral stability
   diagrams
Statics, 2, 3, 22, 27–47. *See also* Chemical
   thermodynamics
   extent of reaction, 33, 34, 47, 66–68
   irreversible thermodynamics, 65–68
     DeDonder's inequality, 67
     entropy production, 66
     linear free energy relationship, 179–180
     principle of microscopic reversibility, 65
     reaction affinity, 66
Suction lysimeters, see tension lysimeters
Sulfate
   anthropogenic, 195, 197, 198, 199, 240
   soil solution, 195–198, 199
Syringe-pressure displacement, *see* Column
   displacement

Technetium, 271–273
Tension lysimeters, 130–135, 142
   construction, 135

dynamic sampling, 133–135
tension effects, 130, 131–133
   constant versus transient, 131–133, 134
   radius of influence, *see* Zone of
     influence
   zone of influence, 130–131
Theory of electrolytic solutions
Trace elements, 246–274
   adsorption by lysimeter packing materials,
     126
   complexation reactions, 250–260, 281–285
     Irving-Wallace order, 254–255
   contamination of lysimetry solutions,
     124–125
   soil solution:
     concentrations, 153, 247, 248, 249
     from waste disposal, 247–250
     weak acid oxyanions, 268–269
Trough lysimeters, *see* Ebermayer lysimeters

Vacuum displacement, *see* Column
   displacement
Varicite, 198–201

Water characteristic function, 16, 131, 142
Water potential, 16, 143, 217–218
Weathering, *see* Mineral weathering

Xenobiotics, 293–310
   analysis, 295–300
     by lysimetry, 125–126, 128, 300–301, 309
     in soil solution, 112, 113, 296, 297
   availability, 219, 303–306
   chemisorption, 296–300, 305, 307–308
     free energy relationship, 180
   intensity response curve, 210, 304, 306
   speciation and complexation, 294–295